CAMBRIDGE TRACTS IN MATHEMATICS
General Editors
B. BOLLOBAS, F. KIRWAN, P. SARNAK, C.T.C. WALL

115

Introduction to H_p Spaces

Paul Koosis
McGill University in Montreal

Introduction to H_p Spaces

Second edition, corrected and augmented

With two appendices by

V. P. Havin
St. Petersburg (Leningrad) State and McGill Universities

CAMBRIDGE
UNIVERSITY PRESS

CAMBRIDGE UNIVERSITY PRESS
Cambridge, New York, Melbourne, Madrid, Cape Town, Singapore, São Paulo

Cambridge University Press
The Edinburgh Building, Cambridge CB2 8RU, UK

Published in the United States of America by Cambridge University Press, New York

www.cambridge.org
Information on this title: www.cambridge.org/9780521455213

First edition published 1980
Second edition published 1998
This digitally printed version 2008

A catalogue record for this publication is available from the British Library

ISBN 978-0-521-45521-3 hardback
ISBN 978-0-521-05681-6 paperback

Contents

Preface to Second Edition

The first edition of this book was published in 1980 in the LMS Lecture Note Series, and a Russian translation by V.V. Peller and A.G. Tumarkin, made under the direction of V.P. Havin, the editor, appeared in 1984.

Both versions of the book are now out of print, and for the past couple of years people have been asking me how they might procure a copy of it. The Cambridge University Press has therefore decided to put out a second edition, and I am grateful to Dr. David Tranah, the Press' senior mathematics editor, for his having arranged to issue it in somewhat improved typographical format as a Cambridge Tract.

In preparing the first edition I had tried to make the exposition as accessible as I could by concentrating on what I thought were the main ideas in the subject rather than on including as many results as possible. The readers I had in mind were those with some training in analysis who were trying to gain a secure foothold in the theory of H_p spaces, whether with the aim of eventually doing serious work in that subject or for the purpose of understanding its applications in other areas (e.g. in operator theory – some of the material is now even used in electrical engineering). I have been guided by the same concern while working on the second edition and have for that reason tried to preserve the book's original character. That has especially meant refraining from attempting to turn it into what it was never intended to be – an all-encompassing treatise.

My first and main preoccupation has been to put to rights the first edition's many troublesome misprints, oversights and actual mistakes. I am grateful to the people who have called some of these to my attention, and their names are cited in the appropriate places. There were at least two serious errors in mathematical reasoning: at the end of §F in Chapter IV and in §G.1 of Chapter X. The proof of the lemma in Chapter II, §C.2, while not actually wrong, was certainly incomplete. An effort has been made to remedy these defects and several other less serious ones; I hope it has been successful.

Some new material that I consider really important has been added. Lindelöf's second theorem on conformal mapping can now be found in the new §C.3 of Chapter II. That result is one of the two main ingredients in the proof of Kellogg's theorem, included in the new §F of Chapter V. The simple geometric construction of Chapter VIII, §D.1 is used to obtain the atomic decomposition for $\Re H_1$ given in a new §E of that chapter, the old §E having become §F. This decomposition is then applied to give an alternative proof of the hard part of Fefferman's duality theorem in a new §G of Chapter X; the former §G is now §H.

The old appendix on Wolff's proof of the corona theorem is now Chapter XI. Where it used to stand there are two new appendices by V.P. Havin, on Peter Jones' interpolation formula and on Havin's proof of the weak sequential completeness of $L_1/H_1(0)$. The first edition of this book had already gone to press when I learned about the former topic, and time had not permitted my inclusion of the other one. Professor Havin was kind enough to write appendices on both for the Russian edition, and it is his appendices that are reproduced here, in my translation. These are included with his permission; he has read them and made certain suggestions that have been adopted. In making the translations I have tried to hew as closely as I could to Havin's own style, considerably different from mine.

The reader will probably also notice some differences in style between the new passages written by me and the older parts of the book. I cannot help that, for I am no longer able to write as I did in 1980.

I have been encouraged over the years by people with whom I was in large part unacquainted, who let me know in various ways that they liked one of the earlier versions of the book. I thank all of them for that encouragement, which strengthened my motivation to go on writing about mathematics.

Argenteuil County, Québec, near Boileau. October 10, 1993

Preface to First Edition

These are the lecture notes for a course I gave on the elementary theory of H_p spaces at the Stockholm Institute of Technology (tekniska högskolan) during the academic year 1977–78. The course concentrated almost exclusively on concrete aspects of the theory in its simplest cases; little time was spent on the more abstract general approach followed, for instance, in Gamelin's book. The idea was to give students knowing basic real and complex variable theory and a little functional analysis enough background to read current research papers about H_p spaces or on other work making use of their theory. For this reason, more attention was given to techniques and to what I believed were the ideas behind them than to the accumulation of a great number of results.

The lectures, about H_p spaces for the unit circle and the upper half plane, went far enough to include interpolation theory and *BMO*, but not as far as the corona theorem. That omission has, however, been put to rights in an appendix, thanks to T. Wolff's recent work. His proof of the corona theorem given there is a beautiful application of some of the methods developed for the study of *BMO*.

For Carleson's original proof of the corona theorem the reader may consult Duren's book. I have not included the more recent applications of the geometric construction Carleson devised for that proof, such as Ziskind's. Work of Douglas, Sarason, S-Y. Chang and Marshall on the algebras lying between H_∞ and L_∞ is not treated either.

Time did not allow me to cover the work of Hunt, Muckenhoupt and Wheeden on weighted mean value inequalities for harmonic conjugation. I did, however, give the proof of the Helson-Szegö theorem. Marshall's theorem (on the uniformly closed convex hull of the set of Blaschke products) is included although I did not lecture on it, and my lecture treatment of Lindelöf's theorem (on behaviour of the conformal mapping function near a point of tangency of the boundary) has been expanded.

In general, the notes stay quite close to the lectures as they were given. The style is loose and informal. Precise bibliographical references are not given in the text, nor the historical outlines at the end of each chapter that one has come to expect. A very partial bibliography is included; its purpose is to suggest further reading rather than to cover the subject thoroughly or to give due credit to all the workers in the field.

Topics not covered here, as well as the further ramifications of those I do cover, are treated in Garnett's extensive monograph now in the process of final revision. That book is recommended to the reader who wishes to go further.

I want to thank Harold Shapiro, Mats Essén and Magnus Giertz of the Stockholm Institute of Technology mathematics department for having helped me get an appointment to give this course. I want to thank the students and auditors for having successfully supported an extension of the course's length from the one semester originally planned to a full academic year. These were the students: Jockum Aniansson, Mats Lindberg, Lars Svensson and Anders Östrand. Björn Gustafsson audited most of the lectures and Dr. Stormark attended many. I was honoured by Dr. G.O. Thorin's presence at all of them. To all these people, my best wishes and warmest regards.

Los Angeles May 26, 1979

I

Functions Harmonic in $|z| < 1$. Rudiments

A. Power series representation

Let $U(z)$ be real and harmonic in $|z| < R$. (This means $U(z)$ is infinitely differentiable there and satisfies

$$\frac{\partial^2 U}{\partial x^2} + \frac{\partial^2 U}{\partial y^2} = 0.$$

We write throughout $z = x + iy$.) Then we can construct another real function $V(z)$, harmonic in $|z| < R$, such that

$$F(z) = U(z) + iV(z)$$

is analytic there. This function V is frequently called a harmonic conjugate of U. The construction of V is completely elementary; one way of doing it as follows:

We want a function V, infinitely differentiable in $|z| < R$ which, with U, will satisfy the Cauchy–Riemann equations*

$$\frac{\partial V}{\partial x} = -\frac{\partial U}{\partial y}$$
$$\frac{\partial V}{\partial y} = \frac{\partial U}{\partial x}.$$

(Then we will automatically have

$$\frac{\partial^2 V}{\partial x^2} + \frac{\partial^2 V}{\partial y^2} = 0.)$$

Such a function V can be found by second year calculus if the differential

$$\frac{\partial U}{\partial y}\,dx - \frac{\partial U}{\partial x}\,dy$$

* E. Trubowitz observed that these were incorrectly written in the first edition!

is exact in $|z| < R$. But it *is* since

$$\frac{\partial^2 U}{\partial x^2} + \frac{\partial^2 U}{\partial y^2} = 0\,!$$

Again by second year calculus, any two functions V which we can find will differ by a constant. Frequently the constant is chosen so as to make $V(0) = 0$.

Once V is found, we have, for $|z| < R$,

$$U(z) = \Re F(z)$$

with $F(z) = \sum_0^\infty a_n z^n$, the power series expansion being uniformly convergent on compact subsets of $|z| < R$. That's because any function analytic in $|z| < R$ has such a power series development.

Writing $z = re^{i\theta}$, we easily find

$$U(re^{i\theta}) = \sum_{-\infty}^{\infty} A_n r^{|n|} e^{in\theta},$$

with

$$\begin{cases} A_n = \tfrac{1}{2}a_n, & n > 0 \\ A_0 = \Re a_0 \\ A_n = \tfrac{1}{2}\bar{a}_{-n}, & n < 0. \end{cases}$$

Thus, *any function $U(z)$ harmonic in $|z| < R$ has a series representation*

$$U(re^{i\theta}) = \sum_{-\infty}^{\infty} A_n r^{|n|} e^{in\theta}$$

uniformly convergent on compact subsets of $|z| < R$.

B. Poisson's formula

The formula derived in the last section can be put in closed form. If $R > 1$ we easily find, for $r < 1$,

$$U(re^{i\theta}) = \frac{1}{2\pi} \int_{-\pi}^{\pi} U(e^{it}) \sum_{-\infty}^{\infty} r^{|n|} e^{in(\theta - t)}\, dt.$$

By summing two geometric series we get

$$\sum_{-\infty}^{\infty} r^{|n|} e^{in\phi} = \frac{1 - r^2}{1 + r^2 - 2r\cos\phi} \quad \text{if } 0 \le r < 1.$$

Thus we have derived *Poisson's representation*: *If $U(z)$ is harmonic for $|z| < R$, if $R > 1$, and if $0 \le r < 1$,*

$$U(re^{i\theta}) = \frac{1}{2\pi} \int_{-\pi}^{\pi} \frac{(1 - r^2) U(e^{it})}{1 + r^2 - 2r\cos(\theta - t)}\, dt.$$

This formula is basic for the whole course – we shall soon see that it holds under much more general conditions than the one stated above. We call

$$P_r(\theta) = \frac{1 - r^2}{1 + r^2 - 2r\cos\theta}$$

the *Poisson kernel* for $|z| < 1$.

C. Poisson representation of harmonic functions in various classes

Suppose we merely know that $U(z)$ is harmonic in $|z| < 1$. It is remarkable that some version of the Poisson representation will frequently hold for U in that circle.

Theorem *Let $p > 1$, let $U(z)$ be harmonic in $|z| < 1$, and suppose the means*

$$\int_{-\pi}^{\pi} |U(re^{i\theta})|^p \, d\theta$$

are bounded for $r < 1$. Then there is an $F \in L_p(-\pi, \pi)$ with

$$U(re^{i\theta}) = \frac{1}{2\pi} \int_{-\pi}^{\pi} \frac{1 - r^2}{1 + r^2 - 2r\cos(\theta - t)} F(t) \, dt$$

for $r < 1$.

Proof For $p > 1$, L_p is the dual of L_q, where $1/p + 1/q = 1$. The functions

$$U_n(\theta) = U\left(\left(1 - \frac{1}{n}\right) e^{i\theta}\right)$$

(instead of $1 - 1/n$, any sequence of r_n tending to 1 from below will do!) have $\|U_n\|_p \le C$ ($\| \ \|_p$ is here taken over $[-\pi, \pi]$, of course!), so, by the Cantor diagonal process, we can extract a subsequence U_{n_j} of them such that

$$LG = \lim_{j \to \infty} \int_{-\pi}^{\pi} G(\theta) U_{n_j}(\theta) \, d\theta$$

exists for G ranging over a *countable dense subset* of L_q. Since $\|U_{n_j}\|_p \le C$, this limit, LG, will actually exist for *all* $G \in L_q$ (easy exercise), and LG is then a bounded linear functional on L_q. So, since L_p is the dual of L_q, there *is* an $F \in L_p$ with

$$LG = \int_{-\pi}^{\pi} F(\theta) G(\theta) \, d\theta$$

for all $G \in L_q$.

Now for each n,

$$u_n(z) = U\left(\left(1 - \frac{1}{n}\right) z\right)$$

is harmonic for

$$|z| < \frac{1}{1 - (1/n)},$$

so, if $r < 1$,

$$u_{n_j}(re^{i\theta}) = \frac{1}{2\pi}\int_{-\pi}^{\pi} P_r(\theta - t)u_{n_j}(e^{it})\,dt = \frac{1}{2\pi}\int_{-\pi}^{\pi} P_r(\theta - t)U_{n_j}(t)\,dt.$$

Fix any $r < 1$ and any θ, and use $G(t) = P_r(\theta - t)$; $G \in L_q$. Then

$$\lim_{j\to\infty}\int_{-\pi}^{\pi} P_r(\theta - t)U_{n_j}(t)\,dt = LG = \int_{-\pi}^{\pi} G(t)F(t)\,dt = \int_{-\pi}^{\pi} P_r(\theta - t)F(t)\,dt.$$

The leftmost member is

$$\lim_{j\to\infty} 2\pi u_{n_j}(re^{i\theta}) = 2\pi U(re^{i\theta}).$$

Thus,

$$U(re^{i\theta}) = \frac{1}{2\pi}\int_{-\pi}^{\pi} P_r(\theta - t)F(t)\,dt,$$

where $F \in L_p$. Q.E.D.

Remark The same result holds, with the same proof, for $p = \infty$, if we change the statement slightly:

Theorem *If $U(z)$ is harmonic and bounded in $|z| < 1$, there is an $F \in L_\infty(-\pi, \pi)$ with*

$$U(re^{i\theta}) = \frac{1}{2\pi}\int_{-\pi}^{\pi} \frac{1 - r^2}{1 + r^2 - 2r\cos(\theta - t)}F(t)\,dt.$$

If I am not mistaken, this result was proved by Fatou, in his famous thesis, *Séries trigonométriques et séries de Taylor*, published before the First World War. The result is indeed the *starting point* of the whole subject treated here. Many of the ideas in the first half of this book have their origin in Fatou's thesis.

What if $p = 1$? $L_1(-\pi, \pi)$ is, unfortunately not the dual of *anything*. But M – the space of finite signed measures μ on $[-\pi, \pi]$ – with $\|\mu\|$ = total variation of μ – *is* the dual of $\mathscr{C}[-\pi, \pi]$ – the space of continuous functions on $[-\pi, \pi]$. If $p \in L_1[-\pi, \pi]$, we can associate to p a signed measure μ_p by putting

$$\int_{-\pi}^{\pi} G(t)\,d\mu_p(t) = \int_{-\pi}^{\pi} G(t)p(t)\,dt;$$

then $\|\mu_p\| = \|p\|_1$.

Here, then, the argument used in proving the first theorem of this section gives:

Theorem *If $U(z)$ is harmonic in $|z| < 1$ and the means*

$$\int_{-\pi}^{\pi} |U(re^{i\theta})|\,d\theta$$

are bounded for $r < 1$, there is a finite signed measure μ on $[-\pi, \pi]$ with

$$U(re^{i\theta}) = \frac{1}{2\pi} \int_{-\pi}^{\pi} P_r(\theta - t)\, d\mu(t), \quad 0 \leq r < 1.$$

Corollary (Evans) *Let $U(z)$ be harmonic in $|z| < 1$ and* positive *there (here and henceforth 'positive' just means 'non-negative'). Then there is a finite* positive *measure μ on $[-\pi, \pi]$ with*

$$U(re^{i\theta}) = \frac{1}{2\pi} \int_{-\infty}^{\infty} P_r(\theta - t)\, d\mu(t), \quad 0 \leq r < 1.$$

Proof For $r < 1$ (using, e.g., the expansion

$$U(re^{i\theta}) = \sum_{-\infty}^{\infty} a_n r^{|n|} e^{in\theta}$$

valid in $|z| < 1$), we have

$$2\pi U(0) = \int_{-\pi}^{\pi} U(re^{i\theta})\, d\theta = \int_{-\pi}^{\pi} |U(re^{i\theta})|\, d\theta,$$

since $U \geq 0$. Now just apply the theorem. The measure μ is positive because here (look again at the proof of the first theorem in this section)

$$\int_{-\pi}^{\pi} G(t)\, d\mu(t)$$

comes out *positive* for each *positive* $G \in \mathscr{C}$ – it's the *limit* of positive things!

D. Boundary behaviour

If we have one of the representations

$$U(re^{i\theta}) = \frac{1}{2\pi} \int_{-\pi}^{\pi} \frac{1 - r^2}{1 + r^2 - 2r\cos(\theta - t)} F(t)\, dt$$

$$U(re^{i\theta}) = \frac{1}{2\pi} \int_{-\pi}^{\pi} \frac{1 - r^2}{1 + r^2 - 2r\cos(\theta - t)}\, d\mu(t)$$

derived in the previous section, we should examine the connection between $U(z)$ and the function $F(t)$ or the measure $d\mu(t)$.

1. Integrability properties; functions given by Poisson's formula

We first obtain some crude results which are sufficient for many investigations.

The Poisson kernel

$$P_r(\phi) = \frac{1 - r^2}{1 + r^2 - 2r\cos\phi} = \sum_{-\infty}^{\infty} r^{|n|} e^{in\phi}$$

has the following properties:

(a) $P_r(\phi) > 0, \quad r < 1$

(b) $P_r(\phi + 2\pi) = P_r(\phi)$

(c) For each $r < 1$,

$$\int_{-\pi}^{\pi} P_r(t)\, dt = 2\pi.$$

Of these, (a) and (b) are evident, and (c) follows from the series development for $P_r(\phi)$.

If $F \in L_p[-\pi, \pi]$, it is convenient to suppose F *defined on all of* $\mathbb{R}$ *by periodicity*, $F(t + 2\pi) = F(t)$. *We henceforth assume this*. First we have converses to the representation theorems given in Section C.

Theorem *If $p \geq 1$ and $F \in L_p[-\pi, \pi]$ and*

$$U(re^{i\theta}) = \frac{1}{2\pi} \int_{-\pi}^{\pi} P_r(\theta - t) F(t)\, dt,$$

then $U(z)$ is harmonic in $|z| < 1$ and

$$\int_{-\pi}^{\pi} |U(re^{i\theta})|^p \, d\theta \leq \text{const.}, \quad r < 1.$$

Proof Let

$$\frac{1}{2\pi} \int_{-\pi}^{\pi} e^{-int} F(t)\, dt = A_n.$$

Then, for $0 \leq r < 1$,

$$U(re^{i\theta}) = \sum_{-\infty}^{\infty} A_n r^{|n|} e^{in\theta},$$

which is harmonic in $|z| < 1$ *by inspection*, because the series converges uniformly in the interior (meaning *uniformly on compact subsets* – complex variable language!) of that region. (If F is *real*, the series is clearly the real part of an analytic function which can be easily written down.)

Given $r < 1$, by property (b) and 2π-periodicity of F we can also write

$$U(re^{i\theta}) = \frac{1}{2\pi} \int_{-\pi}^{\pi} F(\theta - s) P_r(s)\, ds.$$

Now take $G \in L_q[-\pi, \pi]$, $\|G\|_q = 1$, so that (with any given fixed r – of course G will depend on r)

$$\left[\int_{-\pi}^{\pi} |U(re^{i\theta})|^p \, d\theta \right]^{1/p} = \int_{-\pi}^{\pi} U(re^{i\theta}) G(\theta)\, d\theta.$$

By Fubini's theorem, the integral on the right is

$$\frac{1}{2\pi} \int_{-\pi}^{\pi} \int_{-\pi}^{\pi} P_r(s) F(\theta - s) G(\theta)\, d\theta\, ds$$

which is in modulus

$$\leq \frac{1}{2\pi}\int_{-\pi}^{\pi} P_r(s)\,||F||_p\,||G||_q\,\mathrm{d}s = ||F||_p$$

by choice of G and property (c).

In fine,

$$\int_{-\pi}^{\pi} |U(re^{i\theta})|^p\,\mathrm{d}\theta \leq ||F||_p^p$$

and we are done.

Theorem *Let μ be a finite signed measure on $[-\pi, \pi]$. Then*

$$U(re^{i\theta}) = \frac{1}{2\pi}\int_{-\pi}^{\pi} P_r(\theta - t)\,\mathrm{d}\mu(t)$$

is harmonic in $|z| < 1$ and

$$\int_{-\pi}^{\pi} |U(re^{i\theta})|\,\mathrm{d}\theta \leq \text{const.}, \quad r < 1.$$

Proof *Harmonicity* is established as above. Given $r < 1$, let $G \in L_\infty$, $||G||_\infty = 1$, be such that

$$\int_{-\pi}^{\pi} |U(re^{i\theta})|\,\mathrm{d}\theta = \int_{-\pi}^{\pi} G(\theta)U(re^{i\theta})\,\mathrm{d}\theta.$$

The right-hand integral is, by Fubini's theorem, equal to

$$\frac{1}{2\pi}\int_{-\pi}^{\pi}\int_{-\pi}^{\pi} P_r(\theta - t)G(\theta)\,\mathrm{d}\theta\,\mathrm{d}\mu(t)$$

which, by properties (a), (b) and (c) is modulus

$$\leq \frac{1}{2\pi}\int_{-\pi}^{\pi}\int_{-\pi}^{\pi} P_r(\theta - t)\,||G||_\infty\,\mathrm{d}\theta\,|\mathrm{d}\mu(t)| = ||G||_\infty \int_{-\pi}^{\pi} |\mathrm{d}\mu(t)| = \int_{-\pi}^{\pi} |\mathrm{d}\mu(t)|.$$

We are done.

2. Elementary study of boundary behaviour

The Poisson kernel $P_r(\theta)$ has a *fourth property*:

(d) Given any $\delta > 0$,

$$P_r(\theta) \to 0 \quad \text{uniformly for} \quad \delta \leq |\theta| \leq \pi \quad \text{as } r \to 1.$$

This is obvious from the formula for $P_r(\theta)$.

Theorem *Let F be continuous on $\mathbb{R}$ and $F(t + 2\pi) = F(t)$. Let*

$$U(re^{i\theta}) = \frac{1}{2\pi}\int_{-\pi}^{\pi} P_r(\theta - t)F(t)\,\mathrm{d}t.$$

Then $U(z) \to F(\phi)$ *as* $z \to e^{i\phi}$, *and the convergence is uniform in* ϕ.

Proof The result goes back to Poisson himself, who *thought* it showed that the Fourier series of a function *converges* to that function (it doesn't show that!). Write

$$U(re^{i\theta}) = \frac{1}{2\pi}\int_{-\pi}^{\pi} F(\theta - t)P_r(t)\,dt.$$

Given any ϕ, we have, by property (c),

$$F(\phi) = \frac{1}{2\pi}\int_{-\pi}^{\pi} F(\phi)P_r(t)\,dt.$$

Therefore

$$U(re^{i\theta}) - F(\phi) = \frac{1}{2\pi}\int_{-\pi}^{\pi} (F(\theta - t) - F(\phi))\,P_r(t)\,dt.$$

Let $\delta < \pi/2$ be such that $|F(s) - F(\phi)| < \epsilon$ for $|s - \phi| < 2\delta$; δ depends only on ϵ and not on ϕ here by (uniform!) continuity of F.

Write the last right hand integral as a sum of two:

$$|U(re^{i\theta}) - F(\phi)| \le \frac{1}{2\pi}\int_{|t|\le\delta} |F(\theta - t) - F(\phi)|P_r(t)\,dt$$
$$+ \frac{1}{2\pi}\int_{\delta\leqslant|t|\leqslant\pi} |F(\theta - t) - F(\phi)|P_r(t)\,dt.$$

If $|\theta - \phi| < \delta$, the first integral on the right is

$$\le \frac{\epsilon}{2\pi}\int_{|t|\le\delta} P_r(t)\,dt < \epsilon.$$

Let M be a bound on $|F(t)|$. Then the second integral is

$$\le \frac{M}{\pi}\int_{\delta\leqslant|t|\leqslant\pi} P_r(t)\,dt$$

which is $< \epsilon$, say, if r is close enough to 1, by property (d).

So $|U(re^{i\theta}) - F(\phi)| < 2\epsilon$ if $|\theta - \phi| < \delta$ and r is close enough to 1. Q.E.D.

Remark Properties (a), (b), (c) and (d) together constitute the so-called *approximate identity property* of $(1/2\pi)P_r(\theta)$. The above theorem holds good *because of them* – all kinds of other kernels besides the Poisson kernel would work to yield similar theorems.

Theorem *Let* $F \in L_1(-\pi, \pi)$, *and suppose* $F(t)$ *is continuous at* θ_0. *Then*

$$U(re^{i\theta}) = \frac{1}{2\pi}\int_{-\pi}^{\pi} P_r(\theta - t)F(t)\,dt$$

tends to $F(\theta_0)$ *as* $re^{i\theta}$ *tends to* $e^{i\theta_0}$.

Proof Similar to that of the above theorem.

Theorem *Let $F \in L_p$, $1 \le p < \infty$ (sic!) and let*

$$U(re^{i\theta}) = \frac{1}{2\pi}\int_{-\pi}^{\pi} P_r(\theta - t)F(t)\,dt.$$

Then

$$\int_{-\pi}^{\pi} |U(re^{i\theta}) - F(\theta)|^p\,d\theta \to 0$$

as $r \to 1$; i.e., $U(re^{i\theta})$ approaches $F(\theta)$ in L_p norm as $r \to 1$.

Proof Write $F_r(\theta) = U(re^{i\theta})$; then

$$F_r(\theta) - F(\theta) = \frac{1}{2\pi}\int_{-\pi}^{\pi} [F(\theta - t) - F(\theta)]\,P_r(t)\,dt.$$

Using properties (a) and (c) (we think of $F_r(\theta) - F(\theta)$ as a limit of convex combinations of the functions $F(\theta - t) - F(\theta)$, taking t as a parameter and θ as the variable), we have, by an evident generalization of the triangle inequality

$$\left(\int_{-\pi}^{\pi} |F_r(\theta) - F(\theta)|^p\,d\theta\right)^{1/p} \le \frac{1}{2\pi}\int_{-\pi}^{\pi}\left(\int_{-\pi}^{\pi} |F(\theta - t) - F(\theta)|^p\,d\theta\right)^{1/p} \cdot P_r(t)\,dt.$$

That is, if we write

$$\Phi(t) = \left(\int_{-\pi}^{\pi} |F(\theta - t) - F(\theta)|^p\,d\theta\right)^{1/p},$$

$$\|F_r - F\|_p \le \frac{1}{2\pi}\int_{-\pi}^{\pi} \Phi(t)P_r(t)\,dt.$$

But $\Phi(t) \to 0$ as $t \to 0$! *That is because translation is continuous in the L_p norm for* $1 \le p < \infty$. This, in turn, follows from the rudiments of real variable theory as follows: given $F \in L_p(-\pi, \pi)$ and $\epsilon > 0$ take a continuous G, periodic of period 2π, with $\|F - G\|_p < \epsilon$. Then

$$\int_{-\pi}^{\pi} |G(\theta - t) - G(\theta)|^p\,d\theta$$

is obviously $< \epsilon^p$ for $|t| < \delta$, say, by uniform continuity, so

$$\|F(\theta - t) - F(\theta)\|_p < 3\epsilon \quad \text{for} \quad |t| < \delta.$$

In particular, $\Phi(t)$ is *continuous at* 0 *where it equals zero.*

So, by a previous result, $\|F_r - F\|_p \to 0$ as $r \to 1$.

If $p = \infty$ all we have is weak* convergence:

Theorem *If $F \in L_\infty$, and*

$$U(re^{i\theta}) = \frac{1}{2\pi}\int_{-\pi}^{\pi} P_r(\theta - t)F(t)\,dt,$$

then $U(re^{i\theta}) \to F(\theta)$ w* *as* $r \to 1$.

Proof Take any $G \in L_1(-\pi, \pi)$. We are to prove that

$$\int_{-\pi}^{\pi} U(re^{i\theta})G(\theta)\,d\theta \to \int_{-\pi}^{\pi} F(\theta)G(\theta)\,d\theta$$

for $r \to 1$. But this is true, because ($P_r(\phi)$ being even!)

$$\int_{-\pi}^{\pi} G(\theta)P_r(\theta - t)\,d\theta = \int_{-\pi}^{\pi} P_r(t - \theta)G(\theta)\,d\theta$$

tends in L_1 norm to $G(t)$ as $r \to 1$ by the preceding theorem. We need then only apply Fubini's theorem.

Similarly

Theorem *Let*

$$U(re^{i\theta}) = \frac{1}{2\pi}\int_{-\pi}^{\pi} P_r(\theta - t)\,d\mu(t)$$

with μ *a finite signed measure on* $[-\pi, \pi]$. *Then* $U(re^{i\theta})\,d\theta \to d\mu(\theta)$ w* *as* $r \to 1$, *i.e., for any continuous* $G(\theta)$, *periodic and of period* 2π,

$$\int_{-\pi}^{\pi} U(re^{i\theta})G(\theta)\,d\theta \to \int_{-\pi}^{\pi} G(\theta)\,d\mu(\theta)$$

as $r \to 1$.

Proof Use Fubini's theorem together with the first result of this subsection.

3. Deeper study of boundary behaviour; Fatou's theorem

If $U(z)$, harmonic in $|z| < 1$, has one of the representations

$$U(re^{i\theta}) = \frac{1}{2\pi}\int_{-\pi}^{\pi} P_r(\theta - t)F(t)\,dt, \quad F \in L_p, \qquad U(re^{i\theta}) = \frac{1}{2\pi}\int_{-\pi}^{\pi} P_r(\theta - t)\,d\mu(t),$$

we have still to discuss the pointwise behaviour of $U(z)$ as z tends to points $e^{i\theta}$ on the boundary of the unit circle. Study of such behaviour cannot proceed on the basis of the approximate identity properties (a)-(d) alone, but requires a more detailed examination of $P_r(\theta)$.

Both representations written above for $U(re^{i\theta})$ are subsumed in the second one, for if $F \in L_p(-\pi, \pi)$, and we take $d\mu(\theta) = F(\theta)\,d\theta$, then μ *is* in fact a finite signed measure on $[-\pi, \pi]$. In dealing with such a measure, it is convenient to introduce the function $\mu(\theta)$ of bounded variation on $[-\pi, \pi]$ given by

$$\mu(\theta) = \int_{0}^{\theta} d\mu(t)$$

(with usual interpretation of the integral if $\theta < 0$). Then we have the

Theorem (Fatou) *Let $-\pi < \phi_0 < \pi$ and suppose that the derivative $\mu'(\phi_0)$ exists and is finite. Then*

$$U(re^{i\theta}) = \frac{1}{2\pi}\int_{-\pi}^{\pi} P_r(\theta - t)\,d\mu(t)$$

tends to $\mu'(\phi_0)$ for $re^{i\theta}$ tending to $e^{i\phi_0}$, from within any region of the form $|\theta - \phi_0| \leq c(1-r)$.*

Remark 1 Thus, $z = re^{i\theta}$ is required to tend to $e^{i\phi_0}$ *from within sectors of opening* $< 180°$ having *vertex* at $e^{i\phi_0}$, and symmetric about the radius from 0 out to $e^{i\phi_0}$. We frequently say that $U(re^{i\theta}) \to \mu'(\phi_0)$ *as $re^{i\theta}$ tends to $e^{i\phi_0}$ non-tangentially*, and write

$$U(re^{i\theta}) \to \mu'(\phi_0) \quad \text{for} \quad re^{i\theta} \underset{\angle}{\longrightarrow} e^{i\phi_0}.$$

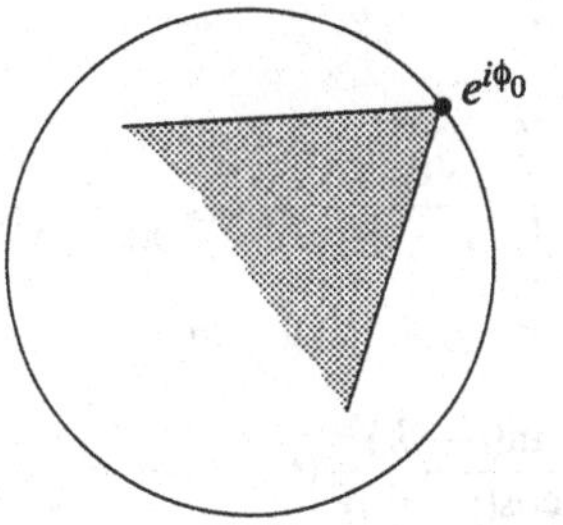

Remark 2 A similar result holds for $\phi_0 = \pm\pi$ provided $\mu'(\phi_0)$ exists in a properly defined sense there. Let the reader figure out what this sense should be.

Proof To simplify the notation, take $\phi_0 = 0$. Then, if $\mu'(0)$ exists and is finite, and $|\theta_r| \leq c(1-r)$, let us show that

$$\frac{1}{2\pi}\int_{-\pi}^{\pi} \frac{1-r^2}{1+r^2-2r\cos(t-\theta_r)}\,d\mu(t) \to \mu'(0)$$

as $r \to 1$. Without loss of generality, assume $\mu'(0) = 0$; otherwise work with $d\mu(t) - \mu'(0)\,dt$ instead of $d\mu(t)$;

$$\frac{1}{2\pi}\int_{-\pi}^{\pi} P_r(\theta_r - t)\mu'(0)\,dt$$

is equal to $\mu'(0)$.

Let $\delta > 0$ be such that $|\mu(t)| \leq \epsilon|t|$ for $|t| \leq \delta$. If $1-r$ is very close to 0, so that $2|\theta_r|$ is much smaller than δ,

$$\frac{1}{2\pi}\int_{-\pi}^{\pi} \frac{1-r^2}{1+r^2-2r\cos(t-\theta_r)}\,d\mu(t) = o(1) + \frac{1}{2\pi}\int_{-\delta}^{\delta} \frac{1-r^2}{1+r^2-2r\cos(t-\theta_r)}\,d\mu(t)$$

* Strictly speaking, such a region is not a sector with *straight* sides and vertex at $e^{i\phi_0}$. But it *becomes* asymptotic to such a sector – of opening $< 180°$ – near the point $e^{i\phi_0}$.

where $o(1) \to 0$ as $r \to 1$. Integrate

$$\frac{1}{2\pi}\int_{-\delta}^{\delta}\frac{1-r^2}{1+r^2-2r\cos(t-\theta_r)}\,d\mu(t)$$

by parts to get an integrated term (is $o(1)$) plus

$$\frac{1}{2\pi}\int_{-\delta}^{\delta}\frac{2(1-r^2)r\sin(t-\theta_r)}{(1+r^2-2r\cos(t-\theta_r))^2}\mu(t)\,dt.$$

Assuming, without loss of generality, that $\theta_r > 0$, we break up the last integral as

$$\frac{1}{2\pi}\left[\int_{-\delta}^{0}+\int_{0}^{2\theta_r}+\int_{2\theta_r}^{\delta}\right]\frac{2r(1-r^2)\sin(t-\theta_r)\mu(t)}{(1+r^2-2r\cos(t-\theta_r))^2}\,dt = \mathrm{I}+\mathrm{II}+\mathrm{III},$$

say. Then

$$|\mathrm{II}| \le \frac{1}{2\pi}\int_{0}^{2\theta_r}\frac{4\theta_r}{(1-r)^3}\cdot\epsilon t\,dt \le \frac{4\epsilon\theta_r^3}{\pi(1-r)^3} \le \frac{4}{\pi}c^3\epsilon,$$

since $0 \le \theta_r \le c(1-r)$. For $2\theta_r \le t \le \delta$, $|\mu(t)| \le \epsilon t \le 2\epsilon(t-\theta_r)$, so

$$|\mathrm{III}| \le \frac{\epsilon}{\pi}\int_{2\theta_r}^{\delta}\frac{2(1-r^2)r\sin(t-\theta_r)}{(1+r^2-2r\cos(t-\theta_r))^2}(t-\theta_r)\,dt = \frac{\epsilon}{\pi}\int_{\theta_r}^{\delta-\theta_r}\frac{2r(1-r^2)\sin t}{(1+r^2-2r\cos t)^2}t\,dt$$

$$\le \frac{\epsilon}{\pi}\int_{0}^{\pi}\frac{2r(1-r^2)\sin t}{(1+r^2-2r\cos t)^2}t\,dt.$$

This last is integrated by parts (in the *opposite direction* to our first integration by parts!) to give

$$\frac{\epsilon}{\pi}\left[o(1)+\int_{0}^{\pi}\frac{(1-r^2)\,dt}{1+r^2-2r\cos t}\right] = \epsilon + o(1).$$

Similarly $|\mathrm{I}| \le \epsilon/2+o(1)$. So $|\mathrm{I}+\mathrm{II}+\mathrm{III}| \le (4c^3/\pi+3/2)\epsilon+o(1)$ as $r \to 1$, and since $\epsilon > 0$ is arbitrary, we are done.

Remark What makes the above proof work is the monotoneity of $P_r(\theta)$ on each of the intervals $[-\pi, 0]$ and $[0, \pi]$.

Theorem (Fatou) *If $-\pi < \phi_0 < \pi$ and if $\mu'(\phi_0)$ exists and is infinite, then, for*

$$U(re^{i\theta}) = \frac{1}{2\pi}\int_{-\pi}^{\pi}P_r(\theta-t)\,d\mu(t),$$

$U(re^{i\phi_0}) \to \mu'(\phi_0)$ as $r \to 1$.

Remark Thus, even when $\mu'(\phi_0)$ is infinite, we still have

$$U(z) \to \mu'(\phi_0)$$

when z goes out radially to $e^{i\phi_0}$.

Proof Take $\phi_0 = 0$ and assume $\mu'(0) = +\infty$. Choose $\delta > 0$ so small that $\mu(t)\operatorname{sgn} t \geq M|t|$ for $|t| \leq \delta$. Then, computing as in the proof of the preceding theorem,

$$U(r) = o(1) + \frac{1}{2\pi}\int_{-\delta}^{\delta} \frac{2(1-r)r\sin t}{(1+r^2-2r\cos t)^2}\mu(t)\,dt \geq o(1) + \frac{1}{\pi}M\int_{0}^{\delta} \frac{2(1-r)r\sin t}{(1+r^2-2r\cos t)^2}t\,dt.$$

Doing the reverse integration by parts, this last is seen to be $o(1) + M$ for r close enough to 1.

Scholium Can we replace '$z \to e^{i\phi_0}$ radially' by '$z \underset{\angle}{\longrightarrow} e^{i\phi_0}$' in the theorem just proved? We *can*, if $U(z) \geq 0$ for $|z| < 1$, i.e. if μ is a *positive* measure.

Lemma (Harnack's Theorem) *Let $U(z)$ be positive and harmonic in $|z| < 1$. Then*

$$U(z) \geq \frac{1-|z|}{1+|z|}U(0).$$

Proof

$$U(re^{i\theta}) = \frac{1}{2\pi}\int_{-\pi}^{\pi} \frac{1-r^2}{1+r^2-2\cos(\theta-t)}d\mu(t) \geq \frac{1-r}{1+r}\cdot\frac{1}{2\pi}\int_{-\pi}^{\pi} d\mu(t) = \frac{1-r}{1+r}U(0),$$

since here $d\mu(t) \geq 0$.

Now, assume $U(z) \geq 0$ and $U(r) \to \infty$ for $r \to 1$. If S is any sector with vertex at 1, symmetric about the positive real axis, of opening $< 180°$, let S' be a similar but *slightly larger* sector:

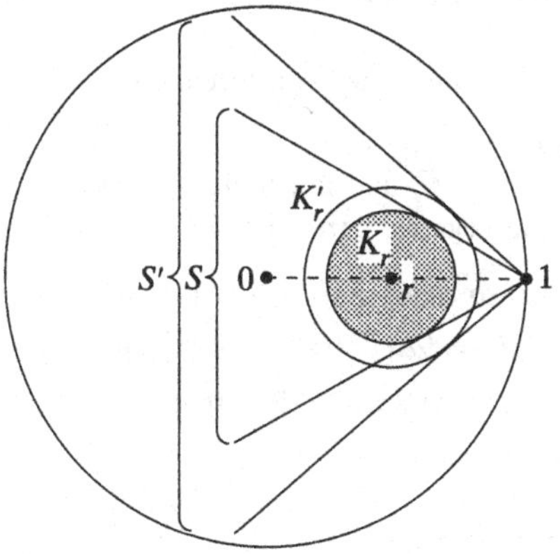

If $0 < r < 1$, let K_r be the circle about r tangent to the sides of S, and K_r' be the circle about r tangent to the sides of S'. Clearly γ = radius of K_r/radius of K_r' is *independent* of r and < 1. Then, $U(z)$ is positive in K_r', so, if z is in K_r', by the lemma,

$$U(z) \geq \frac{1-\gamma}{1+\gamma}U(r),$$

proving $U(z) \to \infty$ as $z \to 1$ from within S.

However, a complete generalization of the second Fatou theorem for non-tangential movement of z out to the boundary is false.

That is the content of the following problem.

In order to simplify the computations, we have made a change of variable corresponding to the conformal representation

$$w = re^{i\theta} = \frac{i-z}{i+z}, \quad z = x + iy,$$

which makes the upper half plane $y > 0$ correspond to the circle $|w| < 1$ and $z = 0$ correspond to $w = 1$.

Then, if $e^{i\tau} = (i-t)/(i+t)$, it is easily checked (see pp. 106–108) that

$$\frac{1-r^2}{1+r^2-2\cos(\theta-\tau)}\, d\nu(\tau) = \frac{y}{(x-t)^2+y^2}\, d\mu(t),$$

where

$$\frac{2\, d\mu(t)}{1+t^2} = d\nu(\tau).$$

Problem 1 To construct a signed measure μ on $[0,1]$ such that $\mu'(0+) = \infty$, but if

$$U(z) = \int_0^1 \frac{y}{(x-t)^2+y^2}\, d\mu(t),$$

$U(x+ix)$ *does not* tend to ∞ as $x \to 0+$.

Procedure Obtain inductively the positive numbers $t_0 = 1$, $x_0 = 1/2$, $t_1 < x_0$, $x_1 < t_1$, $t_2 < x_1$, etc., and construct μ on each of the intervals $[t_1, t_0]$, $[t_2, t_1)$, $[t_3, t_2)$, ... one after the other.

μ is to be a discrete measure, and when constructed will satisfy

(1) $\mu(t) \geq \sqrt{t}$, $0 \leq t \leq 1$;
(2) $\mu_+([t_k, t_{k-1})) = 9(\sqrt{t_{k-1}} - \sqrt{t_k})$ and $\mu_-([t_k, t_{k-1})) = 8(\sqrt{t_{k-1}} - \sqrt{t_k})$, where μ_+ and μ_- denote the positive and negative parts of μ.

Start by taking $t_1 < x_0/10$ so small that $9\sqrt{t_1} < 1 - \sqrt{t_1} = \Delta_1$, say, and, on $[t_1, t_0]$, put

$$d\mu(t) = 9\Delta_1\, d\delta(t-t_1) - 8\Delta_1\, d\delta(t-x_0),$$

where δ denotes the unit point mass at 0. *Show that*, if, at the end of the construction, μ satisfies (2), then

$$\int_0^1 \frac{x_0}{(x_0-t)^2+x_0^2}\, d\mu(t) \leq -\frac{3\Delta_1}{x_0} \leq -\frac{3}{2\sqrt{x_0}}.$$

Next, take $x_1 < t_1$ so small that

$$\int_{t_1}^1 \frac{x_1^2}{(t-x_1)^2+x_1^2}\, d\mu(t) \leq \sqrt{t_1} - \sqrt{x_1}$$

(on $(t_1, 1]$, μ is already constructed!). Assuming that μ will satisfy (2) at the end of the construction, *how should we choose* $t_2 < x_1$, and *how should we define* μ on $[t_2, t_1)$ so as to ensure

$$\int_0^1 \frac{x_1}{(x_1-t)^2+x_1^2}\,d\mu(t) \leq -\frac{2\Delta_2}{x_1} \leq -\frac{1}{\sqrt{x_1}},$$

writing $\Delta_2 = \sqrt{t_1} - \sqrt{t_2}$?

Show how the construction can be carried out so as to get a measure μ on $[0, 1]$ satisfying (1) and (2) with $\mu'(0+) = \infty$ but

$$\int_0^1 \frac{x_k}{(t-x_k)^2+x_k^2}\,d\mu(t) \leq -\frac{1}{\sqrt{x_k}}.$$

Let $F \in L_p[-\pi, \pi]$, $p \geq 1$ and let

$$U(re^{i\theta}) = \frac{1}{2\pi}\int_{-\pi}^{\pi} P_r(\theta - t)F(t)\,dt.$$

A classical theorem of Lebesgue* says that

$$\frac{d}{d\theta}\int_0^{\theta} F(t)\,dt \quad \textit{exists a.e. and equals } F(\theta).$$

In conjunction with the first theorem of this subsection we thus see that, a.e. in ϕ,

$$U(re^{i\theta}) \to F(\phi) \quad \text{as } re^{i\theta} \underset{\angle}{\longrightarrow} e^{i\phi}.$$

Combined with a theorem of Section C, we thus have

Theorem *Let $1 < p \leq \infty$ (sic!) and let $U(z)$ be harmonic in $|z| < 1$, with*

$$\left(\int_{-\pi}^{\pi} |U(re^{i\theta})|^p\,d\theta\right)^{1/p} \leq C$$

for $0 \leq r < 1$. Then, for almost all θ, $U(z)$ tends to a finite limit, say $U(e^{i\theta})$, as $z \underset{\angle}{\longrightarrow} e^{i\theta}$, $U(e^{i\theta}) \in L_p(-\pi, \pi)$, and, for $0 \leq r < 1$,

$$U(e^{i\theta}) = \frac{1}{2\pi}\int_{-\pi}^{\pi} \frac{1-r^2}{1+r^2-2r\cos(\theta-t)}U(e^{it})\,dt.$$

Notation $U(e^{i\theta})$ is called the (*non-tangential* or $\angle$) *boundary value function* for $U(z)$; we frequently write

$$U(e^{i\theta}) = \lim_{z \underset{\angle}{\rightarrow} e^{i\theta}} U(z) \quad \text{a.e.}.$$

In future, whenever we have a function U harmonic in $|z| < 1$, satisfying the hypothesis of

---

* For a *proof*, see the scholium at the end of Subsection B.1 in Chapter VIII.

the above theorem (for $p > 1$), *we assume it to be automatically extended* a.e. *to* $|z| = 1$ *in the manner described*.

In case $p = 1$ the theorem is not completely true. In that case we have a *measure* $d\mu(\theta)$. A decomposition theorem of Lebesgue says that then the derivative $\mu'(\theta)$ *still* exists and is finite a.e., that $\mu'(\theta) \in L_1(-\pi, \pi)$, but that $d\mu(\theta)$ is not in general $\mu'(\theta)\,d\theta$. Instead,

$$d\mu(\theta) = \mu'(\theta)\,d\theta + d\sigma(\theta)$$

where σ is a singular measure, i.e., one supported on a set of Lebesgue measure zero.

Thus, if we *only know* that the means

$$\int_{-\pi}^{\pi} |U(re^{i\theta})|\,d\theta$$

are bounded for $r < 1$, we *still* have a.e. existence of the finite non-tangential

$$\lim_{z \underset{\angle}{\to} e^{i\theta}} U(z) = \mu'(\theta),$$

but we cannot recover $U(z)$ *from this boundary value function*. Instead, we have

$$U(re^{i\theta}) = \frac{1}{2\pi}\int_{-\pi}^{\pi} P_r(\theta - t)\mu'(t)\,dt + \frac{1}{2\pi}\int_{-\pi}^{\pi} P_r(\theta - t)\,d\sigma(t)$$

with some *singular measure* σ.

The simplest cases show that a representation with nonzero σ can actually occur; one such is the *ordinary Poisson kernel*

$$U(re^{i\theta}) = \frac{1 - r^2}{1 + r^2 - 2r\cos\theta}\ !$$

Indeed,

$$\lim_{z \underset{\angle}{\to} e^{i\theta}} U(z) = 0$$

save for $\theta = 0$, and

$$U(re^{i\theta}) = \frac{1}{2\pi}\int_{-\pi}^{\pi} P_r(\theta - t)\cdot 2\pi\,d\delta_0(t)$$

where δ_0 is the unit point mass at 0.

This *distinction between* the cases $p = 1$ and $p > 1$ is one of the *fundamental complications of the theory*, and will be seen to have deep and far-reaching implications in its further development.

E. The harmonic conjugate

Given a function $U(z)$ harmonic in $|z| < 1$, having one of the representations studied in this chapter, we proceed to investigate the pointwise boundary behaviour of its harmonic conjugate. At the beginning of this chapter, we said that a harmonic function $V(z)$ is a *harmonic conjugate* of $U(z)$ if $U(z) + iV(z)$ is analytic in $|z| < 1$. Harmonic conjugates are only defined to within an additive constant; working in the unit circle, it is customary to require $V(0) = 0$; *the resulting harmonic conjugate* $V(z)$ *of* U *is denoted by* $\tilde{U}(z)$. The tilde notation is customary in the designation of harmonic conjugates.

1. Formula for the harmonic conjugate

Suppose that

$$U(re^{i\theta}) = \sum_{-\infty}^{\infty} A_n r^{|n|} e^{in\theta}, \quad 0 \le r < 1,$$

then

$$\tilde{U}(re^{i\theta}) = -\sum_{-\infty}^{\infty} i \operatorname{sgn} n\, A_n r^{|n|} e^{in\theta},$$

where $\operatorname{sgn} 0$ means 0.

Indeed, $\tilde{U}(re^{i\theta})$ is harmonic in $|z| < 1$ *by inspection* (the series converges absolutely there), and $\tilde{U}(0) = 0$. And also

$$U(re^{i\theta}) + i\tilde{U}(re^{i\theta}) = A_0 + \sum_{1}^{\infty} 2A_n r^n e^{in\theta}$$

is *analytic* (by inspection!) in $|z| < 1$.

Now if

$$U(re^{i\theta}) = \frac{1}{2\pi}\int_{-\pi}^{\pi} P_r(\theta - t)\, d\mu(t)$$

with a measure μ on $[-\pi, \pi]$, then the above series development for U is valid with coefficients

$$A_n = \frac{1}{2\pi}\int_{-\pi}^{\pi} e^{-int}\, d\mu(t).$$

Looking at the series development for $\tilde{U}$, we see that

$$\tilde{U}(re^{i\theta}) = -\frac{1}{2\pi}\int_{-\pi}^{\pi}\sum_{-\infty}^{\infty} i \operatorname{sgn} n\, r^{|n|} e^{in(\theta - t)}\, d\mu(t).$$

Call

$$-\sum_{-\infty}^{\infty} i \operatorname{sgn} n\, r^{|n|} e^{in\theta} = Q_r(\theta)$$

the *conjugate Poisson kernel*. By direct summation of two geometric series, we find

$$Q_r(\theta) = \frac{2r\sin\theta}{1 + r^2 - 2r\cos\theta}.$$

Thus:

Theorem *If*

$$U(re^{i\theta}) = \frac{1}{2\pi}\int_{-\pi}^{\pi} \frac{1 - r^2}{1 + r^2 - 2r\cos(\theta - t)}\, d\mu(t)$$

with a measure μ, then the harmonic conjugate $\tilde{U}$ of U is given by

$$\tilde{U}(re^{i\theta}) = \frac{1}{2\pi}\int_{-\pi}^{\pi} \frac{2r\sin(\theta - t)}{1 + r^2 - 2r\cos(\theta - t)}\, d\mu(t).$$

2. Harmonic conjugate near an arc where original boundary function has a continuous derivative

We are especially interested in the boundary behaviour of $\tilde{U}(z)$ when

$$U(re^{i\theta}) = \frac{1}{2\pi}\int_{-\pi}^{\pi} P_r(\theta - t)F(t)\, dt$$

with a *function* F (belonging, say, to $L_p(-\pi,\pi)$, $p \geq 1$).

Assuming, as usual the definition of F to be extended to $\mathbb{R}$ so as to make F 2π-periodic, we have

$$\tilde{U}(re^{i\theta}) = \frac{1}{2\pi}\int_{-\pi}^{\pi} \frac{2r\sin(\theta - t)}{1 + r^2 - 2r\cos(\theta - t)} F(t)\, dt$$
$$= \frac{1}{\pi}\int_{0}^{\pi} \frac{r\sin s}{1 + r^2 - 2r\cos s}(F(\theta - s) - F(\theta + s))\, ds.$$

We have

$$\frac{r\sin s}{1 + r^2 - 2r\cos s} = \frac{2r\sin(s/2)\cos(s/2)}{(1-r)^2 + 4r\sin^2(s/2)},$$

so if

$$\int_{0}^{\pi} \frac{|F(\theta - s) - F(\theta + s)|}{s}\, ds < \infty,$$

we clearly have, for $r \to 1$,

$$\tilde{U}(re^{i\theta}) \to \frac{1}{2\pi}\int_{0}^{\pi} \frac{F(\theta - s) - F(\theta + s)}{\tan(s/2)}\, ds,$$

the integral on the right being absolutely convergent.

This certainly happens if $F'(\theta)$ exists and is finite.

Indeed, *if $F'(\theta)$ is continuous for $\alpha < \theta < \beta$, say, then $\tilde{U}(z)$ has a continuous extension up to any closed subarc of the open arc $\{e^{i\theta}; \alpha < \theta < \beta\}$ on $|z| = 1$, and for such θ,*

$$\tilde{U}(e^{i\theta}) = \lim_{r \to 1} \tilde{U}(re^{i\theta}) = \frac{1}{\pi}\int_{0}^{\pi} \frac{F(\theta - t) - F(\theta + t)}{2\tan(t/2)}\, dt,$$

the integral being absolutely convergent.

Problem 2 Prove the statement just made. (Hint: The mean value theorem for derivatives comes in here.) Also prove the following:

If $F \geq 0$ is 2π-periodic and continuous as a mapping into $[0, \infty]$ (that is, F is allowed to *take the value* ∞, *but if* $F(\theta_0)=\infty$, *then* $F(\theta)\to\infty$ *for* $\theta\to\theta_0$), *and if* $F\in L_1(-\pi,\pi)$, *then*

$$U(re^{i\theta})=\frac{1}{2\pi}\int_{-\pi}^{\pi}P_r(\theta-t)F(t)\,dt$$

extends continuously up to $|z|=1$ *as a mapping to* $[0,\infty]$.

3. Behaviour near points of original boundary function's Lebesgue set
It turns out that the radial boundary value

$$\tilde{U}(e^{i\theta})=\frac{1}{\pi}\int_{0}^{\pi}\frac{F(\theta-t)-F(\theta+t)}{2\tan(t/2)}\,dt$$

of

$$\tilde{U}(re^{i\theta})=\frac{1}{2\pi}\int_{-\pi}^{\pi}Q_r(\theta-t)F(t)\,dt$$

exists a.e. in θ for *very general* functions F. No differentiability is really required for F. Of course, the integral

$$\int_{0}^{\pi}\frac{F(\theta-t)-F(\theta+t)}{2\tan(t/2)}\,dt$$

must be interpreted properly. Just taking it as the limit of

$$\int_{\epsilon}^{\pi}\frac{F(\theta-t)-F(\theta+t)}{2\tan(t/2)}\,dt$$

for $\epsilon\to 0+$ will, as we shall see, be enough.

Definition Let F be 2π-periodic, and in $L_1(-\pi,\pi)$. We say θ is *in the Lebesgue set for* F if

$$\frac{1}{h}\int_{-h}^{h}|F(\theta+t)-F(\theta)|\,dt\to 0$$

as $h\to 0$.

Theorem (Lebesgue!) *Almost every* θ *is in the Lebesgue set for* F.

Proof Given any rational number r, the function $|F(t)-r|$ is in L_1, hence equal to the derivative of its indefinite integral a.e. That is, for almost all θ,

$$\lim_{h\to 0}\frac{1}{h}\int_{\theta}^{\theta+h}|F(t)-r|\,dt=|F(\theta)-r|$$

holds *simultaneously for all rational* r as long as $\theta\notin E$, say, E being a set of measure zero.
Let $\theta\notin E$ and $\epsilon>0$ be given; if r is a rational number with $|F(\theta)-r|<\epsilon$, then

$$\frac{1}{h}\int_{0}^{h}|F(\theta+t)-F(\theta)|\,dt\le\frac{1}{h}\int_{\theta}^{\theta+h}|F(t)-r|\,dt+|F(\theta)-r|\le\epsilon+\frac{1}{h}\int_{\theta}^{\theta+h}|F(t)-r|\,dt.$$

The thing on the right tends to $\epsilon + |F(\theta) - r| \le 2\epsilon$ as $h \to 0$, so

$$\limsup_{h\to 0} \frac{1}{h}\int_0^h |F(\theta+t) - F(\theta)|\,dt \le 2\epsilon$$

if $\theta \notin E$. Since ϵ is arbitrary > 0,

$$\frac{1}{h}\int_0^h |F(\theta+t) - F(\theta)|\,dt \to 0$$

as $h \to 0$, and this holds for all $\theta \notin E$, i.e., for almost all θ.

Now we have the basic

Theorem *Let $F(t+2\pi) = F(t)$, $F \in L_1(-\pi,\pi)$. Then*

$$\frac{1}{\pi}\int_{-\pi}^{\pi} \frac{r\sin t}{1+r^2-2r\cos t}F(\theta-t)\,dt - \frac{1}{\pi}\int_{1-r}^{\pi} \frac{F(\theta-t)-F(\theta+t)}{2\tan(t/2)}\,dt$$

tends to zero as $r \to 1$ for all θ in the Lebesgue set for F, i.e., almost everywhere.

Remark The idea is that

$$\frac{1}{\pi}\int_{\epsilon}^{\pi} \frac{F(\theta-t)-F(\theta+t)}{2\tan(t/2)}\,dt$$

can, for almost all θ, be compared with $\tilde{U}((1-\epsilon)e^{i\theta})$ – a value of the harmonic conjugate inside the unit circle – when ϵ is small.

Proof $r\sin t/(1+r^2-2r\cos t)$ is an odd function of t, so the difference in question *is unchanged if* F is everywhere replaced by $F - F(\theta)$.

Do this replacement. Then the difference breaks up into two parts, of which the first is

$$\mathrm{I} = \frac{1}{\pi}\int_{-(1-r)}^{1-r} \frac{r\sin t}{1+r^2-2r\cos t}(F(\theta-t)-F(\theta))\,dt.$$

We have

$$\left|\frac{r\sin t}{1+r^2-2r\cos t}\right| \le \frac{|\sin t|}{(1-r)^2} \le \frac{1}{1-r}$$

for $|t| \le 1-r$, therefore

$$|\mathrm{I}| \le \frac{1}{\pi(1-r)}\int_{-(1-r)}^{1-r} |F(\theta-t)-F(\theta)|\,dt\,,$$

which $\to 0$ as $r \to 1$ if θ is in the Lebesgue set for F.

It is convenient to write $\Delta = 1-r$. Then the rest of our difference is

$$\mathrm{II} = \frac{1}{\pi}\int_{\Delta\le|t|\le\pi} \left\{\frac{r\sin t}{\Delta^2+4r\sin^2(t/2)} - \frac{\sin t}{4\sin^2(t/2)}\right\}(F(\theta-t)-F(\theta))\,dt.$$

The expression in braces works out to

$$\frac{-\Delta^2\sin t}{4\left(\Delta^2+4r\sin^2(t/2)\right)\sin^2(t/2)}$$

which, for $r \geq 1/2$, say, is in absolute value $\leq C(1-r)^2/|t|^3$, C being a numerical constant.

Thus,

$$|\mathrm{II}| \leq \frac{C\Delta^2}{\pi} \int\limits_{\Delta \leq |t| \leq \pi} \frac{|F(\theta - t) - F(\theta)|}{|t|^3} \, dt.$$

To evaluate, for instance,

$$\Delta^2 \int_{\Delta}^{\pi} \frac{|F(\theta - t) - F(\theta)|}{t^3} \, dt,$$

integrate by parts, getting

$$\frac{\Delta^2}{\pi^3} \int_0^{\pi} |F(\theta - t) - F(\theta)| \, dt - \frac{1}{\Delta} \int_0^{\Delta} |F(\theta - t) - F(\theta)| \, dt +$$

$$+ 3\Delta^2 \int_{\Delta}^{\pi} \frac{\int_0^s |F(\theta - t) - F(\theta)| \, dt}{s^4} \, ds.$$

The first two terms obviously tend to 0 as $\Delta \to 0$, θ being in the Lebesgue set for F.

Given $\epsilon > 0$, pick a *fixed* $\eta > 0$ such that

$$\frac{1}{s} \int_0^s |F(\theta - t) - F(\theta)| \, dt < \epsilon$$

for $0 < s < \eta$; the last term is then

$$\leq 3\Delta^2 \int_{\Delta}^{\eta} \frac{\epsilon}{s^3} \, ds + 3\Delta^2 \int_{\eta}^{\pi} \frac{1}{\eta^4} \int_0^s |F(\theta - t) - F(\theta)| \, dt \, ds$$

$$< \frac{2\Delta^2}{\Delta^2}\epsilon + \frac{3\pi\Delta^2}{\eta^4} \int_0^{\pi} |F(\theta - t) - F(\theta)| \, dt$$

$$= 2\epsilon + \Delta^2 \frac{M}{\eta^4}$$

when $0 < \Delta < \eta$. This is $< 4\epsilon$ if Δ is small enough, so, since ϵ is arbitrary, we see that

$$\Delta^2 \int_{\Delta}^{\pi} \frac{|F(\theta - t) - F(\theta)|}{t^3} \, dt,$$

and hence II, tends to 0 as $\Delta \to 0$.

The theorem is completely proved.

4. Existence of $\tilde{f}(\theta)$ for f in L_2. Discussion

Now we shall use material from Sections C and D together with the comparison theorem of Subsection 3 above in order to show that

$$\lim_{\epsilon \to 0+} \int_{\epsilon}^{\pi} \frac{f(\theta - t) - f(\theta + t)}{2\tan(t/2)} \, dt$$

exists a.e. whenever $f \in L_2(-\pi, \pi)$! This is the first really *deep* theorem in the whole chapter.

First, a quick remark. The relations

$$\frac{1}{2\pi}\int_{-\pi}^{\pi} e^{in\theta}e^{-im\theta}\,d\theta = \begin{cases} 0, & n \neq m \\ 1, & n = m \end{cases}$$

show, together with absolute convergence, that if

$$U(re^{i\theta}) = \sum_{-\infty}^{\infty} A_n r^{|n|} e^{in\theta}$$

is harmonic in $|z| < 1$,

$$\int_{-\pi}^{\pi} |U(re^{i\theta})|^2\,d\theta = 2\pi \sum_{-\infty}^{\infty} |A_n|^2 r^{2|n|}$$

for $0 \leq r < 1$. From this relation together with material of Sections C, D we immediately get the

Lemma *If $U(z)$ is harmonic in $|z| < 1$, one has*

$$U(re^{i\theta}) = \frac{1}{2\pi}\int_{-\pi}^{\pi} \frac{1 - r^2}{1 + r^2 - 2r\cos(\theta - t)} F(t)\,dt$$

with $F \in L_2(-\pi, \pi)$ iff

$$U(re^{i\theta}) = \sum_{-\infty}^{\infty} A_n r^{|n|} e^{in\theta}$$

with

$$\sum_{-\infty}^{\infty} |A_n|^2 < \infty.$$

Theorem *Let $F \in L_2(-\pi, \pi)$ be 2π-periodic. Then*

$$\tilde{F}(\theta) = \lim_{\epsilon \to 0} \frac{1}{\pi}\int_{\epsilon}^{\pi} \frac{F(\theta - t) - F(\theta + t)}{2\tan(t/2)}\,dt$$

exists for almost all θ, $\tilde{F} \in L_2(-\pi, \pi)$, and $\|\tilde{F}\|_2 \leq \|F\|_2$.

If

$$U(re^{i\theta}) = \frac{1}{2\pi}\int_{-\pi}^{\pi} P_r(\theta - t)F(t)\,dt,$$

then

$$\tilde{U}(re^{i\theta}) = \frac{1}{2\pi}\int_{-\pi}^{\pi} Q_r(\theta - t)F(t)\,dt$$

is also equal to

$$\frac{1}{2\pi}\int_{-\pi}^{\pi} P_r(\theta - t)\tilde{F}(t)\,dt.$$

Proof By the lemma,

$$U(re^{i\theta}) = \sum_{-\infty}^{\infty} A_n r^{|n|} e^{in\theta}$$

with

$$\sum_{-\infty}^{\infty} |A_n|^2 < \infty,$$

so therefore, *again* by the lemma,

$$\tilde{U}(re^{i\theta}) = -i \sum_{-\infty}^{\infty} \operatorname{sgn} n\, A_n r^{|n|} e^{in\theta}$$

must in fact equal

$$\frac{1}{2\pi} \int_{-\pi}^{\pi} P_r(\theta - t) G(t)\, dt$$

for some $G \in L_2(-\pi, \pi)$.

By Section D,

$$\tilde{U}(z) \to G(\theta) \quad \text{as } z \underset{L}{\longrightarrow} e^{i\theta}$$

for almost all θ, in particular,

$$\lim_{r\to 1} \tilde{U}(re^{i\theta}) = G(\theta) \quad \text{a.e.}$$

By the last theorem of Subsection 3 we now see that

$$\frac{1}{\pi} \int_{1-r}^{\pi} \frac{F(\theta - t) - F(\theta + t)}{2\tan(t/2)}\, dt$$

must also tend to $G(\theta)$ for almost all θ as $r \to 1$. So most of the theorem holds already if we just write $\tilde{F}(\theta) = G(\theta)$!

It remains to check the norm inequalities. But that's easy. By work in Section D,

$$\|F\|_2^2 = \lim_{r\to 1} \int_{-\pi}^{\pi} |U(re^{i\theta})|^2\, d\theta$$

which equals $2\pi \sum_{-\infty}^{\infty} |A_n|^2$ according to the computation done above; at the same time,

$$\|\tilde{F}\|_2^2 = \lim_{r\to 1} \int_{-\pi}^{\pi} |\tilde{U}(re^{i\theta})|^2\, d\theta = 2\pi \sum_{n\neq 0} |A_n|^2$$

(remember $\operatorname{sgn} 0 = 0$!) That does it.

Discussion Where it exists, the limit

$$\tilde{F}(\theta) = \lim_{\epsilon\to 0} \frac{1}{\pi} \int_{\varepsilon}^{\pi} \frac{F(\theta - t) - F(\theta + t)}{2\tan(t/2)}\, dt$$

is the same as

$$\lim_{\epsilon\to 0} \frac{1}{\pi} \left\{ \int_{-\pi}^{\theta-\epsilon} + \int_{\theta+\epsilon}^{\pi} \right\} \frac{F(t)\, dt}{2\tan((\theta - t)/2)}.$$

This exhibits it as a *Cauchy principal value*. We frequently write

$$\tilde{F}(\theta) = \frac{1}{\pi} \int_{-\pi}^{\pi} \frac{F(t)\,dt}{2\tan((\theta - t)/2)}$$

to emphasize that the expression is *evaluated by omitting a small interval having θ as its midpoint, and then having the width of that interval shrink to zero. The symmetry is crucial here;*

$$\frac{1}{\pi} \int_{-\pi}^{\pi} \frac{F(t)\,dt}{2\tan((\theta - t)/2)}$$

is usually not an integral in any ordinary sense.

We also write

$$\tilde{F}(\theta) = \frac{1}{\pi} \int_{0+}^{\pi} \frac{F(\theta - t) - F(\theta + t)}{2\tan(t/2)}\,dt.$$

Although $\tilde{F}(\theta)$ is obtained from F by such a *delicate* limiting process (the *symmetry* of the omitted interval which shrinks down to θ !) *see how strongly it is bound to F, in the metric sense*! For $\tilde{F}$ depends linearly on F, and $\|\tilde{F}\|_2 \leq \|F\|_2$!

Taken by themselves, the statements about $\tilde{F}(\theta)$ constitute a *purely real variable result*. Yet we used a lot of complex function theory (harmonic functions and their conjugates) to establish it, for we made the passage

F	$\rightarrow$	$U(z)$	$\rightarrow$	$\tilde{U}(z)$	$\rightarrow$	$\tilde{F}$
		harmonic in unit circle		harmonic in unit circle		boundary function

during its proof. And we used a differentiability property of something related to F in order to compare $\tilde{U}((1-\epsilon)e^{i\theta})$ with

$$\frac{1}{\pi} \int_{\epsilon}^{\pi} \frac{F(\theta - t) - F(\theta + t)}{2\tan(t/2)}\,dt.$$

Altogether, a most intricate business. At the beginning of this century, Lusin wanted to find a more direct proof – one which didn't bring in complex variable theory. He thought that doing this would lay bare more of the real mechanism of the interference of translates of functions on the real line which must be operating in order to make

$$\int_{0+}^{\pi} \frac{F(\theta - t) - F(\theta + t)}{2\tan(t/2)}\,dt$$

exist a.e. For it is indeed a real process of interference which is taking place here (see next subsection!).

The nth partial sum of the Fourier series of $F(\theta)$ is essentially given by

$$\int_{-\pi}^{\pi} F(\theta - t) \frac{\sin(nt)}{t}\,dt$$

– an expression very like the one for $\tilde{F}(\theta)$, except for the factor $\sin(nt)$ in the integrand.

Lusin thought that if one could *really see why* the limit defining $\tilde{F}(0)$ exists a.e., one *might stand a chance* of *proving* that the *Fourier series* of an L_2 function *converges* a.e. In a way, he was right. Carleson's celebrated proof of this convergence (published in 1966) depends greatly on delicate properties of the operation taking F to $\tilde{F}$.

Long before Carleson did this, real variable proofs of the existence of $\tilde{F}(\theta)$ *were* found. Lusin himself obtained one. They are *harder* than the classical one given above.

5. Existence of $\tilde{f}(\theta)$ really due to cancellation

It was said in the last subsection that the existence a.e. of the limit

$$\tilde{f}(\theta) = \frac{1}{\pi}\int_{0+}^{\pi} \frac{f(\theta - t) - f(\theta + t)}{2\tan(t/2)}\,dt$$

is *deep*, and comes from some complicated interference phenomenon. The existence of the limit really comes from *cancellation of positive and negative contributions*, and not from the smallness of $|f(\theta - t) - f(\theta + t)|$, *even when f is continuous*. This is shown by the following

Theorem *There exists a continuous function G, periodic of period 2π, such that*

$$\int_0^{\pi} \frac{|G(\theta + t) - G(\theta - t)|}{t}\,dt = \infty$$

for every θ.

Proof By contradiction, using the Baire category theorem.

Denote by $\mathscr{C}$ the space of continuous functions of period 2π. Using the usual norm $\|f\| = \sup_\theta |f(\theta)|$, $\mathscr{C}$ becomes a *complete* normed space.

Assume that for every $f \in \mathscr{C}$ there is *at least one* θ with

$$\int_0^{\pi} \frac{|f(\theta + t) - f(\theta - t)|}{t}\,dt < \infty;$$

we shall arrive at a contradiction. For each $n = 1, 2, 3, \ldots$ let

$$E_n = \left\{ f \in \mathscr{C};\ \text{for at least one } \theta,\ \int_0^{\pi} \frac{|f(\theta + t) - f(\theta - t)|}{t}\,dt \le n \right\}.$$

By hypothesis, $\bigcup_1^\infty E_n = \mathscr{C}$. Now each E_n is *closed* in the norm topology of $\mathscr{C}$. Indeed, suppose $f_k \in E_n$ and $\|f_k - f\| \xrightarrow[k]{} 0$ with $f \in \mathscr{C}$. We prove that $f \in E_n$. For each f_k there is a θ_k with

$$\int_0^{\pi} \frac{|f_k(\theta_k + t) - f_k(\theta_k - t)|}{t}\,dt \le n.$$

By 2π-periodicity, we may take $0 \le \theta_k \le 2\pi$, and then there is no loss of generality in assuming $\theta_k \xrightarrow[k]{} \theta$, $0 \le \theta \le 2\pi$ (otherwise just go to a subsequence). We have

$$\begin{aligned} &|(f(\theta + t) - f(\theta - t)) - (f_k(\theta_k + t) - f_k(\theta_k - t))| \\ &\quad \le |(f(\theta_k + t) - f(\theta_k - t)) - (f_k(\theta_k + t) - f_k(\theta_k - t))| \\ &\qquad + |f(\theta + t) - f(\theta_k + t) - f(\theta - t) + f(\theta_k + t)| \end{aligned}$$

which $\xrightarrow[k]{} 0$ uniformly in t, since $f \in \mathscr{C}$. In particular,

$$|f(\theta + t) - f(\theta - t)| = \lim_{k\to\infty} |f_k(\theta_k + t) - f_k(\theta_k - t)|$$

for all t, so, by Fatou's Lemma,

$$\int_0^\pi \frac{|f(\theta+t)-f(\theta-t)|}{t}\,dt \le \liminf_{k\to\infty} \int_0^\pi \frac{|f_k(\theta_k+t)-f_k(\theta_k-t)|}{t}\,dt$$

which is $\le n$. So $f \in E_n$.

Because $\bigcup_1^\infty E_n = \mathscr{C}$, closure of the E_n implies, by the *Baire category theorem*, that at least one E_n *contains an entire sphere*. That is, there is an $F \in \mathscr{C}$ and a $\rho > 0$ such that $f \in \mathscr{C}$ and $\|f - F\| < 2\rho$ imply $f \in E_n$. Pick an F_0 (e.g., a *trigonometric polynomial*) with $\|F_0 - F\| < \rho$ such that $F_0'(\theta)$ exists everywhere and is finite. If, say, $\|F_0'\| \le K$ (K may of course be *enormous*!) we have

$$\int_0^\pi \frac{|F_0(\theta+t)-F_0(\theta-t)|}{t}\,dt \le 2\pi K$$

for *all* θ by the mean value theorem.

Now let $g \in \mathscr{C}$, and let $\|g\| < \rho$. Then $\|F_0 + g - F\| < 2\rho$, so $F_0 + g \in E_n$. In view of the previous inequality, this means that there must be at least one θ with

$$\int_0^\pi \frac{|g(\theta+t)-g(\theta-t)|}{t}\,dt \le n + 2\pi K,$$

i.e., $\|g\| < \rho$ *implies that for at least one* θ,

$$\int_0^\pi \frac{|g(\theta+t)-g(\theta-t)|}{t}\,dt \le \text{ some fixed number, say } M.$$

But this cannot be. Take any continuous function h *of period* π such that $h(\theta+t)-h(\theta-t)$ is not identically zero in t for *any* θ. Here is an example of such a function:

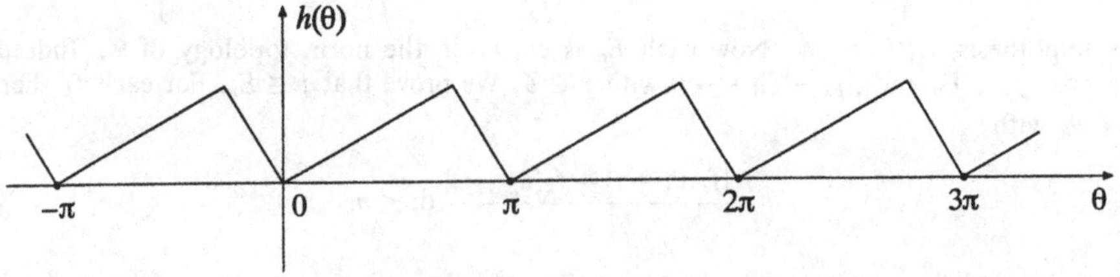

Then for all θ,

$$\int_0^\pi |h(\theta+t)-h(\theta-t)|\,dt$$

is $\ge$ say $\alpha > 0$. Given $m = 2, 3, 4, \ldots$, let $\theta_m \in [0, \pi]$ be such that $m\theta - \theta_m$ is a multiple of

π. Then

$$\int_0^\pi \frac{|h(m(\theta+t)) - h(m(\theta-t))|}{t}\,dt = \int_0^{m\pi} \frac{|h(\theta_m+s) - h(\theta_m-s)|}{s}\,ds$$

$$\geq \int_0^\pi |h(\theta_m+s) - h(\theta_m-s)|\,ds + \frac{1}{2}\int_\pi^{2\pi} |h(\theta_m+s) - h(\theta_m-s)|\,ds + \ldots$$

$$\ldots + \frac{1}{m}\int_{(m-1)\pi}^{m\pi} |h(\theta_m+s) - h(\theta_m-s)|\,ds$$

$$\geq \alpha\left(1+\frac{1}{2}+\ldots+\frac{1}{m}\right) \sim \alpha\log m\,.$$

Taking such an h with $\|h\| < \rho$, we see that for $g(t) = h(mt)$ there is *no* θ with

$$\int_0^\pi \frac{|g(\theta+t) - g(\theta-t)|}{t}\,dt \leq M,$$

provided m is large enough. This contradicts our previous statement, and proves the theorem.

II

Theorem of the Brothers Riesz. Introduction to the Space H_1

A. The F. and M. Riesz theorem

Let μ be a (signed or in general complex-valued) Borel measure on $[-\pi, \pi]$. In 1917, in the proceedings of the fourth Scandinavian mathematical congress (this is a paper every analyst should read!), F. and M. Riesz published the celebrated

Theorem *If*

$$\int_{-\pi}^{\pi} e^{in\theta}\, d\mu(\theta) = 0 \text{ for } n = 1, 2, 3, \ldots,$$

then μ is absolutely continuous with respect to Lebesgue measure.

The theorem is not only deep, but is also the basis for much of the following work. We shall go through three proofs of this theorem, two here, and a third in Chapter IV.

1. Original proof

The original proof of F. and M. Riesz is perhaps the simplest, and goes as follows.

Suppose μ is *not* absolutely continuous. Then there is some set $E \subseteq [-\pi, \pi]$ *of Lebesgue measure zero* but such that

$$\int_E e^{i\theta}\, d\mu(\theta) \neq 0.$$

Let, indeed, ν be the complex valued Borel measure given by

$$\nu(E) = \int_E e^{i\theta}\, d\mu(\theta)$$

for Borel sets E. Then

$$\mu(E) = \int_E e^{-i\theta}\, d\nu(\theta)$$

for such E – to check this, approximate $e^{i\theta}$ uniformly by Borel functions of modulus 1 taking only a finite number of values. But then, if $\nu(E) = 0$ for every Borel set E with $|E| = 0$, we also have $\mu(E) = 0$ for each such E.

Notation As is customary, we henceforth denote the Lebesgue measure of a set E by $|E|$.

From elementary measure theory, we know that we can take E to also be *closed*.

This being granted, we proceed thus. Denote the (disjoint) open intervals complementary to E – the so-called *contiguous intervals* – by (α_n, β_n) – there are at most countably many of them. We mean by this that *the arcs* $\{e^{i\theta};\ \alpha_n < \theta < \beta_n\}$ *are disjoint, and that, together with* $\{e^{i\theta};\ \theta \in E\}$, *they precisely fill up the unit circumference* $|z| = 1$.

Because $|E| = 0$, $\sum_n(\beta_n - \alpha_n) = 2\pi$. Let $p_n \xrightarrow[n]{} \infty$, $p_n > 0$, but in such fashion that $\sum_n p_n(\beta_n - \alpha_n) < \infty$. Such a sequence can always be found. *Then define a function* $F(\theta)$ *as follows*:

(i) $F(\theta + 2\pi) = F(\theta)$;

(ii) $F(\theta) = \infty$, $\theta \in E$;

(iii) If $\alpha_n < \theta < \beta_n$,

$$F(\theta) = p_n \frac{l_n}{\sqrt{l_n^2 - (\theta - \gamma_n)^2}},$$

where $l_n = (\beta_n - \alpha_n)/2$ and $\gamma_n = (\beta_n + \alpha_n)/2$.

Then

$$\int_{\alpha_n}^{\beta_n} F(\theta)\,\mathrm{d}\theta = \pi p_n l_n,$$

so $F \in L_1(-\pi, \pi)$ by choice of the p_n, since $|E| = 0$. $F(\theta)$ is also *continuous from* $\mathbb{R}$ *to* $[0, \infty]$ (with ∞ *included*). Indeed, if $\theta_0 \in (\alpha_n, \beta_n)$ for some n, continuity of F at θ_0 is obvious. If $\theta_0 \in E$, and a sequence of θ's tends to θ_0, then we can *split* the sequence into *two parts*:

One part lying in *finitely many* of the complementary intervals (α_n, β_n) – ultimately, in at most *two* of them – and *tending to some common endpoint of them*. For this part of the sequence it is clear by the formula in (iii) that $F(\theta) \to \infty$.

Another part running through *infinitely many* of the (α_n, β_n). Since, by construction, $F(\theta) \geq p_n$ on (α_n, β_n) and $p_n \to \infty$, we again have $F(\theta) \to \infty$. (That's the *reason* for the *introduction* of the p_n !)

We now put

$$U(re^{i\theta}) = \frac{1}{2\pi}\int_{-\pi}^{\pi} \frac{1 - r^2}{1 + r^2 - 2r\cos(\theta - t)} F(t)\,dt,$$

which we can *do* since $F \in L_1$. By part of Problem 2 above (Chapter I, Section E), $U(z) \to F(\theta)$ for $z \to e^{i\theta}$, i.e., U is continuous from $|z| < 1$ into $[0, \infty]$. We also take

$$\tilde{U}(re^{i\theta}) = \frac{1}{2\pi}\int_{-\pi}^{\pi} \frac{2r\sin(\theta - t)}{1 + r^2 - 2r\cos(\theta - t)} F(t)\,dt.$$

By the *same* Problem 2, $\tilde{U}(z)$ *extends continuously up to each open arc* $\{e^{i\theta}; \alpha_n < \theta < \beta_n\}$, for $F(\theta)$ is $\mathscr{C}_1$ (and even $\mathscr{C}_\infty$) on each such open arc.

Now put

$$\phi(z) = \frac{U(z) + i\tilde{U}(z)}{U(z) + i\tilde{U}(z) + 1}, \qquad |z| < 1.$$

Since $F(\theta) > 0$, $U(z) > 0$, so $\phi(z)$ is *analytic* in $|z| < 1$, and $|\phi(z)| < 1$ there. *It even extends continuously up to* $|z| = 1$. For if $\theta_0 \notin E$, we have just seen that $U(z)$ and $\tilde{U}(z)$ tend to *continuous* limits as z tends to points in some small arc about $e^{i\theta_0}$. And if $\theta_0 \in E$, $U(z) \to \infty$ as $z \to e^{i\theta_0}$, so, *independently of how* $\tilde{U}(z)$ *behaves*,

$$\phi(z) = \frac{U(z) + i\tilde{U}(z)}{U(z) + i\tilde{U}(z) + 1} \to 1$$

for $z \to e^{i\theta_0}$. For $\theta \notin E$ but tending to a point θ_0 of E we also clearly have for the boundary limit $\phi(e^{i\theta})$,

$$\phi(e^{i\theta}) = \frac{F(\theta) + i\tilde{F}(\theta)}{F(\theta) + i\tilde{F}(\theta) + 1} \to 1,$$

since $F(\theta) \to \infty$. Taking $\phi(z)$ to be thus extended by continuity up to $|z| = 1$, we now see that $|\phi(z)| < 1$ *everywhere* for $|z| \le 1$ (sic!), *save when* $z = e^{i\theta}$ *with* $\theta \in E$, *where* $\phi(e^{i\theta}) = 1$. And this function ϕ is *continuous on* $|z| \le 1$ *and analytic in* $|z| < 1$. The first construction of such a function was essentially made by Fatou in his thesis – the Riesz brothers just used it. (But Fatou never thought of the theorem they proved by its help!)

Let $k = 1, 2, \ldots$. Then $[\phi(z)]^k$ is also analytic in $|z| < 1$, continuous in $|z| \le 1$, so

$$[\phi(re^{i\theta})]^k \to [\phi(e^{i\theta})]^k$$

uniformly for $r \to 1$*. Hence for each k,

$$\int_{-\pi}^{\pi} [\phi(re^{i\theta})]^k e^{i\theta} \, d\mu(\theta) \to \int_{-\pi}^{\pi} [\phi(e^{i\theta})]^k e^{i\theta} \, d\mu(\theta)$$

as $r \to 1$.

Now for $r < 1$, by analyticity,

$$[\phi(re^{i\theta})]^k = \sum_0^\infty a_n r^n e^{in\theta},$$

the series being uniformly convergent in θ.

So, *by the hypothesis of the theorem to be proved*,

$$\int_{-\pi}^{\pi} [\phi(re^{i\theta})]^k e^{i\theta} \, d\mu(\theta) = 0.$$

Making $r \to 1$, we have

$$\int_{-\pi}^{\pi} [\phi(e^{i\theta})]^k e^{i\theta} \, d\mu(\theta) = 0, \quad k = 1, 2, \ldots .$$

But now, *by construction*, $\phi(e^{i\theta}) = 1$ for $\theta \in E$ whereas *elsewhere*, $|\phi(e^{i\theta})| < 1$. Therefore,

* The presentation at this point has been improved thanks to the criticism of M.-Y. Couture.

by bounded convergence,

$$\int_{-\pi}^{\pi} [\phi(e^{i\theta})]^k e^{i\theta} \, d\mu(\theta) \to \int_E e^{i\theta} \, d\mu(\theta)$$

as $k \to \infty$, i.e., $\int_E e^{i\theta} \, d\mu(\theta) = 0$ *for any closed E with* $|E| = 0$. Since, as stated at the beginning, if μ were *not* absolutely continuous we could get such an E with $\int_E e^{i\theta} \, d\mu(\theta) \neq 0$, we have reached a contradiction, proving the theorem.

2. Modern Helson and Lowdenslager proof

Here is a modern proof of the F. and M. Riesz theorem, due to Helson and Lowdenslager, ca. 1958.

Given a finite positive measure ν on $[-\pi, \pi]$, we let $L_2(d\nu)$ be the Hilbert space obtained by defining

$$\langle f, g \rangle_\nu = \int_{-\pi}^{\pi} f(\theta)\overline{g(\theta)} \, d\nu(\theta)$$

for *continuous* 2π-periodic functions f and g, and then forming the completion in the usual way, using the norm $\|f\|_\nu = \sqrt{\langle f, f \rangle_\nu}$.

(i) Write $d\mu(\theta) = h(\theta) \, d\theta + ds(\theta)$ where $h \in L_1$ and s is singular, and suppose

$$\int_{-\pi}^{\pi} e^{in\theta} \, d\mu(\theta) = 0, \quad n = 1, 2, \ldots \, .$$

Our main object is to prove that

$$\int_{-\pi}^{\pi} e^{in\theta} \, ds(\theta) = 0, \quad n = 1, 2, \ldots \, .$$

Take

$$d\nu(\theta) = (1 + |h(\theta)|) \, d\theta + |\, ds(\theta)|.$$

We consider $\mathfrak{A}$ = the closed subspace of $L_2(d\nu)$ (sic!) spanned by $\{e^{in\theta}; n = 1, 2, \ldots\}$. Let $F \in \mathfrak{A}$ be the element of $\mathfrak{A}$ making

$$\int_{-\pi}^{\pi} |1 - F(\theta)|^2 \, d\nu(\theta)$$

as small as possible. Since $d\nu(\theta) \geq d\theta$,

$$\int_{-\pi}^{\pi} |1 - F(\theta)|^2 \, d\nu(\theta)$$

$$\geq \inf \left\{ \int_{-\pi}^{\pi} |1 - P(\theta)|^2 \, d\theta; \quad P(\theta) = a_1 e^{i\theta} + a_2 e^{2i\theta} + \ldots + a_p e^{pi\theta} \right\} = 2\pi.$$

By elementary Hilbert space geometry, in $L_2(d\nu)$, $1 - F \perp \mathfrak{A}$, in particular

$$1 - F \perp e^{in\theta}(1 - F(\theta)) \quad \text{in } L_2(d\nu)$$

for $n = 1, 2, \ldots$. Therefore

$$\int_{-\pi}^{\pi} |1 - F(\theta)|^2 e^{-in\theta}\, \mathrm{d}\nu(\theta) = 0, \quad n = 1, 2, \ldots,$$

so, since $\mathrm{d}\nu(\theta) \geq 0$, by *conjugation*,

$$\boxed{|1 - F(\theta)|^2\, \mathrm{d}\nu(\theta) = c\, \mathrm{d}\theta.}$$

By the previous, $c > 0$.

(ii) From $|1 - F(\theta)|^2\, \mathrm{d}\nu(\theta) = c\, \mathrm{d}\theta$, $|1 - F(\theta)|^2\, \mathrm{d}\nu(\theta)$ is absolutely continuous, so

$$|1 - F(\theta)|^2 |\, \mathrm{d}s(\theta)| = 0, \quad \text{or} \quad F(\theta) = 1 \text{ a.e. } |\, \mathrm{d}s|.$$

Therefore

$$|1 - F(\theta)|^2\, \mathrm{d}\nu(\theta) = |1 - F(\theta)|^2 (1 + |h(\theta)|)\, \mathrm{d}\theta.$$

Thus $|1 - F|^2 (1 + |h|) = c$, or

$$\frac{1}{|1 - F|^2} = \frac{1 + |h|}{c} \in L_1(\, \mathrm{d}\theta),$$

so

$$\boxed{\frac{1}{1 - F} \in L_2},$$

and

$$\boxed{|1 - F|(1 + |h|) = \frac{c}{|1 - F|} \in L_2}.$$

(iii) Since $F(\theta) = 1$ a.e. $|\, \mathrm{d}s|$,

$$(1 - \overline{F(\theta)})\, \mathrm{d}\nu(\theta) = (1 - \overline{F(\theta)})(1 + |h(\theta)|)\, \mathrm{d}\theta.$$

Now $1 - F \perp \mathfrak{A}$ in $L_2(\, \mathrm{d}\nu)$ makes, *with this*,

$$\int_{-\pi}^{\pi} e^{in\theta} [1 - \overline{F(\theta)}](1 + |h(\theta)|)\, \mathrm{d}\theta = 0, \quad n = 1, 2, \ldots \ .$$

i.e., since $1/(1 - F) = (1 - \overline{F})(1 + |h(\theta)|)$,

$$\int_{-\pi}^{\pi} \frac{1}{1 - F(\theta)} e^{in\theta}\, \mathrm{d}\theta = 0, \quad n = 1, 2, 3, \ldots \ .$$

With $1/(1 - F) \in L_2$, this last implies there is a sequence of *polynomials* $G_n(z)$ with

$$\left\| G_n(e^{i\theta}) - \frac{1}{1 - F(\theta)} \right\|_2 \xrightarrow[n]{} 0.$$

(iv) From $\int_{-\pi}^{\pi} e^{in\theta}\, \mathrm{d}\mu(\theta) = 0$, $n = 1, 2, \ldots$ and $F(\theta)$'s being in the closure of the set of linear combinations of the $e^{in\theta}$, $n \geq 1$, in $L_2(\, \mathrm{d}\nu)$ with $\mathrm{d}\nu(\theta) \geq |\, \mathrm{d}\mu(\theta)|$, we surely have

$$\int_{-\pi}^{\pi} e^{ik\theta} (1 - F(\theta))\, \mathrm{d}\mu(\theta) = 0, \quad k = 1, 2, \ldots \ .$$

But $d\mu(\theta) = h(\theta)\, d\theta + ds(\theta)$, and $1 - F(\theta) = 0$ a.e. $|ds(\theta)|$.

So

$$\boxed{\int_{-\pi}^{\pi} e^{ik\theta}[1 - F(\theta)]h(\theta)\, d\theta = 0, \quad k = 1, 2, \ldots}\ .$$

By (ii), $(1 - F)h \in L_2$, so, if G_n is the sequence of polynomials from (iii), by Schwarz,

$$\int_{-\pi}^{\pi} e^{ik\theta} h(\theta)\, d\theta = \int_{-\pi}^{\pi} e^{ik\theta} \frac{1}{1 - F(\theta)}[1 - F(\theta)]h(\theta)\, d\theta$$

$$= \lim_{n \to \infty} \int_{-\pi}^{\pi} e^{ik\theta} G_n(e^{i\theta})(1 - F(\theta))h(\theta)\, d\theta.$$

But each of these last integrals is *zero* by previous boxed remark, the $G_n(z)$ being *polynomials* in z.

Conclusion:

$$\int_{-\pi}^{\pi} e^{ik\theta} h(\theta)\, d\theta = 0, \quad k = 1, 2, \ldots\ .$$

(v) The result just obtained shows, with the hypothesis, that

$$\int_{-\pi}^{\pi} e^{ik\theta}\, ds(\theta) = 0, \quad k = 1, 2, 3, \ldots\ .$$

Since $F \in \mathfrak{A}$ is in the closure of the set of linear combinations of $e^{i\theta}$, $e^{2i\theta}$, etc. in $L_2(d\nu)$ where $d\nu \geq |ds|$, we have

$$\int_{-\pi}^{\pi} F(\theta)\, ds(\theta) = 0,$$

i.e., $\int_{-\pi}^{\pi} ds(\theta) = 0$, *since* $F(\theta) = 1$, a.e. $|ds|$ *by step* (ii). *We thus see that we even have*

$$\int_{-\pi}^{\pi} e^{ik\theta} \cdot e^{-i\theta}\, ds(\theta) = 0 \quad \text{for } k = 1, 2, 3, \ldots\ .$$

Now apply the argument in steps (i) to (iv) using a *new measure* μ *given by* $d\mu(\theta) = e^{-i\theta}\, ds(\theta)$. Since this μ is *already entirely singular*, we will obtain *for it*

$$\int_{-\pi}^{\pi} e^{ik\theta} e^{-i\theta}\, d\mu(\theta) = 0, \quad k = 1, 2, 3, \ldots\ ,$$

i.e.,

$$\int_{-\pi}^{\pi} e^{ik\theta} e^{-2i\theta}\, ds(\theta) = 0, \quad k = 1, 2, 3, \ldots\ .$$

Taking now a *new* (singular) μ given by $d\mu(\theta) = e^{-2i\theta}\, ds(\theta)$ and *repeating the process*, we see that

$$\int_{-\pi}^{\pi} e^{ik\theta} e^{-3i\theta}\, ds(\theta) = 0, \quad k = 1, 2, 3, \ldots .$$

By *repetition*, we thus have

$$\int_{-\pi}^{\pi} e^{il\theta}\, ds(\theta) = 0$$

for *all* integers l, whence $ds(\theta) = 0$.

We have thus proven that, for our *given original* μ, $d\mu(\theta) = h(\theta)\, d\theta$. Thus μ is absolutely continuous. Q.E.D.

B. Definition and basic properties of H_1

Admission of complex valued harmonic functions into the discussion is rather straightforward – we simply denote by that name any complex linear combination of ordinary (real valued) harmonic functions. (This notion was avoided up to here because we wanted to be able to say that a harmonic function was the real part of an analytic one!) It becomes convenient now to consider complex valued harmonic functions because we can look upon an analytic function as harmonic. The representation theorems of Chapter I, Section C and the boundary behaviour theorems of Chapter I, Section D do, of course, *hold* for complex valued harmonic functions – their extension is trivial.

Definition $F(z)$, analytic for $|z| < 1$ is said *to be in* H_1 if

$$\int_{-\pi}^{\pi} |F(re^{i\theta})|\, d\theta \quad \text{is } \textit{bounded} \text{ for } r < 1.$$

1. Poisson representation for functions in H_1

Let $F \in H_1$. Since $F(z)$ is, in particular, *harmonic* for $|z| < 1$ we have, by a theorem in Chapter I, Section C,

$$F(re^{i\theta}) = \frac{1}{2\pi} \int_{-\pi}^{\pi} \frac{1 - r^2}{1 + r^2 - 2r\cos(\theta - t)}\, d\mu(t)$$

for some measure μ (here *complex valued*!) on $[-\pi, \pi]$. By Chapter I, Section D,

$$F(re^{i\theta})\, d\theta \to d\mu(\theta) \quad w^*$$

as $r \to 1$. Since F is *analytic* in $|z| < 1$, Cauchy's theorem (or straightforward manipulation with a power series!) here yields, for $r < 1$,

$$\int_{-\pi}^{\pi} e^{in\theta} F(re^{i\theta})\, d\theta = 0, \quad n = 1, 2, \ldots .$$

Therefore

$$\int_{-\pi}^{\pi} e^{in\theta}\, d\mu(\theta) = 0, \quad n = 1, 2, 3, \ldots .$$

Now the theorem of the Brothers Riesz, given above in Section A, *guarantees that* μ *is absolutely continuous, i.e.,* $d\mu(\theta) = h(\theta)\, d\theta$ *for some* $h \in L_1(-\pi, \pi)$.

So in this case we really have

$$F(re^{i\theta}) = \frac{1}{2\pi}\int_{-\pi}^{\pi} P_r(\theta - t)h(t)\,dt$$

with a function h. The distinction between this case (for analytic F) and the more general one treated in Chapter I, *Section* C *(F merely harmonic) is very important for the whole development of the theory.*

By Chapter I, Section D we now have $F(z) \to h(\theta)$ a.e. for $z \xrightarrow[\angle]{} e^{i\theta}$, so if, as mentioned at the end of Chapter I, Section D, we call

$$F(e^{i\theta}) = \lim_{z \xrightarrow[\angle]{} e^{i\theta}} F(z),$$

we have

$$\boxed{F(re^{i\theta}) = \frac{1}{2\pi}\int_{-\pi}^{\pi} \frac{1 - r^2}{1 + r^2 - 2r\cos(\theta - t)} F(e^{it})\,dt}$$

for $F \in H_1$.

2. L_1 convergence to boundary data function

From the boxed formula in Subsection 1, and the elementary approximate identity property of $P_r(\theta)$ (Chapter I, Subsection D.2):

$$\boxed{\int_{-\pi}^{\pi} |F(re^{i\theta}) - F(e^{i\theta})|\, d\theta \to 0 \quad \text{as } r \to 1.}$$

3. Cauchy's formula

By Cauchy's formula, for $|z| < R < 1$,

$$F(z) = \frac{1}{2\pi i}\int_{|\zeta| = R} \frac{F(\zeta)}{\zeta - z}\, d\zeta.$$

Now if $|z| < 1$ is *fixed* and we make $R \to 1$, we conclude from Subsection 2 that

$$\boxed{F(z) = \frac{1}{2\pi i}\int_{0}^{2\pi} \frac{F(e^{it})d(e^{it})}{e^{it} - z},}$$

which is Cauchy's formula for functions in H_1.

C. Digression on conformal mapping theory

A simply connected region whose boundary contains more than two points can be mapped conformally on the interior of the unit circle. This is the Riemann mapping theorem, whose proof belongs to the basic course on complex variable theory.

It is also true that if the simply connected region's boundary is a *Jordan curve*, then the function mapping the region onto $\{|z| < 1\}$ has a continuous one-one extension up to the boundary, which it takes onto $\{|z| = 1\}$.

This fact is often given without proof in complex variable courses. Because we use it sometimes in the present book, we give a complete proof in Subsection C.1 below. The proof uses the Jordan curve theorem, here assumed known, but is otherwise self-contained.

1. Carathéodory's theorem

Definition A *Jordan curve* is a *continuous one-one image of* $\{|\zeta| = 1\}$ *in* $\mathbb{C}$. The continuous one-one function mapping $\{|\zeta| = 1\}$ onto the Jordan curve is called a *parametrization* of the curve.

By means of its parametrization, any Jordan curve has a natural order defined on it, in an obvious fashion. We can thus speak of *arcs* on a Jordan curve, and so forth.

Definition A *Jordan arc* is a continuous one-one image of an *interval* of $\mathbb{R}$ (*with* or *without* endpoints) in $\mathbb{C}$.

A Jordan curve can be a very complicated object:

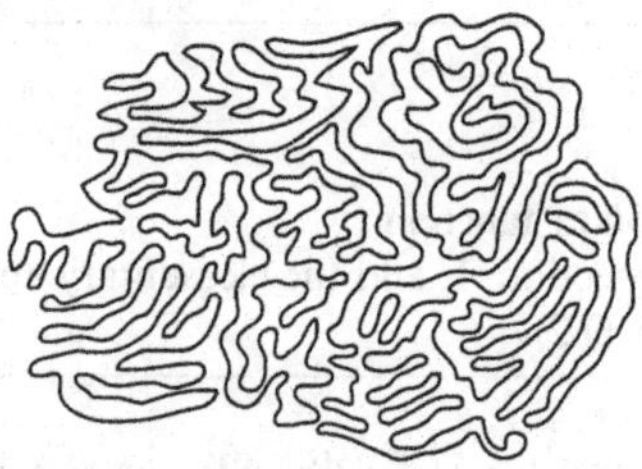

We admit without proof the following Jordan Curve Theorem. *Let* Γ *be a Jordan curve. Then* $\mathbb{C} \sim \Gamma$ *consists of two connected components,* $\mathcal{O}$ *and* Ω, *one of which contains all z with sufficiently large modulus. If* $\omega \in \Gamma$, *any neighbourhood of* ω *has in it points of* $\mathcal{O}$ *and points of* Ω.

Definition Let Ω be the component of $\mathbb{C} \sim \Gamma$ which contains all points of sufficiently large modulus. Ω is called the *outside* of Γ. The other component, $\mathcal{O}$, is called the *inside* of Γ.

In complex variable theory, one uses the following definition:

A *connected* open set in $\mathbb{C}$ is called *simply connected* if, whenever any *Jordan polygon* Π lies in the open set, the *inside* of Π also lies in it.

(A *Jordan polygon* is a Jordan curve made up of straight segments, i.e., a *simple closed broken-line path*. The Jordan curve theorem is elementary for Jordan polygons (although not trivial), and can be proved by induction.)

Lemma *Let* Γ *be a Jordan curve, and* $\mathcal{O}$ *its inside. Then* $\mathcal{O}$ *is simply connected.*

Proof Assume not. Then there is a Jordan polygon $\Pi \subseteq \mathcal{O}$ whose inside, $\mathcal{O}'$, is *not* in $\mathcal{O}$. Therefore there is a $z \in \mathcal{O}'$ with $z \in \Gamma$ or $z \in \Omega$, the *outside* of Γ. If $z \in \Gamma$, let $\mathcal{N}$ be a neighbourhood of z with $\mathcal{N} \subseteq \mathcal{O}'$. By the Jordan curve theorem, $\mathcal{N}$ contains a point z' of Ω. So in any case, if the result is *false*, $\mathcal{O}'$ has in it a point $z' \in \Omega$. Since a connected open set is arcwise connected, we can join z' to ∞ by a path Λ lying in Ω, in particular, *not touching* Γ. Since $z' \in \mathcal{O}'$, *inside* of Π, Λ must touch Π, say in z''. Then $z'' \in \mathcal{O}$. But Λ joins z'' to ∞ without touching Γ, which is impossible. So the lemma is true.

Remark The above argument will be used again in what follows. We shall refer to it as the *Jordan curve argument.*

Let now Γ be a Jordan curve, and let $\mathcal{D}$ be its inside. The Riemann mapping theorem says there is a conformal mapping Φ of $|z| < 1$ onto $\mathcal{D}$ since, by the lemma, $\mathcal{D}$ is *simply connected*. *Our problem is to show that Φ can be continuously extended up to $\{|z| \le 1\}$ so as to take $\{|z| = 1\}$ in continuous one-one fashion onto Γ.*

Lemma *There is a function $\eta(\delta)$ defined for all sufficiently small $\delta > 0$, with $\eta(\delta) \to 0$ as $\delta \to 0$, such that, given $a, b \in \Gamma$ with $|a - b| \le \delta$, there is one and only one arc of Γ having endpoints a and b whose diameter is $\le \eta(\delta)$.*

Proof Let $\psi(e^{it})$ be a parametrization of Γ, and let $\delta_0 > 0$ be so small that whenever $|\psi(\zeta) - \psi(\zeta')| \le \delta_0$, $|\zeta - \zeta'| < 2$. For ζ and ζ' with $|\psi(\zeta) - \psi(\zeta')| \le \delta_0$, let σ be the (unique!) *shorter arc* of $\{|z| = 1\}$ having endpoints ζ and ζ', and call $\gamma = \psi(\sigma)$. By continuity of ψ and its inverse, $\operatorname{diam}\gamma \to 0$ *uniformly* for $|\psi(\zeta) - \psi(\zeta')| \to 0$, whatever ζ and ζ' we take on the unit circle with $|\psi(\zeta) - \psi(\zeta')| \le \delta_0$. So for $\delta < \delta_0$, call

$$\eta(\delta) = \sup\{\operatorname{diam}\gamma;\ |\psi(\zeta) - \psi(\zeta')| \le \delta\}$$

Then $\eta(\delta) \to 0$ as $\delta \to 0$. Let $\delta_1 < \delta_0$ be so small that $\eta(\delta_1) < (\operatorname{diam}\Gamma)/2$. Then the lemma holds for $\delta \le \delta_1$.

Definition Let $a, b \in \Gamma$ with $|a - b|$ sufficiently small. The unique arc of Γ with endpoints a and b having diameter $\le \eta(|a - b|)$ is called *the smaller arc of Γ joining a to b.*

Lemma *Let Φ map $\{|z| < 1\}$ conformally onto $\mathcal{D}$ bounded by the Jordan curve Γ. Let $|\zeta| = 1$. Then $\lim_{z\to\zeta} \Phi(z)$ exists and is on Γ.*

This is the *main part* of the solution of our problem, and will be done in a series of steps.

(i) Without loss of generality, $\zeta = 1$, and, instead of looking at Φ, we look at a conformal mapping $F(z)$ of $\Im z > 0$ onto $\mathcal{D}$. *We are to prove* that $\lim_{z\to 0} F(z)$ exists and is on Γ; this will be enough. With γ_r the half-circle $re^{i\theta}$, $0 < \theta < \pi$, consider the (un-ended) Jordan arc $F(\gamma_r) \subseteq \mathcal{O}$. We have

$$\text{length}\, F(\gamma_r) = \int_0^{\pi} |F'(re^{i\theta})| r \, d\theta.$$

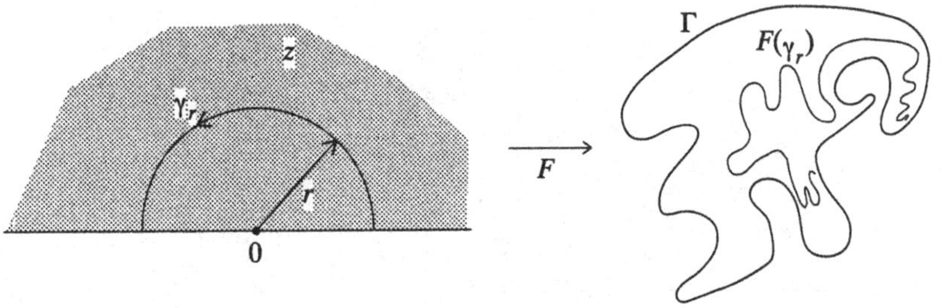

Also,

$$\infty > \text{area}\, \mathcal{D} = \iint_{\Im z > 0} |F'(z)|^2 \, dx\, dy = \int_0^{\infty}\!\!\int_0^{\pi} |F'(re^{i\theta})|^2 r \, d\theta \, dr,$$

and this by Schwarz' inequality is

$$\geq \int_0^\infty \frac{[\text{length}\, F(\gamma_r)]^2}{\pi r}\, dr.$$

So

$$\int_0^\infty \frac{1}{r} [\text{length}\, F(\gamma_r)]^2 dr < \infty.$$

Since for any $p > 0$, $\int_0^p dr/r = \infty$, there must be a sequence $r_n \downarrow 0$ with length $F(\gamma_{r_n}) \to 0$. (N.B. Examples show that *it is not true* that length $F(\gamma_r) \to 0$ for $r \to 0$!) *Choose such a sequence* $\{r_n\}$ *once and for all, and call* $\gamma_{r_n} = \gamma_n$.

(ii) As soon as length $F(\gamma_n) < \infty$, the *limits*

$$a = \lim_{\theta \to 0} F(r_n e^{i\theta}) \quad \text{and} \quad b = \lim_{\theta \to \pi} F(r_n e^{i\theta})$$

exist. They are on Γ. Assume that, e.g., $a \in \mathscr{D}$, and let $\mathscr{N}$ be a neighbourhood of a with $\mathscr{N} \subseteq \mathscr{D}$. Let $z_0 = F^{-1}(a)$; $\Im z_0 > 0$, and choose $\mathscr{N}$ so small that $F^{-1}(\mathscr{N})$ has *compact closure* in $\Im z > 0$. This is possible because F^{-1} is continuous! Then, for $\theta > 0$ sufficiently close to 0, $F(r_n e^{i\theta})$ is in $\mathscr{N}$, so $r_n e^{i\theta}$ is in $F^{-1}(\mathscr{N})$, which is *false* for θ close enough to 0. So a and b *are* on Γ. We adjoin a and b to $F(\gamma_n)$ in an obvious fashion so as to get *either*:

(a) A closed Jordan arc $\overline{F(\gamma_n)}$ if $a \neq b$;
(b) A *Jordan curve* if $a = b$.

In case (a), $|a - b| \leq$ length $F(\gamma_n)$ is small if n is large; therefore by the above lemma there is a *unique smaller arc* Γ_{ab} of Γ going from a to b. *In this case* $F(\gamma_n)$ and Γ_{ab} form, *together*, a Jordan curve. The proof of this consists in *writing down a parametrization* for $F(\gamma_n) \cup \Gamma_{ab}$ in terms of a readily available one for $F(\gamma_n)$ and the one for Γ, and is so straightforward and routine that we omit it.

(iii) Let $\mathcal{O}_n$ be the *inside* of the Jordan curve $F(\gamma_n) \cup \Gamma_{ab}$ in case (a) above, or of $\overline{F(\gamma_n)}$ in case (b) above. *Then* $\mathcal{O}_n \subseteq \mathscr{D}$.

Proof Note that $F(\gamma_n) \subseteq \mathscr{D}$, and *repeat* the Jordan curve argument already used in proving the first lemma.

(iv) Let n be sufficiently large, and let S_n and T_n be the regions shown in $\Im z > 0$:

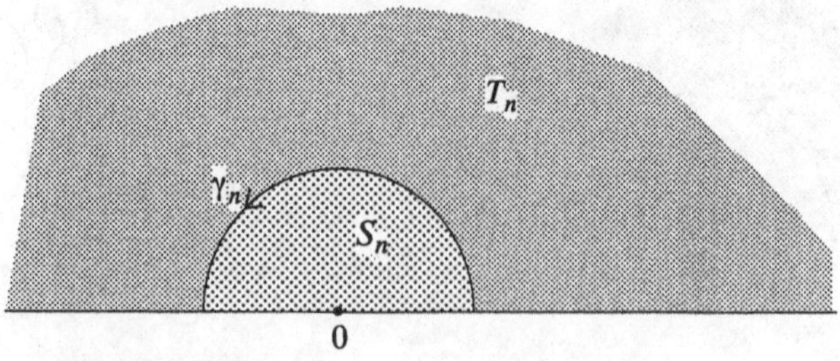

Then $F(S_n) \subseteq \mathcal{O}_n$, the *inside* of $F(\gamma_n) \cup \Gamma_{ab}$ (or of $\overline{F(\gamma_n)}$ if $a = b$).

Proof By (iii), $\mathcal{O}_n \subseteq \mathscr{D}$. Take any $w_0 \in \mathcal{O}_n$, then $w_0 = F(z_0)$ where $\Im z_0 > 0$. Since $w_0 \notin F(\gamma_n)$, $z_0 \in S_n$ or $z_0 \in T_n$.

(a) If $z_0 \in S_n$, $F(S_n) \subseteq \mathcal{O}_n$, and *we're done*. Indeed, S_n is *open and connected*, so $F(S_n)$ is *also*. $F(S_n) \subseteq \mathscr{D}(\subseteq \mathbb{C} \sim \Gamma)$ and $S_n \cap \gamma_n = \emptyset$, so $F(S_n)$ *lies entirely in the complement* of the Jordan curve bounding $\mathcal{O}_n$ ($\overline{F(\gamma_n)}$ or $F(\gamma_n) \cup \Gamma_{ab}$ as the case may be). $F(S_n)$ has a *point in common*, namely w_0, with the *connected component* $\mathcal{O}_n$ of that complement. Therefore $F(S_n)$ *must lie entirely in that component*, i.e., $F(S_n) \subseteq \mathcal{O}_n$.

(b) If $z_0 \in T_n$ then we prove *as in* (a) that $F(T_n) \subseteq \mathcal{O}_n$. However, this we now show to be *impossible* for sufficiently large n. Indeed,

$$\text{area}\, F(T_n) = (\text{area}\, \mathscr{D} - \text{area}\, F(S_n)) \to \text{area}\, \mathscr{D} \text{ as } n \to \infty$$

since $\cap_n S_n = \emptyset$. On the other hand, if length $F(\gamma_n) = \delta_n$, we have $\delta_n \to 0$ by (i), and $|a-b| \leq \delta_n$ so surely Γ_{ab} has diameter $\leq \eta(\delta_n)$ which $\to 0$ as $n \to \infty$, by the *second* lemma. (We count Γ_{ab} as $\{a\}$ if $a = b$. *Of course*, a and b *and* Γ_{ab} *depend* on n, but we *do not show* this dependence in the *notation*, in order to have easier symbols to read!) For any given large n, let Δ_n be a disc of radius $\delta_n + \eta(\delta_n)$ about the end a of $\overline{F(\gamma_n)}$; then the *whole Jordan curve* $F(\gamma_n) \cup \Gamma_{ab}$ lies in Δ_n:

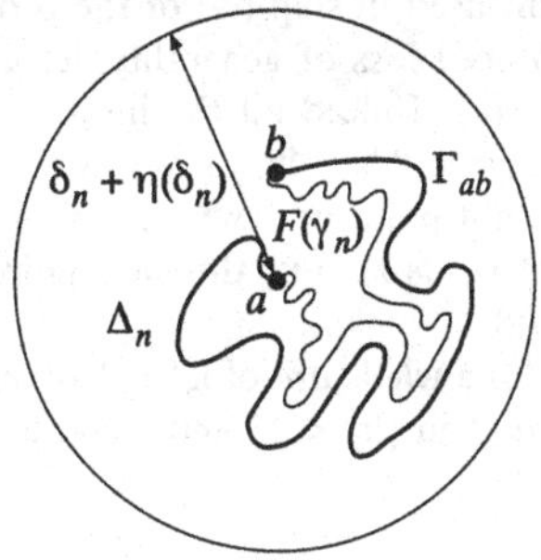

From this it follows that the inside $\mathcal{O}_n$ of $F(\gamma_n) \cup \Gamma_{ab}$ *also* lies in Δ_n – this is proven by a repetition of the *Jordan curve argument*.

So area $\mathcal{O}_n \leq \pi(\delta_n + \eta(\delta_n))^2 \underset{n}{\longrightarrow} 0$ as $n \to \infty$, whilst, as we have seen, area $F(T_n) \underset{n}{\longrightarrow}$ area $\mathscr{D} > 0$. Therefore, for sufficiently large n, $F(T_n) \subseteq \mathcal{O}_n$ is *impossible*, and *we must have* $F(S_n) \subseteq \mathcal{O}_n$.

Q.E.D.

(v) We have $S_n \supseteq S_{n+1} \supseteq \ldots$. By (iv), for sufficiently large n, $F(S_n) \subseteq \mathcal{O}_n$, and as we saw in proving (iv),

$$\text{diam}\, \mathcal{O}_n \leq \delta_n + \eta(\delta_n) \underset{n}{\longrightarrow} 0.$$

Now suppose $\Im z_k > 0$ and $z_k \underset{k}{\longrightarrow} 0$. The sequence $\{F(z_k)\}$ is eventually in *every* $F(S_n)$ and diam $F(S_n) \underset{n}{\longrightarrow} 0$. *Therefore* $F(z_k)$ converges to a definite *limit*, w, say, by *Cauchy's convergence criterion*, and w is *independent* of the particular sequence $\{z_k\}$ chosen of points tending to 0. By the proof in (iv), the distance of any point in $F(S_n) \subseteq \mathcal{O}_n$ to Γ is $\leq \delta_n + \eta(\delta_n)$ which $\underset{n}{\longrightarrow} 0$, so $w \in \Gamma$. (This could also be seen by the argument at the beginning of (ii).)

We have completed the proof of our third lemma.

Go back to $\Phi(z)$ which maps $\{|z| < 1\}$ conformally on $\mathscr{D}$. The lemma just proven shows that for *each* ζ, $|\zeta| = 1$, $\lim_{z \to \zeta} \Phi(z)$ *exists and is on* Γ.

Notation Call $\lim_{z \to \zeta} \Phi(z) = \Phi(\zeta)$ for $|\zeta| = 1$.

Lemma *$\Phi(\zeta)$, as thus defined, is continuous on $\{|\zeta| = 1\}$.*

The proof is easy and routine, so we omit it.

Lemma $\Phi(\zeta)$ *is one-one on* $\{|\zeta| = 1\}$.

Proof Assume not. Then for, say, α and β on the unit circle, $\alpha \neq \beta$, we have $\Phi(\alpha) = \Phi(\beta)$. Let σ be the path shown, and consider $\Phi(\sigma)$.

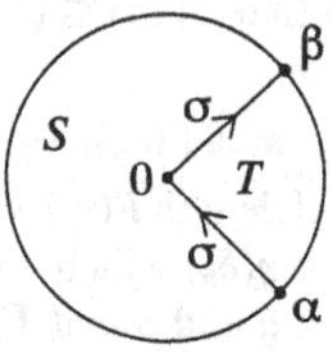

Since Φ is one-one in $\{|z| < 1\}$, it is easy to see that $\Phi(\sigma)$ is a Jordan curve lying entirely in $\mathscr{D}$ except for the one point, $\Phi(\alpha) = \Phi(\beta)$ which lies on Γ. A repetition of the *Jordan curve argument* shows that the *inside*, $\mathscr{O}$, of $\Phi(\sigma)$ lies in $\mathscr{D}$. Let S and T be the two sectors shown of $\{|z| < 1\}$. The argument used in step (iv) of the proof of the *third* lemma shows that $\Phi(S) \subseteq \mathscr{O}$ or $\Phi(T) \subseteq \mathscr{O}$. Without loss of generality, let $\Phi(S) \subseteq \mathscr{O}$. Then, if $z \in S$ and $|z| \to 1$, we must have $\Phi(z) \to \Phi(\alpha)$. Indeed all the limit points of $\Phi(z)$ *must be* on Γ if $|z| \to 1$, by the argument in (ii) above. Also, $\Phi(z)$ stays in $\mathscr{O}$ whose *only limit point* on Γ is $\Phi(\alpha)$. (Proof: $\mathscr{O} \subseteq \mathscr{D}$ so if w is a limit point of $\mathscr{O}$ which is on Γ (hence *not* in $\mathscr{D}$), w is *surely* $\notin \mathscr{O}$, so w is on the *boundary* of $\mathscr{O}$, $\Phi(\sigma)$. But the *only* point of $\Phi(\sigma)$ which lies on Γ is $\Phi(\alpha)$!) So $\Phi(z)$ necessarily $\to \Phi(\alpha)$.

Since the boundary of S includes a *whole arc* of $|\zeta| = 1$ (from α to β), and $\Phi(z)$ is *analytic* in $|z| < 1$, *it must in fact be constant* in $\{|z| < 1\}$ and equal to $\Phi(\alpha)$. This is absurd, and the lemma is proven.

With $\Phi(z)$ extended as indicated above to $\{|z| \leq 1\}$, it becomes *continuous* on that *closed disc* and takes $\{|z| = 1\}$ in *one-one fashion into* Γ.

Lemma Φ *takes* $\{|z| = 1\}$ *onto* Γ.

Proof Let $w_0 \in \Gamma$. By the Jordan curve theorem there is a sequence of points $w_n \in \mathscr{D}$ with $w_n \xrightarrow[n]{} w_0$. Let z_n, $|z_n| < 1$, be such that $\Phi(z_n) = w_n$, then $\Phi(z_n) \xrightarrow[n]{} w_0$. Without loss of generality, $z_n \xrightarrow[n]{} z_0$, $|z_0| \leq 1$. But $|z_0| < 1$ is *impossible*, for *then* $w_0 = \lim_{n\to\infty} \Phi(z_n) = \Phi(z_0)$ would be in $\mathscr{D}$. So $|z_0| = 1$, and, by the above-described extension of Φ, $\Phi(z_0) = w_0$.

Theorem *The conformal mapping* Φ *of* $\{|z| < 1\}$ *onto the inside* $\mathscr{D}$ *of* Γ *has a continuous one-one extension up to* $\{|z| = 1\}$ *and when thus extended takes* $\{|z| = 1\}$ *onto* Γ.

Proof Combine the last 4 lemmas.

Remark The result is due to Carathéodory, ca. 1912 or 1914. The above proof I learned, in part, from my teacher, *Rafael Robinson* – he gave the analytic argument of (i) in his course in Berkeley, in the autumn of 1951. I suspect that the whole argument consists of very well-known ideas.

2. Lindelöf's first theorem on conformal mapping

Suppose $\mathscr{D}$ is a connected region bounded by a Jordan curve Γ, and suppose, without loss of generality, that $0 \in \Gamma$. Suppose $f(z)$ maps $\{|z| < 1\}$ conformally onto $\mathscr{D}$; as we have seen, f has a continuous one-one extension up to $|z| = 1$, and, when thus extended, takes that circumference onto Γ, yielding a *parametrization* of Γ. We suppose, without loss of generality, that $f(1) = 0$:

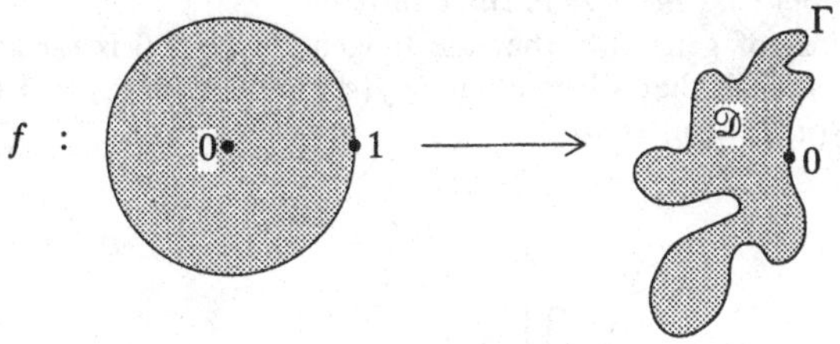

In 1917 (in the same Scandinavian Congress proceedings where the famous paper of F. and M. Riesz appeared), Lindelöf published a theorem about the case where Γ has a *tangent* at 0. This theorem states that for $|z| < 1$, $z \to 1$, we have

$$\arg f(z) - \arg(1 - z) \to \text{constant.}$$

It means that the *conformal images of sectors in* $|z| < 1$ *with their vertices at* 1 *are asymptotically like sectors in* $\mathscr{D}$ *of the same opening with their vertices at* 0:

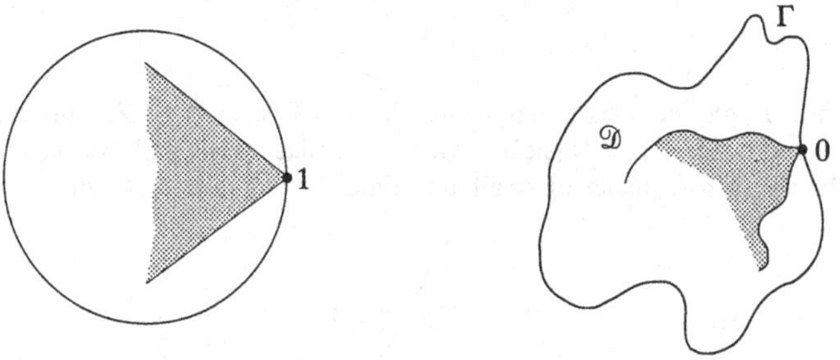

Let us, for the moment, grant the following:

Lemma $\arg f(z)$ *is bounded in* $|z| < 1$.

Then the proof of Lindelöf's theorem runs as follows:

Since $f(e^{i\theta})$ is continuous and $f(e^{i\theta}) \neq 0$ for $e^{i\theta} \neq 1$ (the Jordan curve Γ being *simple*), $\arg f(e^{i\theta})$ (defined as $\lim_{z \to e^{i\theta}} \arg f(z)$) is *continuous* for $e^{i\theta}$ away from 1; the lemma says it's *bounded*.

For Γ to have a *tangent* at 1 *means* that $\arg f(e^{i\theta})$ *tends* to a *certain limit*, say α, as $\theta \to 0+$, and tends to α + an *odd* multiple of π as $\theta \to 0-$. *It is reasonable from considerations of orientation that this multiple be simply* π; *granted this* (which will be proved after the lemma), we see that $\arg f(e^{i\theta}) - \arg(1 - e^{i\theta})$ is continuous at 0, where it equals $\alpha + (\pi/2)$. Since $\arg f(z) - \arg(1 - z)$ is *harmonic* and *bounded* in $|z| < 1$, we have, by Chapter I, Section C,

$$\arg f(re^{i\theta}) - \arg(1 - re^{i\theta}) = \frac{1}{2\pi} \int_{-\pi}^{\pi} \frac{1 - r^2}{1 + r^2 - 2r\cos(\theta - t)} \arg\left[\frac{f(e^{it})}{1 - e^{it}}\right] dt$$

for $|z| < 1$. The above-mentioned continuity now shows that

$$\arg f(z) - \arg(1 - z) \to \alpha + \frac{\pi}{2}$$

as $z \to 1$, by a theorem in Chapter I, Subsection D.2.

To make this argument rigorous, the lemma and the statement about $\arg f(e^{\pm i\theta})$ must be proved. *A rigorous proof uses the Jordan curve theorem.*

Assume without loss of generality that the tangent to Γ at 0 is *vertical*. Then, given $\eta > 0$ there is a $\delta > 0$ such that all points $w = f(e^{i\theta})$ on Γ with $|w| < \delta$ lie in one of the two sectors $|\arg w \pm (\pi/2)| < \eta$:

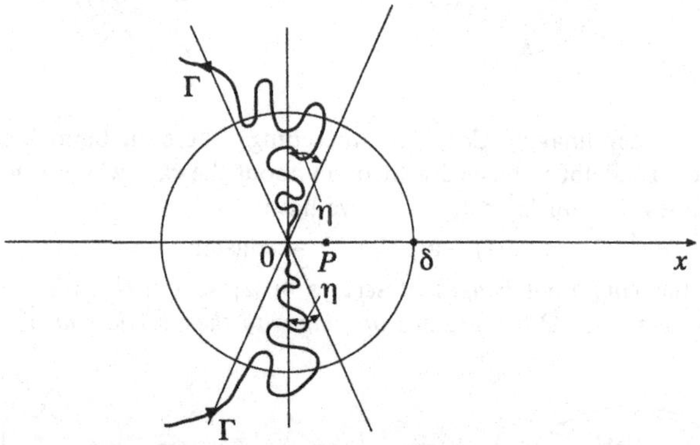

There are points P on the x-axis arbitrarily close to 0 with $P \notin \mathscr{D}$. Indeed, let Φ be a conformal mapping of $|z| > 1$ (sic!) onto the *outside* of Γ, and consider the curves $\gamma_n = \Phi(\lambda_n)$, where λ_n is a sequence of small arcs about 1 lying in $|z| > 1$, whose radius tends to 0

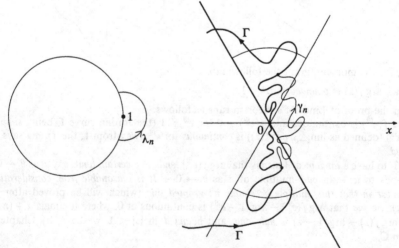

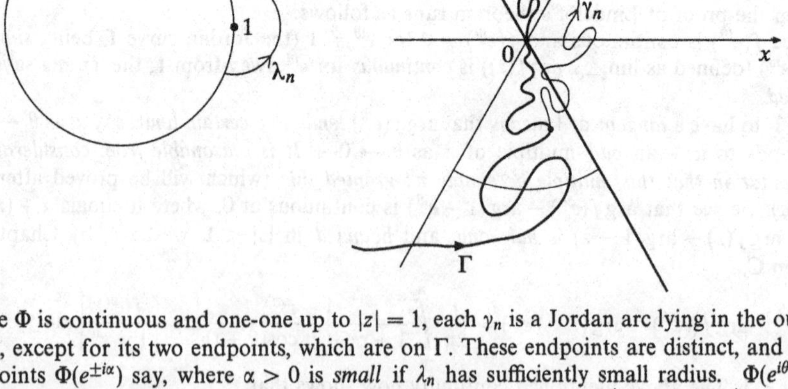

Since Φ is continuous and one-one up to $|z| = 1$, each γ_n is a Jordan arc lying in the outside of Γ, except for its two endpoints, which are on Γ. These endpoints are distinct, and equal to points $\Phi(e^{\pm i\alpha})$ say, where $\alpha > 0$ is *small* if λ_n has sufficiently small radius. $\Phi(e^{i\theta})$ is a parametrization of Γ, so, since Γ has a vertical tangent at 0, $\Phi(e^{i\alpha})$ and $\Phi(e^{-i\alpha})$ must lie on *opposite sides* of 0 in the union of the two *vertical sectors* $|\arg w \pm (\pi/2)| < \eta$, if $\alpha > 0$

is small. That means that γ_n *starts on one side of the* x-*axis and ends on the other side.* Therefore it *crosses* it at, say, P, so P lies in the *outside* of Γ.

Now if we do this construction with radius λ_n *sufficiently small*, we get $0 < |P| < \delta$.

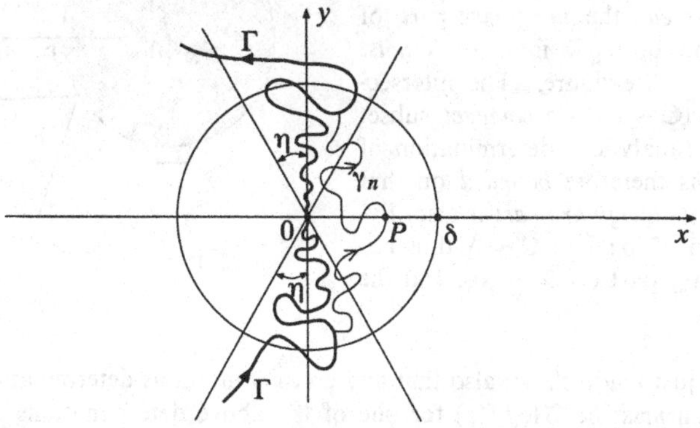

If, then, $P > 0$, the *whole sector* $|\arg w| < (\pi/2) - \eta$ in $|w| < \delta$ *lies in the outside* of Γ, for any w there can be joined to P by a line not touching Γ. A similar statement of course holds if $P < 0$.

Since $\{|z| < 1\}$ is simply connected, and $f(z)$ analytic and $\neq 0$ there, we *do have* single valued branches of $\log f(z)$ in that open disc. It is claimed that any one of those has *bounded imaginary part* there. This imaginary part *is* a determination of $\arg f(z)$ for $|z| < 1$, so establishment of the claim will prove the lemma.

Fix any one of the points $P \neq 0$ just found on the x-axis, with $|P| < \delta$ and $P \notin \mathscr{D}$ (indeed, $P \notin \overline{\mathscr{D}}$ since $P \notin \Gamma$). Without loss of generality we may assume $P > 0$ as in the above figure; then the interval $(0, P]$ on the x-axis is also disjoint from Γ.

Since our point P lies *outside* of Γ, there is a broken line path going out from P towards ∞ (*straight* where sufficiently far out) which never touches Γ. Taking Λ' to consist of the segment $[0, P]$ followed by that broken line path, we get a polygonal curve from 0 out towards ∞, touching Γ *only* at 0. Let us now *cut off* (one by one) any loops Λ' may have at the points where it crosses itself. The result is a *new* path Λ without self-intersections consisting of a segment $[0, Q]$, $0 < Q \leq P$, followed by a broken line σ from Q (sic!) out towards ∞ which never touches Γ. The whole path Λ, except for 0, lies in the *outside* of Γ. Therefore $\overline{\mathscr{D}} \sim \{0\}$ lies in the complement of $\mathbb{C} \sim \Lambda$.

That complement is, however, a *simply connected* domain not containing 0, so we have analytic single-valued branches of $\log w$ defined therein. Since $\overline{\mathscr{D}} \sim \{0\} \subseteq \mathbb{C} \sim \Lambda$, *any* single-valued analytic determination of $\log w$ in $\mathscr{D}$ extends to one in $\mathbb{C} \sim \Lambda$ by analytic continuation. For $|z| \leq 1$ and $z \neq 1$, we can thus substitute $w = f(z)$ ($\in \overline{\mathscr{D}}$ and $\neq 0$) into one of the functions $\log w$ just described and get a continuous single-valued determination of $\log f(z)$ for such z, analytic for $|z| < 1$. And *any* such determination of $\log f(z)$ in $\{|z| < 1\}$ *is obtained* in this manner from a suitable branch of $\log w$ defined in $\mathbb{C} \sim \Lambda$. *The lemma* will therefore follow if we show that *each* such determination of $\log w$ *has bounded imaginary part* on $\overline{\mathscr{D}} \sim \{0\} \subseteq \mathbb{C} \sim \Lambda$.

The portion σ of Λ has *positive distance* ε from 0, $0 < \varepsilon \leq Q$. Let S be the slit disk consisting of the w, $0 < |w| < \varepsilon$, with $w \notin (0, \varepsilon)$; then $(\mathbb{C} \sim \Lambda) \cap \{|w| < \varepsilon\} = S$. Since S is *connected*, the *imaginary part* of any determination of $\log w$ in $\mathbb{C} \sim \Lambda$ *must be bounded in* S – see figure. The intersection $(\bar{\mathscr{D}} \sim \{0\}) \cap (\mathbb{C} \sim S)$ is a *compact* subset of $\mathbb{C} \sim \Lambda$. Any (analytic!) determination of $\log w$ in $\mathbb{C} \sim \Lambda$ is therefore *bounded* on that intersection, and its *imaginary part* is also. Every determination of $\log w$ in $\mathbb{C} \sim \Lambda$ thus *has* bounded imaginary part on $\bar{\mathscr{D}} \sim \{0\}$, and the lemma is proved.

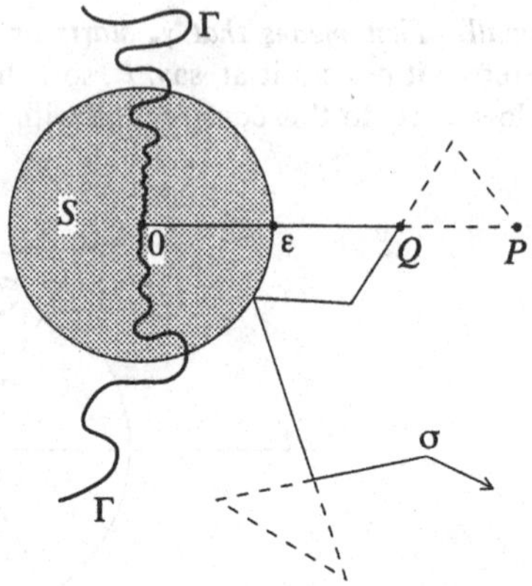

The argument just made shows also that any given continuous determination of $\arg f(z)$ for $|z| < 1$ (which *must be* $\Im \log f(z)$ for one of the above determinations of $\log f(z)$) is not only *bounded there*, but in fact has a *continuous extension* to $\{|z| = 1\} \sim \{1\}$. Such an extension gives us values of $\arg f(e^{i\theta})$ for $0 < |\theta| \leq \pi$.

Supposing always a vertical tangent, let us see why $\arg f(e^{-i\theta}) \cong \arg f(e^{i\theta}) + \pi$ if $\theta > 0$ *is small*. We have just seen that for $|z| < 1$, $|z - 1| < \epsilon$, say, the *variation* of $\arg f(z)$ is at most $\pi + 2\eta$, because then $|f(z)| < \delta$ and $f(z)$ is confined to a sector of opening $\pi + 2\eta$. So, since $|\arg f(e^{\pm i\theta}) \pm (\pi/2)| < \eta$ (modulo 2π) if θ is small, and Γ *has* a tangent at 0, *the only two possibilities are*

$$\arg f(e^{-i\theta}) \cong \arg f(e^{i\theta}) + \pi$$

and

$$\arg f(e^{-i\theta}) \cong \arg f(e^{i\theta}) - \pi.$$

We must confirm the first possibility and invalidate the second.

$f(z)$ is a *conformal* mapping of $|z| < 1$ onto $\mathscr{D}$, therefore, if $P \notin \mathscr{D}$ and $r < 1$, the variation of $\arg(f(z) - P)$ around the circle $z = re^{i\theta}$ is *zero*. Since $f(z)$ is *continuous* up to $|z| = 1$, we can make $r \to 1$, and we see that

$$\Delta_\Gamma \arg(w - P) = 0, \quad P \notin \mathscr{D}.$$

Since, on the other hand, $f(z)$ is conformal, by the principle of argument, if $P' \in \mathscr{D}$, for $r < 1$ sufficiently close to 1 the variation of $\arg(f(z) - P')$ around $z = re^{i\theta}$ is 2π. Making $r \to 1$, we get

$$\Delta_\Gamma \arg(w - P') = 2\pi$$

if $P' \in \mathscr{D}$.

We just saw in the proof of the lemma that there are $P \notin \mathscr{D}$ with $0 < |P|$ *arbitrarily small* and P on the x-axis; without loss of generality let these points be on the *positive* x-axis. Then, the whole circular sector $|\arg w| < (\pi/2) - \eta$, $|w| < \delta$, lies *outside* $\mathscr{D}$. An entirely similar argument shows that there are points P' *in* $\mathscr{D}$ on the x-axis with $|P'|$ arbitrarily small – *these* must then lie on the *negative* x-axis. We see that a circular sector of the form $(\pi/2) + \eta < \arg w < (3\pi/2) - \eta$, $|w| < \delta$, *lies in* $\mathscr{D}$.

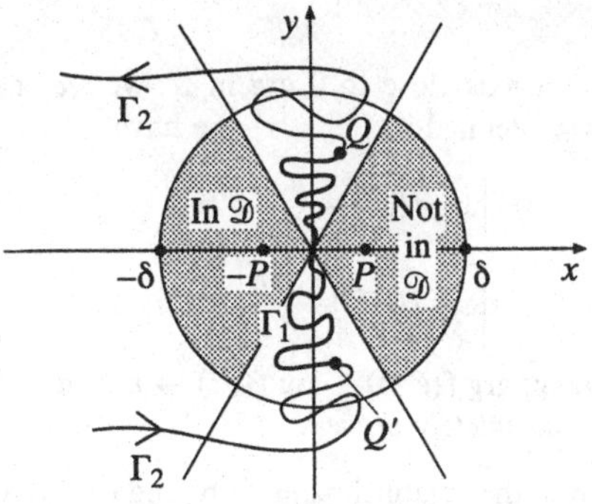

Now take a small fixed $\alpha > 0$, and put $Q = f(e^{i\alpha})$, $Q' = f(e^{-i\alpha})$. Let Γ_1 be the arc of Γ from Q' to Q (corresponding to $-\alpha \le \theta \le \alpha$) and Γ_2 the complementary arc corresponding to $\alpha \le \theta \le 2\pi - \alpha$. Let $P > 0$, *P small*, then

$$\Delta_\Gamma \arg(w - P) = 0,$$
$$\Delta_\Gamma \arg(w + P) = 2\pi,$$

as we have just seen.

So

$$\Delta_{\Gamma_1} \arg(w - P) + \Delta_{\Gamma_2} \arg(w - P) = 0,$$
$$\Delta_{\Gamma_1} \arg(w + P) + \Delta_{\Gamma_2} \arg(w + P) = 2\pi.$$

If P is close to zero, the two *second* terms are sensibly equal – Γ_2 stays away from 0 –, so we find

$$\Delta_{\Gamma_1} \arg(w + P) - \Delta_{\Gamma_1} \arg(w - P) \to 2\pi$$

as $P \to 0+$.

Now if $Q' = f(e^{-i\alpha})$ lies *below* 0 in the double sector

$$\left|\arg w \pm \frac{\pi}{2}\right| < \eta$$

and $Q = f(e^{i\alpha})$ *lies above*, then for P close to 0,

$$|\Delta_{\Gamma_1} \arg(w + P) - \pi| < 2\eta$$

and

$$|\Delta_{\Gamma_1} \arg(w - P) + \pi| < 2\eta.$$

Whereas, if Q' lies *above* 0 and Q lies *below* it,

$$|\Delta_{\Gamma_1} \arg(w + P) + \pi| < 2\eta,$$
$$|\Delta_{\Gamma_1} \arg(w - P) - \pi| < 2\eta.$$

The *first possibility* gives

$$\Delta_{\Gamma_1}(\arg(w + P) - \arg(w - P)) \cong 2\pi,$$

and the *second*

$$\Delta_{\Gamma_1}(\arg(w + P) - \arg(w - P)) \cong -2\pi.$$

So the first is correct:

$$f(e^{i\alpha}) \quad \text{lies } \textit{above } 0$$

and

$$f(e^{-i\alpha}) \quad \textit{below}.$$

Since points on the *negative* x-axis close to 0 *are* in $\mathscr{D}$, we *see* that, corresponding to an appropriate branch of $\log f(z)$ defined in $\{|z| < 1\}$, we have

$$\left|\arg f(e^{i\alpha}) - \frac{\pi}{2}\right| < 2\eta,$$

$$\left|\arg f(e^{-i\alpha}) - \frac{3\pi}{2}\right| < 2\eta,$$

and, since η is finally *arbitrary*, $\arg f(e^{-i\alpha}) - \arg f(e^{i\alpha}) \to \pi$ as $\alpha \to 0+$.

Lindelöf's theorem is thus completely proved.

An important application is the establishment of boundary behaviour results for certain functions analytic in regions bounded by *rectifiable* Jordan curves. *A rectifiable Jordan curve has a tangent at almost every one of its points.*

Therefore, *at almost every boundary point* of such a region (i.e., one bounded by a *rectifiable* Jordan curve), the *notion of non-tangential limit transfers over from the unit circle by conformal mapping*. Such a conformal mapping takes a sector of opening $< 180°$ in the unit circle, with its vertex at the pre-image of a boundary point, to a subdomain which is asymptotically like a sector of the *same* opening having its vertex at the boundary point.

This fact, together with the important theorem of F. and M. Riesz to be proved in Section D below, enables us to extend many of the results proven in the next chapter about boundary behaviour of functions analytic in $|z| < 1$ to corresponding results about boundary behaviour of functions analytic in a region bounded by a rectifiable Jordan curve.

3. Lindelöf's second theorem on conformal mapping

We continue with the discussion begun in the last subsection, but assume now that the Jordan curve Γ has a tangent at *each* of its points. The line tangent to Γ at w thereon will be denoted by T_w. As before, $f(z)$ is a fixed conformal mapping of $\{|z| < 1\}$ onto the domain $\mathscr{D}$ bounded by Γ.

Under the present circumstances, $f(z)$ (continuous, as we know, up to $\{|z| = 1\}$ and one-one even there) has, at every point $e^{i\theta_0}$ on the unit circumference, the property established above. This means that for *each* such point, any continuous single valued determination of $\arg(f(z) - f(e^{i\theta_0}))$ in $\{|z| < 1\}$ is *bounded* there, and *extends continuously* up to $\{|z| = 1\} \sim \{e^{i\theta_0}\}$, and that after *making* that extension,

$$\lim_{\theta \to \theta_0 +} \arg(f(e^{i\theta}) - f(e^{i\theta_0})) \quad \text{and} \quad \lim_{\theta \to \theta_0 -} \arg(f(e^{i\theta}) - f(e^{i\theta_0}))$$

both exist, with *the second limit equal to π plus the first one*. *Either* of those two limits can be taken as the argument of a vector parallel to $T_{f(e^{i\theta_0})}$; the two vectors thus specified point in opposite directions.

We wish to specify a tangent vector pointing in the direction 'of increasing θ'. That we do by making the

Definition $p(e^{i\theta_0})$ will denote the complex number of modulus 1 with argument equal to

$$\lim_{\theta \to \theta_0 +} \arg(f(e^{i\theta}) - f(e^{i\theta_0}))$$

(using any one of the single valued determinations of $\arg(f(e^{i\theta}) - f(e^{i\theta_0}))$ just described).

We now *further restrict* our attention to Jordan curves Γ *having continuously turning tangents*. By this we mean curves Γ having a tangent at each of their points such that, for w and w' anywhere on Γ, *the smaller angle made by* T_w *and* $T_{w'}$ *tends to zero whenever* $|w - w'| \to 0$.

Lemma *If* Γ *has a continuously turning tangent,* $p(e^{i\theta_0})$ *is a continuous function of* θ_0.

Proof We may, without loss of generality, take $\theta_0 = 0$ and $f(1) = 0$ with $p(1) = i$ (i.e., T_0 *vertical*) as in the discussion of the preceding subsection. Let us show that then $p(e^{i\theta}) \to i$ for $\theta \to 0+$; the argument for $\theta \to 0-$ is the same.

Fix a small $\eta > 0$; then, as in the passage referred to, $w = f(e^{i\theta})$ will lie in the sector $|\arg w - (\pi/2)| < \eta$ for all sufficiently small $\theta > 0$ (and in the sector $|\arg w - (3\pi/2)| < \eta$ for all $\theta < 0$ close enough to zero). This gives us an $\alpha > 0$ such that (for suitable specification of the argument) $|\arg f(e^{i\theta}) - (\pi/2)| < \eta$ when $0 < \theta < \alpha$. The hypothesis says that if $\alpha > 0$ is small enough, $T_{f(e^{i\theta})}$ *will also make an acute angle* $< \eta$ with the *vertical* (direction of T_0) for $0 < \theta < \alpha$.

The second property shows us at once that $p(e^{i\theta})$ is close to i or $-i$ when $0 < \theta < \alpha$, and the lemma will be proved if we show that the *first* possibility must hold. Assume then, reasoning by contradiction, that there *is* a θ_1, $0 < \theta_1 < \alpha$, with $|\arg p(e^{i\theta_1}) + (\pi/2)| < \eta$ (for suitable specification of the argument).

Using a suitable continuous determination of $\arg(f(z) - f(e^{i\theta_1}))$ on $\{|z| \le 1\} \sim \{e^{i\theta_1}\}$ we have, however,

$$\arg(f(e^{i\theta}) - f(e^{i\theta_1})) \to \arg p(e^{i\theta_1})$$

for $\theta \to \theta_1 +$, and then, by the result proved at the end of the last subsection,

$$\arg(f(e^{i\theta}) - f(e^{i\theta_1})) \to \arg p(e^{i\theta_1}) + \pi$$

for $\theta \to \theta_1 -$. Thence, and by the previous relation,

$$\left|\arg(f(e^{i\theta}) - f(e^{i\theta_1})) - \frac{\pi}{2}\right| < 2\eta$$

for all $\theta < \theta_1$ (sic!) sufficiently close to θ_1. This means that for such θ,

$$\Im f(e^{i\theta}) > \Im f(e^{i\theta_1}),$$

$\eta > 0$ having been chosen *small*.

At the same time, for $\theta = 0$,

$$\Im f(1) = 0 < \Im f(e^{i\theta_1}).$$

since $0 < \theta_1 < \alpha$ (making $|\arg f(e^{i\theta_1}) - (\pi/2)| < \eta$).

There must therefore be a θ_2, $0 < \theta_2 < \theta_1$, where $\Im f(e^{i\theta})$ *attains its maximum* for $0 \le \theta \le \theta_1$. But then, at $w_2 = f(e^{i\theta_2})$, T_{w_2} is *horizontal*, and hence makes angles of 90° with T_0. This contradicts the fact that T_{w_2} can only make an acute angle $< \eta$ with T_0, since $0 < \theta_2 < \alpha$.

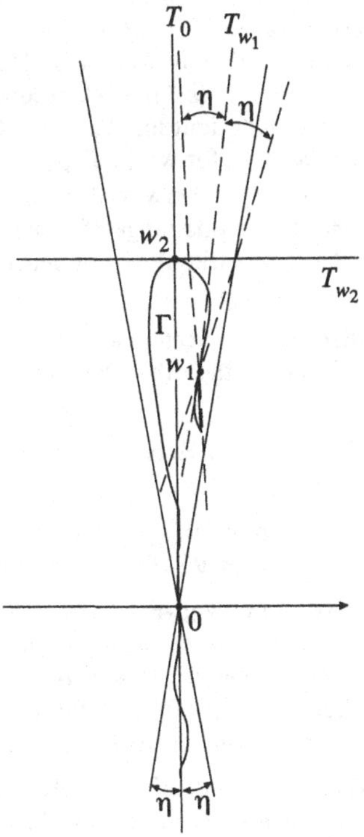

We must therefore have

$$|\arg p(e^{i\theta}) - (\pi/2)| < \eta$$

(for proper specification of the argument) when $0 < \theta < \alpha$, and see that $p(e^{i\theta}) \to i = p(1)$ for $\theta \to 0+$. Done.

Lemma *Suppose that* Γ *has a continuously turning tangent. Then, given any* $\varepsilon > 0$ *there* *is an* $\alpha > 0$ *such that, if we take, for* any θ, *any one of the continuous determinations of* $\arg(f(z) - f(e^{i\theta}))$ *on the set* $\{|z| \leq 1\} \sim \{e^{i\theta}\}$, *we will have, for* $0 < h < \alpha$,

$$\left|\arg(f(e^{i(\theta+h)}) - f(e^{i\theta})) - \lim_{\theta' \to \theta+} \arg(f(e^{i\theta'}) - f(e^{i\theta}))\right| < \varepsilon.$$

Remark Crudely stated, this says that the convergence of $\arg(f(e^{i\theta'}) - f(e^{i\theta}))$ to its limit for $\theta' \to \theta+$ is uniform in θ.

Proof of Lemma The function $p(e^{i\theta})$, continuous on the unit circumference by the preceding lemma, must there be *uniformly* continuous. Given $\varepsilon > 0$, without loss of generality $\varepsilon < \pi/2$, there is then an $\alpha > 0$ such that $|p(e^{i\theta'}) - p(e^{i\theta})| < 2\sin(\varepsilon/2)$ *whenever* $|\theta' - \theta| < \alpha$. It is this α which will do the job.

Fixing any θ_0 and any continuous determination of $\arg(f(z) - f(e^{i\theta_0}))$ for $|z| \leq 1$, $z \neq e^{i\theta_0}$, we write

$$A = \lim_{\theta \to \theta_0+} \arg(f(e^{i\theta}) - f(e^{i\theta_0})),$$

making surely

$$|\arg(f(e^{i\theta}) - f(e^{i\theta_0})) - A| < \varepsilon$$

for all $\theta > \theta_0$ *sufficiently close* to θ_0. What we have to do is prove that this inequality *persists*, as long as $0 < \theta - \theta_0 < \alpha$.

Assume that it *does not*. Then there is a *least* value of θ, $\theta_0 < \theta < \theta_0 + \alpha$, where $|\arg(f(e^{i\theta}) - f(e^{i\theta_0})) - A|$ becomes *equal* to ε, say for $\theta = \theta_1$. Denote then by $\mathscr{L}$ the (infinite!) straight line through $f(e^{i\theta_0})$ and $f(e^{i\theta_1})$, and observe that then there must be a θ_2, $\theta_0 < \theta_2 < \theta_1$, for which $T_{f(e^{i\theta_2})}$ is *parallel* to $\mathscr{L}$. This is indeed obvious if $f(e^{i\theta})$ *lies on* $\mathscr{L}$ for $\theta_0 \leq \theta \leq \theta_1$; otherwise there is a θ, $\theta_0 < \theta < \theta_1$, for which $f(e^{i\theta})$ is *as far away as possible from* $\mathscr{L}$ (for this range of values of θ), and then θ_2 may be taken equal to *that* value of θ.

Suppose without loss of generality that

$$\arg(f(e^{i\theta_1}) - f(e^{i\theta_0})) = A + \varepsilon.$$

Then the unit vectors $e^{i(A+\varepsilon)}$ and $-e^{i(A+\varepsilon)}$ are parallel to the line $\mathscr{L}$, so $p(e^{i\theta_2})$ must coincide with one or the other. We have, however, $p(e^{i\theta_0}) = e^{iA}$ *by definition*. Thence,

$$|p(e^{i\theta_2}) - p(e^{i\theta_0})| = |1 \pm e^{i\varepsilon}| \geq 2\sin(\varepsilon/2),$$

ε being $< \pi/2$. Here,

$$|\theta_2 - \theta_0| < \theta_1 - \theta_0 < \alpha,$$

and the choice of α is contradicted. This means that the inequality in question must remain valid for $\theta_0 < \theta < \theta_0 + \alpha$, and the lemma holds.

We can now give a result that is fundamental for any further study of how the curve Γ's regularity induces regularity properties of the conformal mapping $f(z)$. Since $f'(z) \neq 0$ for $|z| < 1$, single valued analytic branches of $\log f'(z)$ are defined there, and any two of them differ by a constant integral multiple of $2\pi i$. These branches give us the various continuous determinations of the (harmonic) function $\arg f'(z)$ in $\{|z| < 1\}$.

Theorem (Lindelöf) *If the curve* Γ *has a continuously turning tangent, each harmonic determination of* $\arg f'(z)$ *in* $\{|z| < 1\}$ *has a continuous extension to the closed unit disc.*

Proof Fix for the time being any *small* $h > 0$ (so as to ensure the behaviour guaranteed by the last lemma) and look at the ratio

$$g_h(z) = \frac{f(e^{ih}z) - f(z)}{e^{ih}z - z}.$$

When $z \to 0$, this tends to the value $f'(0) \neq 0$. The mapping $f(z)$ is actually continuous and one-one on $\{|z| \leq 1\}$ as we know, so $g_h(z)$ cannot vanish for $0 < |z| \leq 1$. That function is therefore continuous and $\neq 0$ on the *closed* unit disc. It is also *analytic* for $|z| < 1$, so we have analytic branches of $\log g_h(z)$ there. Our first task is to show that any one of those branches has a *continuous* extension up to $\{|z| = 1\}$. That we do by a tiling and patching construction.

Look at the function

$$F_h(s) = f(e^{s+ih}) - f(e^s)$$

for $\Re s \leq 0$ and close to 0 and for $0 \leq \Im s \leq 2\pi$. We will frequently write $s = \sigma + i\tau$ with σ and τ real.

$F_h(s)$ is continuous for $\Re s \leq 0$, and $F_h(s + 2\pi i) = F_h(s)$. $F_h(i\tau)$ is also bounded and *bounded away from* 0 for real τ. We therefore have a $\delta > 0$ such that, on any small square bordering the τ-axis of the form

$$Q_b = \{s;\ -\delta \leq \sigma \leq 0,\ b \leq \tau \leq b + \delta\},$$

single valued continuous determinations of $\log F_h(s)$ *are available.* Two such determinations on the *same* Q_b differ thereon by a constant integral multiple of $2\pi i$.

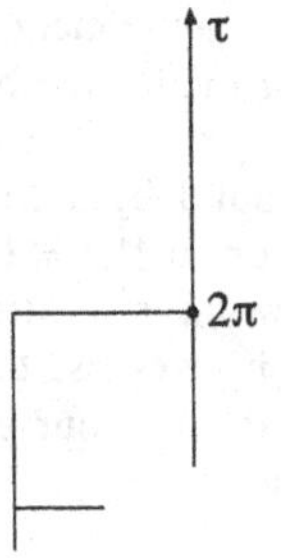

We may just as well take $\delta = 2\pi/N$, with N a large integer. Then the strip $\{s;\ -\delta \leq \sigma \leq 0,\ 0 \leq \tau \leq 2\pi\}$ is precisely covered ('*tiled*') by N non-overlapping squares Q_b, with b taking successively the values $0, 2\pi/N, 4\pi/N, \ldots, 2(N-1)\pi/N$. The squares lie one above the other, and two adjacent ones have a horizontal side in common. Choose now a continuous determination of $\log F_h(s)$ on each of these Q_b in such fashion that the determinations on any two of them that have a horizontal side in common coincide along that horizontal side. One may, for instance, pick a determination on the *lowest* square Q_0 at pleasure, then take on $Q_{2\pi/N}$ the determination matching the former one where $\tau = 2\pi/N$, and so forth. This gives us a continuous determination of $\log F_h(s)$ on the *whole strip*

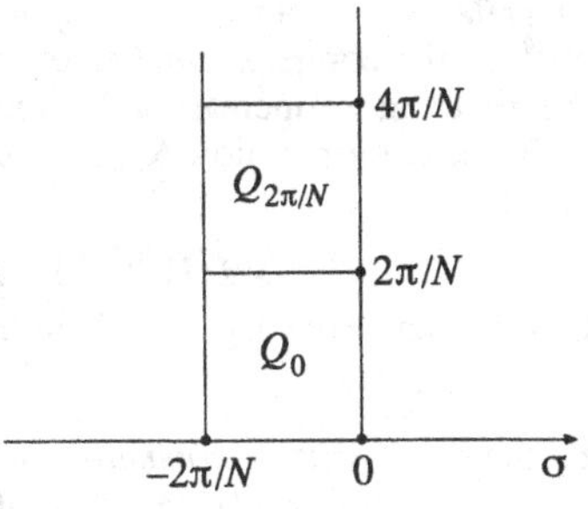

$$\{s;\ -2\pi/N \leq \sigma \leq 0,\ 0 \leq \tau \leq 2\pi\},$$

analytic in the interior of that strip.

For such a determination we have

$$\log F_h(\sigma + 2\pi i) = 2\pi i + \log F_h(\sigma).$$

Indeed, $f(e^{ih}z) - f(z)$ vanishes *precisely once* in $\{|z| < 1\}$, namely, for $z = 0$, so *any* continuous determination of

$$\log F_h(s) = \log(f(e^{s+ih}) - f(e^s))$$

must, *by the principle of argument*, increase by $2\pi i$ when $\tau = \Im s$ does and $\sigma = \Re s$ is held constant < 0. This must remain true for $\sigma = 0$ by continuity.

On account of this property of $\log F_h(s)$, we obtain a *continuous single valued determination of*

$$\log\left(\frac{f(e^{ih}z) - f(z)}{z}\right)$$

on the closed ring $\{e^{-2\pi/N} \leq |z| \leq 1\}$ by putting that logarithm equal to $\log F_h(s) - s$ for $z = e^s$, $-2\pi/N \leq \Re s \leq 0$, $0 \leq \Im s \leq 2\pi$. This gives us a continuous determination of

$$\log g_h(z) = \log\left(\frac{f(e^{ih}z) - f(z)}{z}\right) - \log(e^{ih} - 1)$$

on that closed ring.

We have at the same time the branches of $\log g_h(z)$ analytic for $|z| < 1$, each of which is *continuous on the closed disc* $\{|z| \leq e^{-\pi/N}\}$. Taking such a branch which *agrees* with the preceding determination of $\log g_h(z)$ on the *intersection* of that closed disc with the above closed ring (agreement of the two logarithms at *one point* of the intersection is already enough for this!), we obtain a *continuous single valued determination of* $\log g_h(z)$ *for* $|z| \leq 1$, *analytic in* $\{|z| < 1\}$. Any *other* branch of $\log g_h(z)$ analytic for $|z| < 1$ must differ from the one just found by a constant integral multiple of $2\pi i$, and hence *also* have a continuous extension up to $\{|z| = 1\}$.

We now fix our attention on *one* of the determinations of $\log g_h(z)$ continuous on $\{|z| \leq 1\}$; it gives rise to a determination of $\arg g_h(z) = \Im \log g_h(z)$, *continuous on the closed unit disc* and of course *harmonic* for $|z| < 1$. *This particular determination is henceforth denoted by*

$$\operatorname{Arg} g_h(z).$$

With its help we proceed to obtain a *continuous determination* of $\arg(p(e^{i\theta})/e^{i\theta})$ *on the unit circumference* under the assumption that $h > 0$ was fixed *small enough to begin with*. Here $p(e^{i\theta})$ is the complex number of modulus 1 figuring earlier in this subsection, pointing in the direction 'of increasing θ' along the tangent to Γ at the point $f(e^{i\theta})$.

To the determination $\operatorname{Arg} g_h(z)$ corresponds one of $\arg(f(e^{i(\theta+h)}) - f(e^{i\theta}))$ for $0 \leq \theta \leq 2\pi$, namely

$$\operatorname{Arg}(f(e^{i(\theta+h)}) - f(e^{i\theta})) = \operatorname{Arg} g_h(e^{i\theta}) + \theta + \arg(e^{ih} - 1);$$

here we may take $\arg(e^{ih} - 1) = (\pi + h)/2$. For *each* θ, $0 \leq \theta \leq 2\pi$, we now let

$$\operatorname{Arg}_\theta(f(z) - f(e^{i\theta}))$$

be the *continuous determination of* $\arg(f(z) - f(e^{i\theta}))$ *on* $\{|z| \leq 1\} \sim \{e^{i\theta}\}$ *which agrees with* $\operatorname{Arg}(f(e^{i(\theta+h)}) - f(e^{i\theta}))$ (as just defined) *for* $z = e^{i(\theta+h)}$. Existence of this determination is guaranteed by the work in the last subsection. If ε is any quantity > 0 (and without loss of generality $< \pi/2$), we *know* by the *second* of the above lemmas that *for a suitable initial choice of* $h > 0$,

$$\left|\operatorname{Arg}_\theta(f(e^{i(\theta+h)}) - f(e^{i\theta})) - \lim_{\theta' \to \theta+} \operatorname{Arg}_\theta(f(e^{i\theta'}) - f(e^{i\theta}))\right| < \varepsilon$$

for $0 \leq \theta \leq 2\pi$. Writing

$$A(\theta) = \lim_{\theta' \to \theta+} \operatorname{Arg}_\theta(f(e^{i\theta'}) - f(e^{i\theta}))$$

for $0 \leq \theta \leq 2\pi$, we thus have

$$\left|\operatorname{Arg}(f(e^{i(\theta+h)}) - f(e^{i\theta})) - A(\theta)\right| < \varepsilon$$

on that interval.

However, $p(e^{i\theta}) = e^{iA(\theta)}$ by definition. $A(\theta)$ is thus a determination of $\arg p(e^{i\theta})$ for $0 \leq \theta \leq 2\pi$, and we now put

$$\operatorname{Arg}\left(\frac{p(e^{i\theta})}{e^{i\theta}}\right) = A(\theta) - \theta, \quad 0 \leq \theta \leq 2\pi;$$

it is claimed that the left side is *continuous* (and *single valued*) *on the unit circumference.*

By the preceding relations, we have

$$\left|\operatorname{Arg} g_h(e^{i\theta}) + \frac{\pi+h}{2} - A(\theta) + \theta\right| < \varepsilon$$

for $0 \leq \theta \leq 2\pi$, with $\operatorname{Arg} g_h(e^{i\theta})$ *continuous* (and single valued!) on the unit circumference. Here $A(\theta)$ is at each θ a *possible value* of $\arg p(e^{i\theta})$, where $p(e^{i\theta})$, *of modulus* 1, is a *continuous* function of θ by the *first* of the above lemmas. Therefore, since any two such values must differ by an integral multiple of 2π and $\varepsilon < \pi/2$, *$A(\theta)$ must also be continuous for $0 \leq \theta \leq 2\pi$.* By the same token, $A(\theta) - \theta$ must take *identical values for $\theta = 0$ and for $\theta = 2\pi$*, since that is true of $\operatorname{Arg} g_h(e^{i\theta})$. Continuity of $\operatorname{Arg}(p(e^{i\theta})/e^{i\theta})$ on the unit circumference is thus established, and we have

$$\left|\operatorname{Arg}\left(\frac{p(e^{i\theta})}{e^{i\theta}}\right) - \operatorname{Arg} g_h(e^{i\theta}) - \frac{\pi+h}{2}\right| < \varepsilon$$

thereon.

Because $\operatorname{Arg} g_h(z)$, continuous for $|z| \leq 1$, is *harmonic* in $\{|z| < 1\}$, we have there

$$\operatorname{Arg} g_h(z) = \frac{1}{2\pi}\int_0^{2\pi} \frac{1-|z|^2}{|z-e^{i\theta}|^2} \operatorname{Arg} g_h(e^{i\theta})\, d\theta$$

by Poisson's formula. This and the preceding relation now yield

$$\left|\operatorname{Arg}\left(\frac{f(e^{ih}z)-f(z)}{e^{ih}z-z}\right) + \frac{\pi}{2} - \frac{1}{2\pi}\int_0^{2\pi} \frac{1-|z|^2}{|z-e^{i\theta}|^2} \operatorname{Arg}\left(\frac{p(e^{i\theta})}{e^{i\theta}}\right) d\theta\right| < \varepsilon + \frac{h}{2}$$

for $|z| < 1$.

Consider now what happens if we start with *some other small value* $h' > 0$ of h, corresponding to a value ε', $0 < \varepsilon' < \pi/2$, of the ε figuring in the second of the above lemmas. By going through exactly the same reasoning, we get, corresponding to a continuous determination $\operatorname{Arg}' g_{h'}(z)$ of $\arg g_{h'}(z)$ on the closed unit disc, some *continuous* (single-valued) determination $\operatorname{Arg}'(p(e^{i\theta})/e^{i\theta})$ of $\arg(p(e^{i\theta})/e^{i\theta})$ *on the unit circumference*, with

$$\left|\operatorname{Arg}' g_{h'}(z) + \frac{\pi}{2} - \frac{1}{2\pi}\int_0^{2\pi} \frac{1-|z|^2}{|z-e^{i\theta}|^2} \operatorname{Arg}'\left(\frac{p(e^{i\theta})}{e^{i\theta}}\right) d\theta\right| < \varepsilon' + \frac{h'}{2}$$

for $|z| < 1$. Of course $\operatorname{Arg}'(p(e^{i\theta})/e^{i\theta})$ *need not agree* with $\operatorname{Arg}(p(e^{i\theta})/e^{i\theta})$ on the unit circumference, but since *both* are *continuous* there, they *must differ by a constant integral multiple of* 2π. Adding such a multiple to $\operatorname{Arg}' g_{h'}(z)$ gives us a *new* continuous determination of $\arg g_{h'}(z)$, *which we may denote by* $\operatorname{Arg} g_{h'}(z)$, on $\{|z| \leq 1\}$, and we have then

$$\left|\operatorname{Arg} g_{h'}(z) + \frac{\pi}{2} - \frac{1}{2\pi}\int_0^{2\pi} \frac{1-|z|^2}{|z-e^{i\theta}|^2} \operatorname{Arg}\left(\frac{p(e^{i\theta})}{e^{i\theta}}\right) d\theta\right| < \varepsilon' + \frac{h'}{2}$$

for $|z| < 1$.

We take finally a sequence of values $h_n > 0$ of h tending to zero (corresponding to values of ε tending to 0) and get, in the way just described, determinations

$$\operatorname{Arg} g_{h_n}(z) = \operatorname{Arg}\left(\frac{f(e^{ih_n}z)-f(z)}{e^{ih_n}z-z}\right)$$

of $\arg g_{h_n}(z)$ tending *uniformly* to

$$\frac{1}{2\pi}\int_0^{2\pi} \frac{1-|z|^2}{|z-e^{i\theta}|^2}\operatorname{Arg}\left(\frac{p(e^{i\theta})}{e^{i\theta}}\right)\,d\theta - \frac{\pi}{2}$$

for $|z| < 1$. The limit of the functions $\operatorname{Arg} g_{h_n}(z)$ in $\{|z| < 1\}$ must, being *finite* (and harmonic) there, coincide with some determination of the argument of

$$\lim_{n\to\infty} \frac{f(e^{ih_n}z) - f(z)}{e^{ih_n}z - z} = f'(z)$$

for $|z| < 1$. Denoting that determination by $\operatorname{Arg} f'(z)$, we have, for $|z| < 1$,

$$\boxed{\operatorname{Arg} f'(z) = \frac{1}{2\pi}\int_0^{2\pi} \frac{1-|z|^2}{|z-e^{i\theta}|^2}\operatorname{Arg}\left(\frac{p(e^{i\theta})}{e^{i\theta}}\right)\,d\theta - \frac{\pi}{2},}$$

which *exhibits a continuous extension* of $\operatorname{Arg} f'(z)$ to the *closed unit disc*, equal, for $z = e^{i\theta}$, to $\operatorname{Arg}(p(e^{i\theta})/e^{i\theta}) - \pi/2$. Any *other* harmonic determination of $\arg f'(z)$ in $\{|z| < 1\}$ differs there from the *present* one by a constant integral multiple of 2π, and hence *also* has a continuous extension up to the unit circumference. The theorem is proved.

Remark The boxed formula obtained at the end of the last proof has its uses; it is usually written

$$\operatorname{Arg} f'(z) = \frac{1}{2\pi}\int_0^{2\pi} \frac{1-|z|^2}{|z-e^{i\theta}|^2}\operatorname{Arg}\left(\frac{n(e^{i\theta})}{e^{i\theta}}\right)\,d\theta,$$

with $n(e^{i\theta}) = p(e^{i\theta})/i$ designating the *unit outward normal* to the curve Γ at the point $f(e^{i\theta})$.

The theorem itself (as well as the formula) will be applied in Section F of Chapter V in conjunction with other material of that chapter, so as to obtain some results about the conformal mapping function $f(z)$ depending on the smoothness properties of Γ.

D. Domains bounded by rectifiable Jordan curves

Consider now a domain $\mathscr{D}$ bounded by a rectifiable Jordan curve Γ. Let Φ be a conformal mapping of $\{|z| < 1\}$ onto $\mathscr{D}$:

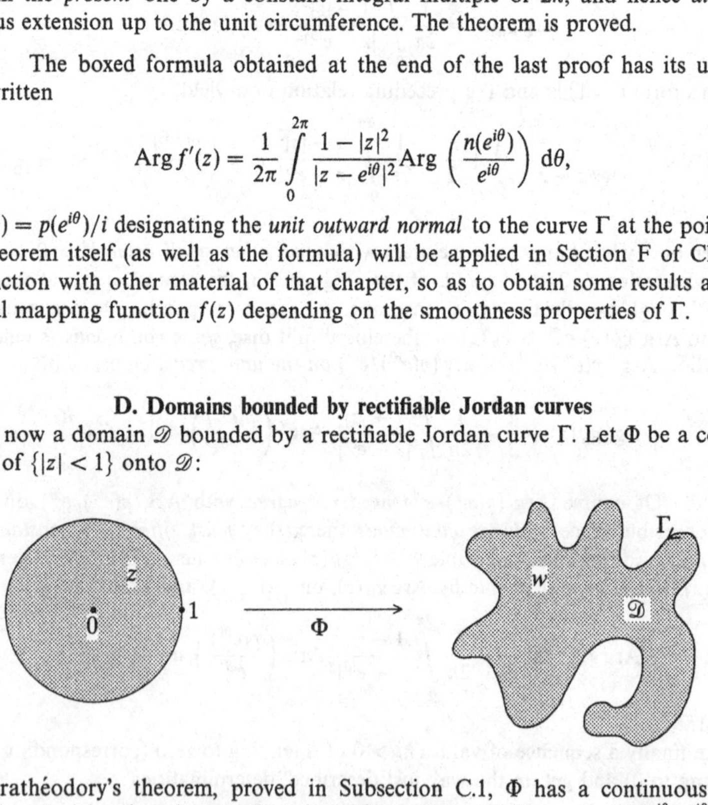

By Carathéodory's theorem, proved in Subsection C.1, Φ has a continuous one-one extension up to $|z| = 1$ and maps that circumference *onto* Γ. So surely, if $[e^{i\theta_0}, e^{i\theta_1}, \ldots, e^{i\theta_p}]$ is a *partition* of $\{|z| = 1\}$, $[\Phi(e^{i\theta_0}), \Phi(e^{i\theta_1}), \ldots, \Phi(e^{i\theta_p})]$ is a *partition* of Γ.

1. Derivative of conformal mapping function is in H_1

Theorem $\Phi'(z) \in H_1$.

Proof (Beckenbach) Let $\varepsilon = e^{2\pi i/n}$. Then, for $|z| < 1$,

$$S(z) = |\Phi(\varepsilon z) - \Phi(z)| + |\Phi(\varepsilon^2 z) - \Phi(\varepsilon z)| + \ldots + |\Phi(\varepsilon^n z) - \Phi(\varepsilon^{n-1} z)|$$

is *subharmonic*, and $S(z)$ is continuous for $|z| \leq 1$ because $\Phi(z)$ is. So by *principle of maximum*, if $|z| < 1$,

$$S(z) \leq \max_{|\zeta|=1} S(\zeta).$$

But if $|\zeta| = 1$, the points

$$[\Phi(\zeta),\ \Phi(\varepsilon\zeta),\ \Phi(\varepsilon^2\zeta), \ldots, \Phi(\varepsilon^n\zeta)]$$

form a partition of Γ, so by *definition of curve length*(!)

$$S(\zeta) \leq \text{length}\,\Gamma < \infty.$$

Therefore if $|z| < 1$, $S(z) \leq \text{length}\,\Gamma$. Now fix $r < 1$. We will have

$$\sum_{k=1}^{n} |\Phi(\varepsilon^k r) - \Phi(\varepsilon^{k-1} r)| = S(r) \leq \text{length}\,\Gamma.$$

Making $n \to \infty$ and using continuity of $\Phi'(re^{i\theta})$ *in* θ for $r < 1$, we get in limit

$$\int_0^{2\pi} |\Phi'(re^{i\theta})| r \,\mathrm{d}\theta \leq \text{length}\,\Gamma,$$

and since this is valid for all $r < 1$, $\Phi'(z) \in H_1$. Q.E.D.

2. Image of a set of measure zero on unit circumference

By Subsection C.1, if Λ is an arc of Γ, Φ maps some arc J of $\{|z| = 1\}$ in one-one bicontinuous fashion onto Λ. The very definition of arc length now gives us

$$\text{length}\,\Lambda = \int_J |\,\mathrm{d}\Phi(e^{i\theta})|.$$

Theorem (F. and M. Riesz) *If J is an arc of the unit circle and* $\Lambda = \Phi(J)$,

$$\text{length}\,\Lambda = \int_J |\Phi'(e^{i\theta})| \,\mathrm{d}\theta.$$

Proof By the theorem of the preceding section $\Phi'(z) \in H_1$, so by Subsection B.2,

$$\int_{-\pi}^{\pi} |\Phi'(e^{i\theta}) - \Phi'(re^{i\theta})| \,\mathrm{d}\theta \to 0$$

for $r \to 1$.

If an open arc J is given, we can find a continuous function $T(\theta)$, $|T(\theta)| \leq 1$, *vanishing identically outside* J, such that

$$\left| \int_J T(\theta) \,\mathrm{d}\Phi(e^{i\theta}) - \int_J |\,\mathrm{d}\Phi(e^{i\theta})| \right| < \varepsilon,$$

$$\left| \int_J T(\theta) \cdot ie^{i\theta}\Phi'(e^{i\theta}) \,\mathrm{d}\theta - \int_J |\Phi'(e^{i\theta})| \,\mathrm{d}\theta \right| < \varepsilon,$$

$\varepsilon > 0$ being arbitrary. To see this, choose $T(\theta)$ so as to have

$$\int_J |T(\theta) - U(\theta)| \, |\,\mathrm{d}\Phi(e^{i\theta})| < \varepsilon,$$

where $U(\theta)$ is the Borel function of modulus 1 making

$$|\,d\Phi(e^{i\theta})| = U(\theta)\,d\Phi(e^{i\theta}).$$

Because J is an *open* arc, we can get such a $T(\theta)$ vanishing outside J. Note that $|\Phi'(e^{i\theta})|\,d\theta$ is just the *absolutely continuous* part of $|\,d\Phi(e^{i\theta})|$, and $ie^{i\theta}\Phi'(e^{i\theta})\,d\theta$ that of $d\Phi(e^{i\theta})$. *Here, we can even take $T(\theta)$ to be continuously differentiable.* Then, integrating by parts,

$$\int_J T(\theta)\,d\Phi(e^{i\theta}) = -\int_J \Phi(e^{i\theta})T'(\theta)\,d\theta = -\lim_{r\to 1}\int_J \Phi(re^{i\theta})T'(\theta)\,d\theta$$

by the theorem of Subsection C.1. But, for each $r < 1$,

$$-\int_J \Phi(re^{i\theta})T'(\theta)\,d\theta = \int_J ire^{i\theta}\Phi'(re^{i\theta})T(\theta)\,d\theta,$$

which, by the remark made at the beginning, must tend to

$$\int_J ie^{i\theta}\Phi'(e^{i\theta})T(\theta)\,d\theta$$

as $r \to 1$. This is, by choice of T, within ε of $\int_J |\Phi'(e^{i\theta})|\,d\theta$, whilst the quantity we started with, which is *equal* to it, is within ε of $\int_J |\,d\Phi(e^{i\theta})|$. Since this last is the same as length Λ, we get the theorem on making $\varepsilon \to 0$.

Arc length on Γ can be used in an evident fashion to define linear measure on Γ. First if $\mathcal{O} \subseteq \Gamma$ is (relatively) open on Γ, $\mathcal{O}$ is a disjoint countable union of open arcs Λ_k of Γ, and we take $|\mathcal{O}| = \sum_k \text{length}\,\Lambda_k$. Then, for $E \subseteq \Gamma$ arbitrary, we take $|E|$ to be

$$\inf\{|\mathcal{O}|;\ \mathcal{O} \supseteq E, \mathcal{O} \text{ open on } \Gamma\}.$$

Using the one-one bicontinuity of Φ as a mapping from $\{|z| = 1\}$ to Γ it is now easy to build upon the above theorem and see that

$$|\Phi(E)| = \int_E |\Phi'(e^{i\theta})|\,d\theta$$

for Borel sets E on the unit circumference.

We have especially the important

Theorem (F. and M. Riesz) *If E is on unit circumference and $|E| = 0$, then $|\Phi(E)| = 0$.*

Proof Let the Ω_n be open sets on $\{|z| = 1\}$, $\Omega_n \supset \Omega_{n+1} \supset \ldots$, $\Omega_n \supset E$, such that $|\Omega_n| \xrightarrow[n]{} 0$. Then $|\Phi(E)| \le |\Phi(\Omega_n)|$ for all n. But by the previous theorem and the discussion immediately following it,

$$|\Phi(\Omega_n)| = \int_{\Omega_n} |\Phi'(e^{i\theta})|\,d\theta,$$

which goes to zero as $n \to \infty$ because $|\Omega_n| \xrightarrow[n]{} 0$ and $\Phi'(e^{i\theta}) \in L_1(-\pi, \pi)$. Q.E.D.

3. Taylor series of mapping function converges absolutely in closed unit disk

Theorem (Hardy) *The power series of $\Phi(z)$ converges absolutely up to $|z| = 1$.*

Proof By Subsection 1, $\Phi'(z) \in H_1$. Also, since Φ is *conformal*, $\Phi'(z)$ never $= 0$ in $|z| < 1$. Therefore we can define an analytic $\Psi(z) = \sqrt{\Phi'(z)}$ in $|z| < 1$. Now write, for $|z| < 1$,

$$\Phi'(z) = \sum_0^\infty a_n z^n.$$

Then we have

$$\Psi(z) = \sum_0^\infty b_n z^n.$$

with

$$b_0^2 = a_0, \quad b_1 b_0 + b_0 b_1 = a_1, \quad b_2 b_0 + b_1 b_1 + b_0 b_2 = a_2, \quad \text{etc.}$$

Since $\Phi'(z) \in H_1$,

$$\int_{-\pi}^{\pi} |\Psi(re^{i\theta})|^2 \, d\theta$$

is *bounded* for $r < 1$. By Parseval, this yields

$$\sum_0^\infty |b_n|^2 < \infty.$$

Now write

$$\psi(z) = \sum_0^\infty |b_n| z^n.$$

By Parseval,

$$\int_{-\pi}^{\pi} |\psi(re^{i\theta})|^2 \, d\theta$$

is bounded for $|z| < 1$.

Let $\Theta(z) = [\psi(z)]^2 = \sum_0^\infty A_n z^n$, say. Then, on the one hand, $\Theta(z) \in H_1$, and on the other,

$$A_n = \sum_0^n |b_k||b_{n-k}| \geq \left| \sum_0^n b_k b_{n-k} \right| = |a_n|.$$

We have

$$\Phi(z) = c_0 + \sum_0^\infty \frac{a_n}{n+1} z^{n+1},$$

so to prove absolute convergence of power series of $\Phi(z)$ up to $|z| = 1$, we need to show

$$\sum_0^\infty \frac{|a_n|}{n+1} < \infty.$$

The above computation guarantees this if $\sum_0^\infty A_n/(n+1) < \infty$.

Now for $|z| < 1$, using the principal determination of the logarithm,

$$-\frac{\pi}{2} < \Im \log(1-z) < \frac{\pi}{2}$$

and

$$\log(1-z) = -\sum_1^\infty \frac{z^n}{n},$$

so that

$$\Im \log(1-z) = \frac{i}{2} \sum_{-\infty}^{\infty}{}' \frac{\operatorname{sgn} n}{|n|} r^{|n|} e^{in\theta}$$

for $z = re^{i\theta}$, $0 \leq r < 1$, whence, with $\Theta(z) = \sum_0^\infty A_n z^n$, using absolute convergence and orthogonality,

$$-\pi i \sum_1^\infty \frac{r^{2n} A_n}{n} = \int_0^{2\pi} \Theta(re^{i\theta}) \Im \log(1 - re^{i\theta})\, d\theta$$

and this is in absolute value

$$\leq \frac{\pi}{2} \int_0^{2\pi} |\Theta(re^{i\theta})|\, d\theta$$

which is $\leq M$ independently of $r < 1$. Since $A_n \geq 0$, we get, making $r \to 1$, $\sum_1^\infty A_n/n \leq M/\pi$, as required.

Problem 3 Let Γ be a rectifiable Jordan curve bounding a domain $\mathscr{D}$. If $f(z)$ is *analytic* in $\mathscr{D}$ and *continuous* on $\overline{\mathscr{D}}$, prove that $\int_\Gamma f(z)\, dz = 0$. (Hint: use conformal mapping.)

III

Elementary Boundary Behaviour Theory for Analytic Functions

A. Existence of non-tangential boundary values a.e.

Recall, from Chapter II Section B (or simply from the general results of Chapter I):

If $F(z) \in H_1$, $\quad F(e^{i\theta}) = \lim_{z \underset{\angle}{\to} e^{i\theta}} F(z)$ exists and is finite for almost all θ.

In particular, we have the

Corollary (Fatou) *If $F(z)$ is regular and* bounded *in $\{|z| < 1\}$, it has non-tangential boundary values almost everywhere on the unit circumference.*

B. Uniqueness theorem for H_1 functions

Theorem *Let $F(z) \in H_1$ and suppose, for a set E of positive measure, that $F(e^{i\theta}) = 0$ for $\theta \in E$. Then $F(z) \equiv 0$.*

Proof Without loss of generality, $0 < |E| < 2\pi$. Put, for $0 \le r < 1$,

$$U(re^{i\theta}) = \frac{1}{2\pi|E|}\int_E P_r(\theta - t)\,dt - \frac{1}{2\pi(2\pi - |E|)}\int_{[0,2\pi]\sim E} P_r(\theta - t)\,dt.$$

Then $U(z)$ is harmonic for $|z| < 1$ and

$$-\frac{1}{2\pi - |E|} \le U(z) \le \frac{1}{|E|}$$

there.

Besides, $U(0) = 0$ and especially

$$U(z) \to -\frac{1}{2\pi - |E|} \quad \text{as} \quad z \underset{L}{\longrightarrow} e^{i\theta}$$

for almost all θ *in* $[0, 2\pi] \sim E$. Using $\tilde{U}(z)$ to denote the harmonic conjugate of $U(z)$, we see that

$$\varphi(x) = \exp[U(z) + i\tilde{U}(z)]$$

is *bounded* in $|z| < 1$, $|\varphi(0)| = 1$, and for almost all $\theta \in [0, 2\pi] \sim E$,

$$|\varphi(e^{i\theta})| = \exp\left(-\frac{1}{2\pi - |E|}\right).$$

(A.e. existence of $\varphi(e^{i\theta})$ is guaranteed by Section A above!)

Since $\varphi(z)$ is bounded, for each k, $[\varphi(z)]^k F(z) \in H_1$. Suppose $F(z) \not\equiv 0$; we may, without loss of generality, suppose that $F(0) \neq 0$; otherwise we just work with the H_1 function $F(z)/z^m$ instead of $F(z)$, m being the order of the zero F has at the origin. By Chapter II, Section B we now get

$$[\varphi(0)]^k F(0) = \frac{1}{2\pi} \int_0^{2\pi} [\varphi(e^{i\theta})]^k F(e^{i\theta})\, d\theta = \frac{1}{2\pi} \int_{[0,2\pi]\sim E} [\varphi(e^{i\theta})]^k F(e^{i\theta})\, d\theta,$$

since $F(e^{i\theta}) = 0$, $\theta \in E$. From this,

$$|F(0)| = |\varphi(0)|^k |F(0)| \leq \frac{1}{2\pi} \exp\left(-\frac{k}{2\pi - |E|}\right) \int_{[0,2\pi]\sim E} |F(e^{i\theta})|\, d\theta$$

which tends to zero as $k \to \infty$ ($F(e^{i\theta})$ is in L_1!). So $F(0) = 0$, a contradiction. We are done.

C. More existence theorems for limits

1. Analytic functions with positive real part

Theorem *If* $f(z)$ *is regular in* $|z| < 1$ *and* $\Re f(z) \geq 0$ *there,* $\lim_{z \underset{L}{\to} e^{i\theta}} f(z)$ *exists and is finite for almost all* θ.

Proof (Titchmarsh) $F(z) = 1/(1 + f(z))$ is regular and bounded in $|z| < 1$ so, by Section A, $F(z) \to$ a finite limit $F(e^{i\theta})$ for almost all θ as $z \underset{L}{\longrightarrow} e^{i\theta}$. Since $f(z) = (1/F(z)) - 1$, $f(z)$ tends to the finite limit $(1/F(e^{i\theta})) - 1$ as $z \underset{L}{\longrightarrow} e^{i\theta}$ unless $F(e^{i\theta}) = 0$. Since, for $F \neq 0$, $|F(e^{i\theta})| > 0$ a.e. by Section B, we are done.

2. A.e. existence of $\tilde{F}(\theta)$ for F in L_1

Recall the notation of Chapter I, Section E, where (Subsection E.4) we proved existence of $\tilde{F}(\theta)$ a.e. for $F \in L_2(-\pi, \pi)$.

Theorem *Let* $F(\theta + 2\pi) = F(\theta)$ *and* $F \in L_1(-\pi, \pi)$. *Then the principal value*

$$\tilde{F}(\theta) = \frac{1}{\pi} \int_{-\pi}^{\pi} \frac{F(t)}{2\tan((\theta - t)/2)}\, dt$$

exists and is finite a.e.

Proof (Titchmarsh) Without loss of generality, $F(\theta) \geq 0$. Then, put

$$U(re^{i\theta}) = \frac{1}{2\pi} \int_{-\pi}^{\pi} P_r(\theta - t) F(t)\, dt,$$

and let $\tilde{U}(z)$ be the harmonic conjugate of $U(z)$.

$f(z) = U(z) + i\tilde{U}(z)$ is regular in $\{|z| < 1\}$ and $\Re f(z) \geq 0$ there. So by Subsection 1 above, $f(z) \to f(e^{i\theta})$ finite for almost all θ as $z \underset{L}{\longrightarrow} e^{i\theta}$. In particular, for almost all θ,

$$\lim_{r \to 1} \tilde{U}(re^{i\theta}) = \Im f(e^{i\theta})$$

exists and is finite. By Chapter I, Subsection E.3,

$$\frac{1}{2\pi} \int_{1-r<|t-\theta|<\pi} \frac{F(t)}{\tan((\theta - t)/2)}\, dt \; - \; \tilde{U}(re^{i\theta}) \; \to \; 0$$

as $r \to 1$ for almost all θ. This does it.

D. Privalov's uniqueness theorem

1. Privalov's ice-cream cone construction

Definition If $|\zeta| = 1$,

$$S_\zeta = \left\{ z;\ |z| > \frac{1}{\sqrt{2}},\ |\arg(\zeta - z)| < \frac{\pi}{4} \right\}.$$

S_1 is the shaded region shown in the following figure:

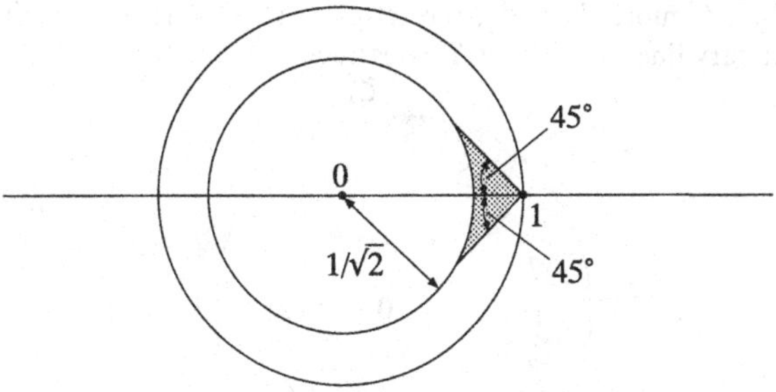

We make a series of obvious remarks:

(a) $\bigcup_{|\zeta|=1} S_\zeta$ is all of $\{1/\sqrt{2} < |z| < 1\}$.

(b) If $1/\sqrt{2} < |z| < 1$ for some z, $\{\zeta;\ |\zeta| = 1\ \&\ z \in S_\zeta\}$ is the (open) arc $\widehat{\zeta_1, \zeta_2}$ of the unit circle constructed as follows:

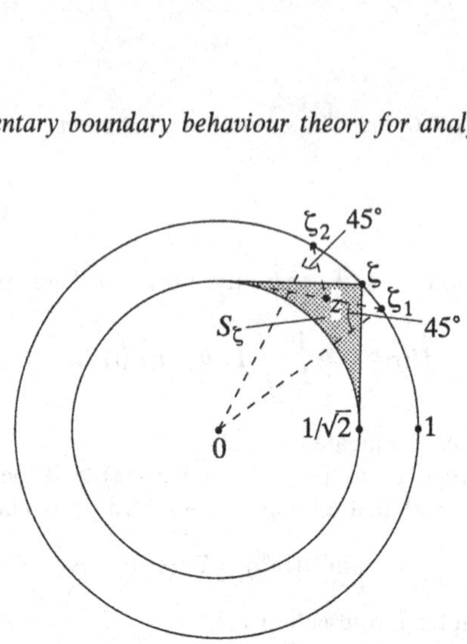

(c) If $J = \widehat{\zeta_1, \zeta_2}$ is an arc of the unit circle with opening $\leq 90°$, the set of z, $1/\sqrt{2} < |z| < 1$, such that an S_ζ contains z *only* for some $\zeta \in J$, consists of the closed curvilinear triangle T constructed upon J in the following manner:

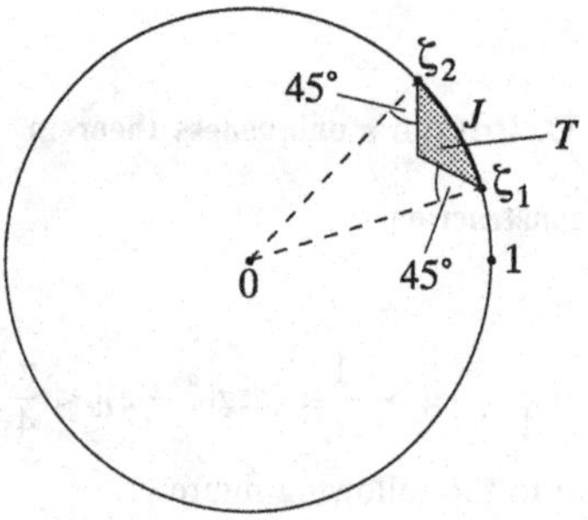

This follows very easily from (a) and (b).

If the arc J subtends more than 90°, the set of z, $1/\sqrt{2} < |z| < 1$, with $z \in S_\zeta$ *only* for $\zeta \in J$ is the closed curvilinear *trapezoid* T constructed thus:

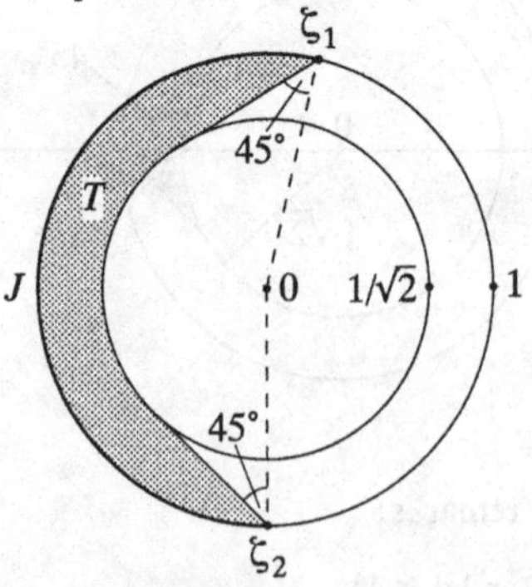

Now we can describe Privalov's construction.

Given a closed set E on the unit circumference, let $\{J_k\}$ be the (at most countable) set of arcs on that circumference *contiguous* (complementary) to E. Using each arc J_k as a base construct upon it the triangle or trapezoid T_k according to the recipe in (c) above. Take the closed domain

$$\overline{\mathscr{D}} = \{|z| \le 1\} \sim \bigcup_k T_k^\circ \sim \bigcup_k J_k.$$

(The superscript $^\circ$ denotes *interior* and the *bar* denotes *closure*.)

Our domain $\overline{\mathscr{D}}$ has the following important property:

> Every $z \in \overline{\mathscr{D}}$ of modulus $> 1/\sqrt{2}$ is in an $\overline{S}_\zeta$ for some $\zeta \in E$.

This follows directly from remarks (a) and (c) above.

This is a picture of $\overline{\mathscr{D}}$:

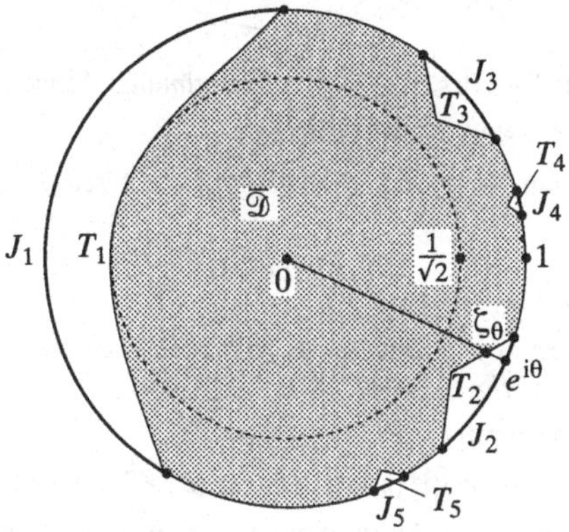

Note that $\partial\overline{\mathscr{D}}$ is a *Jordan curve*. Indeed, if ζ_θ denotes the single point where the ray from 0 to $e^{i\theta}$ hits $\partial\overline{\mathscr{D}}$, ζ_θ is a continuous one-one mapping from the unit circumference onto $\partial\overline{\mathscr{D}}$. This boundary of $\overline{\mathscr{D}}$ is even a *rectifiable* Jordan curve, because, for each k, perimeter $T_k \le C|J_k|$, where C is a geometric constant whose value we need not calculate.

We denote by $\mathscr{D}$ the *interior* of $\overline{\mathscr{D}}$. *We see that the results of Chapter* II, *Sections* C, D *all apply to* $\mathscr{D}$.

Remark Besides being used to prove Privalov's theorem below, $\mathscr{D}$ (or its analogue for the upper half-plane) comes up in Chapter VIII in the study of maximal functions and of Carleson measures.

2. Use of Egoroff's theorem

Theorem *Let $f(z)$ be analytic in $|z| < 1$, and put, for $|\zeta| = 1$,*

$$M_f(\zeta) = \sup\{|f(z)|;\ z \in S_\zeta\}.$$

Then $M_f(\zeta)$ is Lebesgue measurable.

Proof For $n \ge 3$, take $r_n = 1 - (1/n)$ and put, for $|\zeta| = 1$,

$$M_n(\zeta) = \sup\left\{|f(z)|;\ \frac{1}{\sqrt{2}} \le |z| \le r_n \ \&\ |\arg(r_n\zeta - z)| \le \frac{\pi}{4}\right\}.$$

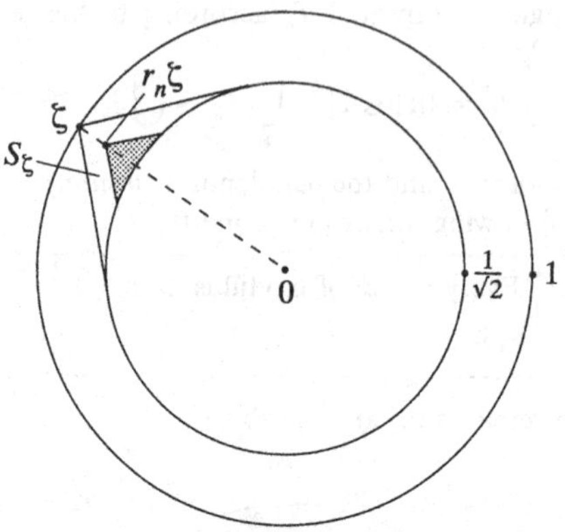

Because $f(z)$ is continuous for $|z| \le r_n$, $M_n(\zeta)$ is *continuous*. Since clearly $M_n(\zeta) \xrightarrow[n]{} M_f(\zeta)$ pointwise in ζ, $M_f(\zeta)$ is Lebesgue measurable.

Theorem *Let $f(z)$ be regular in $|z| < 1$. Suppose there is a set G of positive measure on the unit circumference such that*

$$\lim_{z \xrightarrow[L]{} \zeta} f(z) = 0$$

for each $\zeta \in G$. Then there is a closed *set E, $|E| > 0$, such that $f(z) \to 0$* uniformly *for $|z| \to 1$ and z in the union of the S_ζ with $\zeta \in E$.*

Proof For $n \ge 3$ and $|\zeta| = 1$ put

$$P_n(\zeta) = \sup\{|f(z)|;\ z \in S_\zeta\ \&\ |z| \ge 1 - (1/n)\}.$$

The argument used in the proof of the previous theorem shows that each $P_n(\zeta)$ is Lebesgue measurable; so, then, is the set G' of ζ for which $P_n(\zeta) \xrightarrow[n]{} 0$.

By hypothesis, $P_n(\zeta) \xrightarrow[n]{} 0$ for $\zeta \in G$, and $|G| > 0$. Therefore $|G'| > 0$, and Egoroff's theorem gives us a measurable $E_0 \subseteq G'$, $|E_0| > 0$, with $P_n(\zeta) \xrightarrow[n]{} 0$ *uniformly* for $\zeta \in E_0$. Taking a closed $E \subseteq E_0$ with $|E| > 0$, we have the theorem.

3. Privalov's uniqueness theorem

Theorem (Privalov, ca. 1917) *Let $f(z)$ be analytic in $|z| < 1$ and suppose there is a set G of* positive measure *on the unit circumference with*

$$\lim_{z \xrightarrow[L]{} \zeta} f(z) = 0 \quad \text{for } \zeta \in G.$$

Then f vanishes identically.

Proof By Subsection 2 we can find a closed E, $|E| > 0$, on the unit circumference with $f(z) \to 0$ *uniformly* as $|z| \to 1$ if z is in the union of the S_ζ, $\zeta \in E$. This means that if we make Privalov's construction, described in Subsection 1, starting from E, we will obtain a domain $\mathscr{D} \subset \{|z| < 1\}$ with $f(z) \to 0$ *uniformly* for $|z| \to 1$, $z \in \mathscr{D}$.

Looking at the construction in Subsection 1, we see that $\partial\mathscr{D}$ consists of some segments in $\{|z| < 1\}$ going out to points of E on the unit circumference, together with the set E thereupon. Therefore, if we *define* $f(z)$ to be *zero* on E, we get a function *continuous on* $\overline{\mathscr{D}}$ whose restriction to $\mathscr{D}$ is analytic.

As explained in Subsection 1, $\partial\mathscr{D}$ is a *rectifiable* Jordan curve. Take any conformal mapping φ of $\{|w| < 1\}$ onto $\mathscr{D}$ and, for $|w| < 1$, put $F(w) = f(\varphi(w))$. By Carathéodory's theorem (Chapter II, Section C) φ actually extends continuously up to $\{|w| = 1\}$ and maps that circumference in continuous one-one fashion onto $\partial\mathscr{D}$. This means that $F(w)$ extends continuously up to $\{|w| = 1\}$ since $f(z)$ extends continuously up to $\partial\mathscr{D}$. Let $S = \varphi^{-1}(E)$. Then $F(w) = 0$ for $w \in S$. Since $|E| > 0$, *E has positive linear measure as a subset of the rectifiable curve $\partial\mathscr{D}$. So by a theorem of F. and M. Riesz* (Chapter II, Section D), $|S| > 0$. Since $F(w)$ is regular in $\{|w| < 1\}$, *continuous* on the *closed unit circle, and zero on S, it therefore follows that* $F \equiv 0$ by Section B. So $f(z) \equiv 0$. Q.E.D.

Remark 1 Privalov's theorem is valid for functions $f(z)$ which are *merely supposed meromorphic* in $|z| < 1$. Indeed, the theorems of Subsection 2 hold (with the same proofs) for such functions, and we can proceed as above at the start of the proof of Privalov's theorem, getting a domain $\mathscr{D}$ such that $f(z) \to 0$ *uniformly* for $|z| \to 1$, $z \in \mathscr{D}$. That means we can certainly find an $r < 1$ such that $f(z)$ has *no poles* in $\mathscr{D}$ for $|z| > r$. So the only poles of f *in* $\mathscr{D}$ must be for $|z| \le r$. But, f being meromorphic in the *whole unit circle, its poles in $|z| \le r$ are finite in number*, say $z_1, \dots, z_m$. Then

$$g(z) = (z - z_1)(z - z_2)\cdots(z - z_m)f(z)$$

is regular in $\mathscr{D}$, continuous on $\overline{\mathscr{D}}$, and zero on $E \subset \partial\mathscr{D}$. The remaining part of the proof of Privalov's theorem is now applied to g instead of f, and one sees as before that $g \equiv 0$, hence $f \equiv 0$.

Remark 2 In the statement of Privalov's theorem (where *no* growth restrictions are imposed on $|f(z)|$ for $|z| \to 1$) it is *crucial* that *non-tangential boundary values* (on the set of positive measure where these limits are supposed to vanish) *and not merely radial ones be involved*. One can, indeed, construct non-zero functions $f(z)$ regular for $|z| < 1$, for which*

$$\lim_{r\to 1} f(re^{i\theta}) = 0 \quad \text{a.e.,} \quad 0 \le \theta \le 2\pi.$$

We do not give such a construction here. One can be found in Privalov's book on boundary properties of analytic functions, whose second Russian edition came out around 1950. There is a German translation. More recently, the construction given in that book has been simplified, I believe, by Rudin†.

E. Generalizations of the Schwarz reflection principle

1. Schwarz reflection for H_1 functions

Theorem *Let $F \in H_1$ and suppose that $\Im F(\zeta) = 0$ a.e. for ζ belonging to an* arc *J of the unit circle. Then $F(z)$ can be continued analytically across J into $\{|z| > 1\}$ by putting $F(z^*) = \overline{F(z)}$ for $|z| < 1$, where z^* denotes $1/\bar{z}$.*

Proof By Chapter II, Section B, for $0 \le r < 1$,

$$F(re^{i\theta}) = \frac{1}{2\pi}\int_{-\pi}^{\pi} P_r(\theta - t)F(e^{it})\,dt,$$

* I thank P. J. Rippon for having pointed out to me that 'a.e.' was missing from the following relation in the first edition.

† His simplification is reproduced in the book by Collingwood and Lohwater.

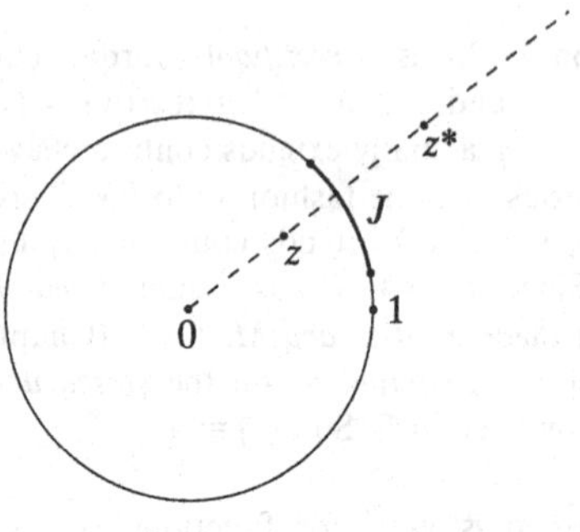

whence

$$\Im F(re^{i\theta}) = \frac{1}{2\pi}\int_{-\pi}^{\pi} P_r(\theta - t)\Im F(e^{it})\,dt,$$

which reduces (with slight abuse of notation!) to

$$\frac{1}{2\pi}\int_{[-\pi,\pi]\sim J} P_r(\theta - t)\Im F(e^{it})\,dt$$

by hypothesis. From Chapter 1, Section D we now see that, as $r \to 1$, $\Im F(re^{i\theta}) \to 0$ *uniformly for* $e^{i\theta}$ *ranging over any proper closed subarc of* J.

The result now follows from the classical Schwarz reflection principle.

2. A theorem of Carleman

Here is a result of Carleman on analytic continuation which has some applications in harmonic analysis.

Theorem *Let F and G be in H_1, let J be an arc of the unit circle, and suppose that $F(\zeta) = \overline{G(\zeta)}$ a.e. for $\zeta \in J$. Then $F(z)$ can be analytically continued across J by putting, for $|z| < 1$, $F(z^*) = \overline{G(z)}$, where $z^* = 1/\bar{z}$.*

Proof For $|z| < 1$, write $S(z) = F(z) + G(z)$ and $D(z) = F(z) - G(z)$; S and D are in H_1, and by hypothesis,

$$\Im S(\zeta) = 0 \quad \text{a.e.,} \quad \zeta \in J,$$

$$\Re D(\zeta) = 0 \quad \text{a.e.,} \quad \zeta \in J.$$

Therefore by Subsection 1 the functions $S(z)$ and $D(z)$ can be analytically continued across J by putting, for $|z| < 1$ and $z^* = 1/\bar{z}$,

$$S(z^*) = \overline{S(z)},$$

$$D(z^*) = -\overline{D(z)}.$$

Therefore $F(z) = (S(z) + D(z))/2$ also continues analytically across J by putting, for $|z| < 1$,

$$F(z^*) = (\overline{S(z)} - \overline{D(z)})/2 = \overline{G(z)}.$$ Q.E.D.

Problem 4 Let $U(z)$ be harmonic in $|z| < 1$, and let $V(z)$ be a harmonic conjugate of $U(z)$ (in $|z| < 1$). Suppose there is a measurable subset E of the unit circumference such that $\lim_{z \underset{L}{\to} \zeta} U(z)$ exists and is finite for every $\zeta \in E$. Show that $\lim_{z \underset{L}{\to} \zeta} V(z)$ *then* exists and is finite for almost every $\zeta \in E$. *A fairly concise solution is required.*

IV

Application of Jensen's Formula. Factorization into a Product of Inner and Outer Functions

A. Blaschke products

If $0 < |z_n| < 1$, $n = 1, 2, 3, \ldots$, and the infinite product

$$\prod_1^\infty \frac{|z_n|}{z_n} \frac{z_n - z}{1 - \bar{z}_n z}$$

*converges absolutely** for $|z| < 1$, it represents a certain function, analytic there, called a *Blaschke product*. We can even allow *a finite number* of the z_n to be *zero*, in which case the factors corresponding to

$$\frac{|z_n|}{z_n} \frac{z_n - z}{1 - \bar{z}_n z}$$

are simply replaced by z.

1. Convergence criterion

We have

$$\frac{|z_n|}{z_n} \frac{z_n - z}{1 - \bar{z}_n z} = |z_n| \frac{1 - (z/z_n)}{1 - \bar{z}_n z} = |z_n| + \frac{(\bar{z}_n - (1/z_n))|z_n| z}{1 - \bar{z}_n z} = |z_n| + \frac{|z_n|^2 - 1}{1 - \bar{z}_n z} \frac{|z_n| z}{z_n}.$$

Therefore

$$\frac{|z_n|}{z_n} \frac{z_n - z}{1 - \bar{z}_n z} = 1 + \{|z_n| - 1\} \left\{ 1 + \frac{(|z_n| + 1)|z_n|}{z_n(1 - \bar{z}_n z)} \cdot z \right\},$$

so the infinite product converges absolutely *for* $z = 0$ iff

$$\sum_n (1 - |z_n|) < \infty.$$

* see Subsections 1 and 2 below

But if $\sum_n(1-|z_n|)<\infty$, *by the same formula just found*,

$$\sum_n\left|1-\frac{|z_n|}{z_n}\frac{z_n-z}{1-\bar{z}_n z}\right| = \sum_n[1-|z_n|]\left|\frac{(|z_n|+1)|z_n|}{z_n(1-\bar{z}_n z)}\cdot z\right| < \infty$$

for $|z|<1$, so the infinite product converges absolutely in $\{|z|<1\}$ (and indeed *uniformly* on compact subsets of that disk) when $\sum_n(1-|z_n|)<\infty$. Thus:

$\prod_n \frac{|z_n|}{z_n}\frac{z_n-z}{1-\bar{z}_n z}$ converges absolutely in $\{|z|<1\}$ iff $\sum_n(1-|z_n|)<\infty$.

2. Boundary values have modulus 1 a.e.

Let $\sum_n(1-|z_n|)<\infty$, so that

$$\prod_n \frac{|z_n|}{z_n}\frac{z_n-z}{1-\bar{z}_n z}$$

converges absolutely and uniformly on compact subsets of $\{|z|<1\}$ and represents a function $B(z)$ *analytic there*. By elementary complex variable theory, *each factor* of the product is in modulus <1 for $|z|<1$, so $|B(z)|<1$ for $\{|z|<1\}$.

Therefore, for almost all ζ, $|\zeta|=1$, $B(\zeta)=\lim B(z)$ for $z \underset{L}{\longrightarrow} \zeta$ *exists* (Fatou's theorem, Chapter III, Section A).

Theorem $|B(e^{i\theta})|=1$ *a.e.*

Proof Without loss of generality, all $|z_n|>0$ (otherwise work with $B(z)/z^k$ instead of $B(z)$). Then $\log|B(0)|=\sum_n\log|z_n|$. Now because $\sum_n(1-|z_n|)<\infty$, $\sum_n\log|z_n|>-\infty$ (N.B. $\log|z_n|<0$ for each n !). Let $0<r<1$, with r not equal to any $|z_n|$. Then, by the most elementary form of Jensen's formula,

$$\log|B(0)| = \sum_{|z_n|<r}\log\left(\frac{|z_n|}{r}\right)+\frac{1}{2\pi}\int_{-\pi}^{\pi}\log|B(re^{i\theta})|\,d\theta,$$

i.e.,

$$\sum_n\log|z_n| = \sum_{|z_n|<r}\log\left(\frac{|z_n|}{r}\right)+\frac{1}{2\pi}\int_{-\pi}^{\pi}\log|B(re^{i\theta})|\,d\theta.$$

That is,

$$\frac{1}{2\pi}\int_{-\pi}^{\pi}\log|B(re^{i\theta})|\,d\theta = \sum_{|z_n|<r}\log\left(\frac{r}{|z_n|}\right)-\sum_n\log\frac{1}{|z_n|}.$$

Pick any fixed p so that $\sum_{n>p}\log(1/|z_n|)<\epsilon$, and take $r<1$ so close to 1 that $|z_n|<r$ if $n=1,2,\ldots,p$. Then, the previous relation yields

$$\frac{1}{2\pi}\int_{-\pi}^{\pi}\log|B(re^{i\theta})|\,d\theta \geq \sum_1^p\log\left(\frac{r}{|z_n|}\right)-\sum_1^p\log\frac{1}{|z_n|}-\epsilon,$$

or, making $r<1$ *close enough* to 1,

$$\frac{1}{2\pi}\int_{-\pi}^{\pi}\log|B(re^{i\theta})|\,d\theta > -2\epsilon.$$

That means

$$\limsup_{r\to 1-} \frac{1}{2\pi}\int_{-\pi}^{\pi} \log|B(re^{i\theta})|\,d\theta \geq 0$$

since $\epsilon > 0$ is arbitrary.

But $B(re^{i\theta}) \to B(e^{i\theta})$ a.e. as $r \to 1$ and

$$\log|B(re^{i\theta})| \leq 0.$$

Therefore, by *Fatou's lemma* (use a *sequence* of r's tending to 1)

$$\frac{1}{2\pi}\int_{-\pi}^{\pi} \log|B(e^{i\theta})|\,d\theta \geq 0.$$

Since $|B(e^{i\theta})| \leq 1$, we have $\log|B(e^{i\theta})| = 0$ a.e. Q.E.D.

B. Factorizing out Blaschke products

1. Formation of a Blaschke product having same zeros as a given analytic function

Theorem *Let $F(z) \not\equiv 0$ be regular in $\{|z| < 1\}$, and let the z_n be its zeros there, $|z_n| < 1$. Suppose that*

$$\int_{-\pi}^{\pi} \log|F(re^{i\theta})|\,d\theta$$

is bounded above for $r < 1$. Then

$$\sum_n (1 - |z_n|) < \infty,$$

so that

$$B(z) = \prod_n \frac{|z_n|}{z_n}\frac{z_n - z}{1 - \bar{z}_n z}$$

converges absolutely for $\{|z| < 1\}$, and we have $F(z) = B(z)G(z)$ where $G(z)$ is regular and has no zeros in $\{|z| < 1\}$.

Proof Without loss of generality, $F(0) \neq 0$ – if not, work with $F(z)/z^k$ instead of $F(z)$. Then, if $0 < r < 1$, and no $|z_n| = r$, by Jensen's formula:

$$\log|F(0)| = \sum_{|z_n|<r} \log\left|\frac{z_n}{r}\right| + \frac{1}{2\pi}\int_{-\pi}^{\pi} \log|F(re^{i\theta})|\,d\theta,$$

i.e., by hypothesis,

$$\sum_{|z_n|<r} \log\left|\frac{r}{z_n}\right| \leq M - \log|F(0)|,$$

where M is independent of r. Making $r \to 1$, we get, for any fixed p,

$$\sum_{1}^{p} \log\frac{1}{|z_n|} \leq M - \log|F(0)|,$$

so

$$\sum_1^\infty \log \frac{1}{|z_n|} < \infty.$$

The existence of $B(z)$ now follows by Subsection A.1. Defining, in $\{|z| < 1\}$, $G(z) = F(z)/B(z)$, we're done.

2. The class H_p. Factorization

Definition If $p > 0$ (sic!), H_p is the set of $F(z)$ analytic in $\{|z| < 1\}$ with

$$\sup_{0 \le r < 1} \int_{-\pi}^{\pi} |F(re^{i\theta})|^p \, d\theta < \infty;$$

(H_1 is a special case of this definition).

Theorem *Let $F(z) \not\equiv 0$ belong to H_p. Then there is a Blaschke product $B(z)$ and a $G(z) \in H_p$ with $F(z) = B(z)G(z)$, $G(z)$ without zeros in $\{|z| < 1\}$.*

Proof If $r < 1$, by the inequality between arithmetic and geometric means,

$$\frac{1}{2\pi} \int_{-\pi}^{\pi} p \log |F(re^{i\theta})| \, d\theta \le \log \frac{1}{2\pi} \int_{-\pi}^{\pi} |F(re^{i\theta})|^p \, d\theta$$

which by hypothesis is $\le \log C$ independent of r. So, by Subsection 1, if the z_n are the zeros of $F(z)$ in $\{|z| < 1\}$, $\sum_n (1 - |z_n|) < \infty$ and the Blaschke product $B(z)$ can be formed. If $G(z) = F(z)/B(z)$, $G(z)$ has no zeros in $\{|z| < 1\}$.

Write

$$B_N(z) = \prod_1^N \frac{|z_n|}{z_n} \frac{z_n - z}{1 - \bar{z}_n z}.$$

The product for $B(z)$ converges absolutely in $\{|z| < 1\}$ and, for $r < 1$, $B_N(z) \underset{N}{\longrightarrow} B(z)$ *uniformly* on $|z| \le r$. Pick such an r for which no $|z_n| = r$.

Then

$$\int_{-\pi}^{\pi} |G(re^{i\theta})|^p \, d\theta = \lim_{N \to \infty} \int_{-\pi}^{\pi} \left| \frac{F(re^{i\theta})}{B_N(re^{i\theta})} \right|^p d\theta.$$

But for each N,

$$G_N(z) = \frac{F(z)}{B_N(z)}$$

is regular in $|z| < 1$, so $|G_N(z)|^p$ is subharmonic there and, for fixed $r < 1$,

$$\int_{-\pi}^{\pi} |G_N(re^{i\theta})|^p \, d\theta \le \limsup_{R \to 1} \int_{-\pi}^{\pi} |G_N(Re^{i\theta})|^p \, d\theta.$$

However, for *fixed* N, as is easily seen, $|B_N(Re^{i\theta})| \to 1$ *uniformly* as $R \to 1$, so

$$\limsup_{R \to 1} \int_{-\pi}^{\pi} |G_N(Re^{i\theta})|^p \, d\theta = \limsup_{R \to 1} \int_{-\pi}^{\pi} |F(Re^{i\theta})|^p \, d\theta$$

which by hypothesis is $\leq 2\pi C$, say. Thus, for each $r < 1$,

$$\int_{-\pi}^{\pi} |G_N(re^{i\theta})|^p \, d\theta \leq 2\pi C$$

for all N, and finally, by the previous,

$$\int_{-\pi}^{\pi} |G(re^{i\theta})|^p \, d\theta \leq 2\pi C.$$

Since $r < 1$ is arbitrary, $G \in H_p$. Q.E.D.

Scholium We see that, in the above factorization,

$$\sup_{r<1} \int_{-\pi}^{\pi} |G(re^{i\theta})|^p \, d\theta = \sup_{r<1} \int_{-\pi}^{\pi} |F(re^{i\theta})|^p \, d\theta.$$

For the argument at the end of the above proof yields

$$\sup_{r<1} \int_{-\pi}^{\pi} |G(re^{i\theta})|^p \, d\theta \leq \sup_{r<1} \int_{-\pi}^{\pi} |F(re^{i\theta})|^p \, d\theta.$$

Since, however, $F = BG$ and $|B(z)| \leq 1$ in $\{|z| < 1\}$, the reverse inequality must also hold.

Corollary *If $f \not\equiv 0$ belongs to H_1, we can find g and $h \in H_1$, $g(z)$ and $h(z)$* without zeros *in $\{|z| < 1\}$, with*

$$\int_{-\pi}^{\pi} |g(e^{i\theta})| \, d\theta \leq \int_{-\pi}^{\pi} |f(e^{i\theta})| \, d\theta,$$

$$\int_{-\pi}^{\pi} |h(e^{i\theta})| \, d\theta \leq \int_{-\pi}^{\pi} |f(e^{i\theta})| \, d\theta,$$

and such that $f = g + h$.

Remark This is a useful technical device, because many inequalities for H_1 are easier to prove for functions without zeros in $\{|z| < 1\}$.

Proof Let $B(z)$ be the Blaschke product formed from the zeros of $f(z)$, then, by the theorem, $f = BF$ where $F(z)$ has no zeros in $\{|z| < 1\}$, $F \in H_1$ and

$$\int_{-\pi}^{\pi} |F(e^{i\theta})| \, d\theta \leq \int_{-\pi}^{\pi} |f(e^{i\theta})| \, d\theta.$$

Now the result follows with

$$g(z) = \tfrac{1}{2}(1 + B(z))F(z)$$
$$h(z) = -\tfrac{1}{2}(1 - B(z))F(z),$$

because, by direct inspection (or the strong maximum principle),

$$|B(z)| < 1 \quad \text{for } |z| < 1.$$

Corollary *Let $f \not\equiv 0$ belong to H_p. Then we can write*

$$f(z) = B(z)[g(z)]^{1/p}$$

where $B(z)$ is a Blaschke product and $g \in H_1$ has no zeros in $\{|z| < 1\}$.

Proof $f = BF$ with $F \in H_p$ having no zeros in $\{|z| < 1\}$. Take $g(z) = [F(z)]^p$, $|z| < 1$; the pth power is defined and regular in $\{|z| < 1\}$ because F never equals 0 there.

C. Functions in H_p

Definition If $F \in H_p$ and $p \geq 1$, we write

$$\|F\|_p = \sup_{r<1} \left(\int_{-\pi}^{\pi} |F(re^{i\theta})|^p \, d\theta \right)^{1/p}$$

If $0 < p < 1$, we write

$$\|F\|_p = \sup_{r<1} \int_{-\pi}^{\pi} |F(re^{i\theta})|^p \, d\theta$$

(*without* the pth root!).

Then $\| \ \|_p$ satisfies the triangle inequality (hence yields a *metric* for H_p) *in all cases*, but is *positive homogeneous* (i.e., a *norm*) *only* for $p \geq 1$.

1. Boundary data function. Norm

Theorem *If $p > 0$ and $F \in H_p$, for almost all $e^{i\theta}$, $\lim F(z)$ for $z \underset{\angle}{\longrightarrow} e^{i\theta}$ exists and is finite, and if we call that limit $F(e^{i\theta})$,*

$$\int_{-\pi}^{\pi} |F(e^{i\theta})|^p \, d\theta = \|F\|_p, \quad 0 < p < 1,$$

$$\int_{-\pi}^{\pi} |F(e^{i\theta})|^p \, d\theta = \|F\|_p^p, \quad p \geq 1.$$

Proof We may as well assume $F \not\equiv 0$. Then, by the scholium of Subsection B.2, $F = BG$ where B is a Blaschke product, $G(z)$ has no zeros in $\{|z| < 1\}$, and

$$\|F\|_p = \|G\|_p .$$

Now write $G(z) = [h(z)]^{1/p}$, as in the second corollary of Subsection B.2. Then $h \in H_1$, and by Chapter II, Subsection B.2,

$$\|h\|_1 = \limsup_{r \to 1} \int_{-\pi}^{\pi} |h(re^{i\theta})| \, d\theta = \int_{-\pi}^{\pi} |h(e^{i\theta})| \, d\theta,$$

where the $\angle$ boundary value $h(e^{i\theta})$ of $h(z)$ exists a.e. (Chapter II Subsection B.1). Since we are assuming that $F \not\equiv 0$, we have, by Chapter III, Section B, $h(e^{i\theta}) \neq 0$ for almost all θ. Therefore, for almost all θ, $\lim G(z) = \lim [h(z)]^{1/p}$ exists as $z \underset{\angle}{\longrightarrow} e^{i\theta}$. Call this limit $G(e^{i\theta})$. Then, since $B(z) \to B(e^{i\theta})$ as $z \underset{\angle}{\longrightarrow} e^{i\theta}$ for almost all θ, $F(z) \to B(e^{i\theta})G(e^{i\theta})$, which call $F(e^{i\theta})$, for almost all θ as $z \underset{\angle}{\longrightarrow} e^{i\theta}$.

Finally, if $0 < p < 1$,

$$\|F\|_p = \|G\|_p = \left\|h^{1/p}\right\|_p = \text{(clearly)} = \|h\|_1 = \int_{-\pi}^{\pi} |h(e^{i\theta})|\,d\theta$$

$$= \int_{-\pi}^{\pi} |G(e^{i\theta})|^p\,d\theta = \int_{-\pi}^{\pi} |F(e^{i\theta})|^p\,d\theta$$

because $B(e^{i\theta}) = 1$ a.e.

If $p \geq 1$, we get the same result, but the integrals in the above chain of equalities, and $\|h\|_1$, are affected with p^{th} roots. We're done.

2. Convergence in mean to boundary data function

Theorem *If $f \in H_p$,*

$$\int_{-\pi}^{\pi} |f(re^{i\theta}) - f(e^{i\theta})|^p\,d\theta \to 0$$

as $r \to 1$, where $f(e^{i\theta})$ is the $\angle$ boundary value of $f(z)$ which, by the previous theorem, exists a.e.

Proof If $f \not\equiv 0$ and $B(z)$ is the Blaschke product formed from the zeros of $f(z)$, $f(z) = B(z)F(z)$, where $F \in H_p$ and $F(z)$ is without zeros in $\{|z| < 1\}$. If $r < 1$, and $0 < p < 1$,

$$\int_{-\pi}^{\pi} |f(e^{i\theta}) - f(re^{i\theta})|^p\,d\theta \leq \int_{-\pi}^{\pi} |B(re^{i\theta})|^p |F(e^{i\theta}) - F(re^{i\theta})|^p\,d\theta$$

$$+ \int_{-\pi}^{\pi} |B(e^{i\theta}) - B(re^{i\theta})|^p |F(e^{i\theta})|^p\,d\theta$$

$$\leq \int_{-\pi}^{\pi} |F(e^{i\theta}) - F(re^{i\theta})|^p\,d\theta + \int_{-\pi}^{\pi} |B(re^{i\theta}) - B(e^{i\theta})|^p |F(e^{i\theta})|^p\,d\theta.$$

For $p \geq 1$, similar inequalities hold, save that the integrals are affected with p^{th} roots.

Now $|B(z)| \leq 1$, $B(re^{i\theta}) \to B(e^{i\theta})$ a.e. as $r \to 1$, and $|F(e^{i\theta})|^p \in L_1$.

Therefore

$$\int_{-\pi}^{\pi} |B(re^{i\theta}) - B(e^{i\theta})|^p |F(e^{i\theta})|^p\,d\theta \to 0$$

as $r \to 1$ by dominated convergence.

It is thus enough to prove that

$$\int_{-\pi}^{\pi} |F(e^{i\theta}) - F(re^{i\theta})|^p\,d\theta \to 0$$

as $r \to 1$ if $F \in H_p$ is *without zeros* in $\{|z| < 1\}$.

Now, if $p \geq 1$, this is easy, because, by Chapter II, Subsection B.1,

$$F(re^{i\theta}) = \frac{1}{2\pi} \int_{-\pi}^{\pi} P_r(\theta - t) F(e^{it})\,dt,$$

since, in particular, $F \in H_1$. By Subsection 1, $F(e^{it}) \in L_p$ if $F \in H_p$, so, by the approximate identity property of P_r,

$$\int_{-\pi}^{\pi} |F(e^{i\theta}) - F(re^{i\theta})|^p \, d\theta \to 0$$

as $r \to 1$.

Suppose $p \geq 1/2$. Using a clever trick of Zygmund, we put $G(z) = \sqrt{F(z)}$; the square root is well defined here because $F(z)$ never $= 0$ in $\{|z| < 1\}$. Then $G \in H_{2p}$ with $2p \geq 1$. We have:

$$\int_{-\pi}^{\pi} |F(re^{i\theta}) - F(e^{i\theta})|^p \, d\theta = \int_{-\pi}^{\pi} |[G(re^{i\theta}) - G(e^{i\theta})][G(re^{i\theta}) + G(e^{i\theta})]|^p \, d\theta$$

which, by Schwarz, is

$$\leq \sqrt{\int_{-\pi}^{\pi} |G(re^{i\theta}) - G(e^{i\theta})|^{2p} \, d\theta} \sqrt{\int_{-\pi}^{\pi} |G(re^{i\theta}) + G(e^{i\theta})|^{2p} \, d\theta}$$

$$\leq C \sqrt{\int_{-\pi}^{\pi} |G(re^{i\theta}) - G(e^{i\theta})|^{2p} \, d\theta}$$

with C independent of r since $G \in H_{2p}$. But now, since $2p \geq 1$, this last expression $\to 0$ as $r \to 1$ by what has been proved above. So the result holds for $p \geq 1/2$.

If, now, $p \geq 1/4$, we make again the substitution $G = \sqrt{F}$ and argue as above, using the result just proven.

In this way we prove the theorem successively for $p \geq 1/4$, $p \geq 1/8$, $p \geq 1/16$, etc., and we GET DONE.

Scholium The ideas of the above proof can be modified so as to obtain a *new proof* of the celebrated *theorem of the brothers Riesz* (Chapter II, Section A).

First of all, the theorem just given is elementary for $p = 2$. Indeed, if $f(z) = \sum_0^\infty A_n z^n$ is in H_2, we have

$$\int_{-\pi}^{\pi} |f(re^{i\theta})|^2 \, d\theta = \pi \sum_0^\infty |A_n|^2 r^{2n}$$

by absolute convergence and orthogonality, so $\sum_0^\infty |A_n|^2 < \infty$.

By the Riesz-Fischer theorem, there is a function $f(e^{i\theta}) \in L_2(-\pi, \pi)$ having the Fourier series $\sum_0^\infty A_n e^{in\theta}$. It is now easy to check by direct calculation that, for $r < 1$,

$$f(re^{i\theta}) = \frac{1}{2\pi} \int_{-\pi}^{\pi} P_r(\theta - t) f(e^{it}) \, dt$$

(work with the series!), so that, *in fact*,

$$f(z) \to f(e^{i\theta}) \quad \text{a.e. for } z \underset{\angle}{\longrightarrow} e^{i\theta}$$

and

$$\int_{-\pi}^{\pi} |f(re^{i\theta}) - f(e^{i\theta})|^2 \, d\theta \to 0$$

as $r \to 1$, by the basic material in Chapter I, Section D.

Granted this result, we can prove the F. and M. Riesz theorem as follows. Take any (non-zero) measure μ on $[-\pi, \pi]$ with

$$\int_{-\pi}^{\pi} e^{in\theta} \, d\mu(\theta) = 0, \quad n = 1, 2, 3, \ldots$$

and put

$$F(re^{i\theta}) = \frac{1}{2\pi} \int_{-\pi}^{\pi} P_r(\theta - t) \, d\mu(t)$$

for $0 \le r < 1$. As we easily see,

$$F(re^{i\theta}) = \sum_0^\infty a_n r^n e^{in\theta},$$

where

$$a_n = \frac{1}{2\pi} \int_{-\pi}^{\pi} e^{-in\theta} \, d\mu(\theta),$$

so $F(z)$ is *regular* in $\{|z| < 1\}$ – by Chapter I, Section C, $F \in H_1$. As at the beginning of the proof of the the theorem given above, we can now write $F(z) = B(z)G(z)$ where $B(z)$ is a Blaschke product and $G \in H_1$, $G(z)$ *without zeros* in $\{|z| < 1\}$. *Therefore we can write* $G = f^2$ for an $f(z)$ analytic in $\{|z| < 1\}$; it is practically evident that $f \in H_2$. So use the special case of the theorem just proved; put $F(e^{i\theta}) = B(e^{i\theta})[f(e^{i\theta})]^2$; then $F(e^{i\theta}) \in L_1(-\pi, \pi)$. Now

$$\int_{-\pi}^{\pi} |F(re^{i\theta}) - F(e^{i\theta})| \, d\theta \le \int_{-\pi}^{\pi} |B(re^{i\theta}) - B(e^{i\theta})| \, |f(e^{i\theta})|^2 \, d\theta$$

$$+ \int_{-\pi}^{\pi} |B(re^{i\theta})| \cdot |(f(re^{i\theta}))^2 - (f(e^{i\theta}))^2| \, d\theta.$$

As $r \to 1$, the *first* integral on the right goes to zero by dominated convergence. The *second* is

$$\le \int_{-\pi}^{\pi} |[f(re^{i\theta})]^2 - [f(e^{i\theta})]^2| \, d\theta;$$

this is now shown to $\to 0$ as $r \to 1$ by the argument of Zygmund's trick, using the already known fact that

$$\int_{-\pi}^{\pi} |f(re^{i\theta}) - f(e^{i\theta})|^2 \, d\theta \to 0, \quad r \to 1.$$

In fine,

$$\int_{-\pi}^{\pi} |F(re^{i\theta}) - F(e^{i\theta})| \, d\theta \to 0, \quad r \to 1.$$

But by Chapter I, Section D, as $r \to 1$,

$$F(re^{i\theta}) \, d\theta \to d\mu(\theta) \quad w^*.$$

So $d\mu(\theta) = F(e^{i\theta}) \, d\theta$ and μ is absolutely continuous, proving the F. and M. Riesz theorem.

3. Smirnov's theorem

Theorem (Smirnov) *Let $f \in H_p$. If $p' > p$ and $f(e^{i\theta}) \in L_{p'}$, then $f \in H_{p'}$.*

Proof Taking f to be $\not\equiv 0$, we have $f = BF$ where B is a Blaschke product and $F(z)$ without zeros in $\{|z| < 1\}$. Since $|B(z)| \le 1$, it is enough to prove that $F \in H_{p'}$, if $F(e^{i\theta}) \in L_{p'}$. F being without zeros, write

$$G(z) = [F(z)]^p.$$

Then $G \in H_1$ and $G(e^{i\theta}) \in L_{p'/p}$ with $p'/p > 1$. It is enough to prove that $G(z) \in H_{p'/p}$. But now use the result of Chapter II, Subsection B.1:

$$G(re^{i\theta}) = \frac{1}{2\pi} \int_{-\pi}^{\pi} P_r(\theta - t) G(e^{i\theta}) \, dt,$$

and the fact that $G(e^{i\theta}) \in L_{p'/p}$ with $p'/p > 1$ to show that

$$\int_{-\pi}^{\pi} |G(re^{i\theta})|^{p'/p} \, d\theta \le \int_{-\pi}^{\pi} |G(e^{i\theta})|^{p'/p} \, d\theta$$

for $r < 1$. We're done.

D. Inner and outer factors

1. A condition guaranteeing absolute continuity

Lemma *Let $\Phi(x)/x \to \infty$ for $x \to \infty$, where $\Phi(x) \ge 0$. Let $f_n(t) \ge 0$,*

$$\int_{-\pi}^{\pi} \Phi(f_n(t)) \, dt \le C,$$

and $f_n(t) \, dt \xrightarrow[n]{} d\mu(t)$ w where μ is a measure on $[-\pi, \pi]$. Then μ is absolutely continuous.*

Proof For $K > 0$, denote by η_K the supremum of $x/\Phi(x)$ for $x \ge K$. Then $\eta_K \to 0$ as $K \to \infty$. Let $E \subseteq [-\pi, \pi]$ be compact and $|E| = 0$: to show that $\mu(E) = 0$. With $\epsilon > 0$ arbitrary, take $\mathcal{O}$ open $\supset E$ with $|\mathcal{O}| < \epsilon$, and let $\psi(t)$ be continuous on $[-\pi, \pi]$, $0 \le \psi(t) \le 1$, $\psi(t) \equiv 0$ outside $\mathcal{O}$, and $\psi(t) \equiv 1$ on E. Since $f_n(t) \, dt \xrightarrow[n]{} d\mu(t)$ w*,

$$\mu(E) \le \int_{-\pi}^{\pi} \psi(t) \, d\mu(t) = \lim_{n \to \infty} \int_{-\pi}^{\pi} \psi(t) f_n(t) \, dt$$

$$\le \limsup_{n \to \infty} \int_{\mathcal{O}} f_n(t) \, dt.$$

Let, for given large K, and any n,

$$\mathcal{O}(K,n) = \{t \in \mathcal{O};\ f_n(t) \le K\}.$$

Then

$$\int_{\mathcal{O}} f_n(t)\,dt \le K|\mathcal{O}(K,n)| + \int_{\mathcal{O}\sim\mathcal{O}(K,n)} f_n(t)\,dt \le K|\mathcal{O}| + \eta_K \int_{-\pi}^{\pi} \Phi(f_n(t))\,dt$$

$$\le K|\mathcal{O}| + \eta_K C \le K\epsilon + C\eta_K.$$

Given $\delta > 0$ choose first K so large that $C\eta_K < \delta/2$, then take $\epsilon > 0$ so small that $K\epsilon < \delta/2$. One finds $\mu(E) < \delta$. Since $\delta > 0$ is arbitrary, $\mu(E) = 0$. Q.E.D.

2. Boundedness of means of $\left|\log|F(re^{i\theta})|\right|$

Lemma *If $F(z)$ is regular in $\{|z| < 1\}$ and*

$$\int_{-\pi}^{\pi} \log^+ |F(re^{i\theta})|\,d\theta$$

is bounded for $0 \le r < 1$, then, if $F \not\equiv 0$,

$$\int_{-\pi}^{\pi} \left|\log|F(re^{i\theta})|\right| r\,d\theta$$

*is bounded for $0 < r < 1$.**

Remark This is an important result in the theory of functions.

Proof Without loss of generality, $F(0) \ne 0$ (otherwise work with $F(z)/z^k$ instead of $F(z)$). Then, by Jensen's inequality,

$$-\infty < \log|F(0)| \le \frac{1}{2\pi}\int_{-\pi}^{\pi} \log|F(re^{i\theta})|\,d\theta$$

$$= \frac{1}{2\pi}\int_{-\pi}^{\pi} 2\log^+|F(re^{i\theta})|\,d\theta - \frac{1}{2\pi}\int_{-\pi}^{\pi} \left|\log|F(re^{i\theta})|\right|\,d\theta$$

for $0 \le r < 1$. Therefore

$$\int_{-\pi}^{\pi} \left|\log|F(re^{i\theta})|\right|\,d\theta \le 2\int_{-\pi}^{\pi} \log^+|F(re^{i\theta})|\,d\theta - 2\pi\log|F(0)|.$$

The result follows.

3. Expression of an analytic function in terms of its real part

First, an identity:

$$\frac{1-r^2}{1+r^2-2r\cos\theta} + \frac{2ir\sin\theta}{1+r^2-2r\cos\theta} = \frac{(1+re^{i\theta})(1-re^{-i\theta})}{(1-re^{i\theta})(1-re^{-i\theta})} = \frac{1+re^{i\theta}}{1-re^{i\theta}}.$$

* An oversight in the first edition's statement of this lemma was observed by Kanghui Guo. When $F(0) = 0$, a factor of r in second integral is *necessary* in order to keep it bounded for $r \to 0$.

Thence, and from Chapter I, if $F(z)$ is regular *in a region including* $\{|z| \leq 1\}$ *in its interior* and $0 \leq r < 1$,

$$F(re^{i\theta}) = \frac{1}{2\pi} \int_{-\pi}^{\pi} \frac{1 + re^{i(\theta-t)}}{1 - re^{i(\theta-t)}} \cdot \Re F(e^{it})\, dt + ic,$$

where c is a real constant ($= \Im F(0)$!).

We usually write:

$$\boxed{F(re^{i\theta}) = \frac{1}{2\pi} \int_{-\pi}^{\pi} \frac{e^{it} + re^{i\theta}}{e^{it} - re^{i\theta}} \Re F(e^{it})\, dt + i\Im F(0).}$$

4. Factorization into inner and outer factors

Theorem *Let $F(z) \not\equiv 0$ belong to H_p, $p > 0$. Let $B(z)$ be the Blaschke product consisting of the zeros of $F(z)$. Then there is a singular measure $\sigma \geq 0$ on $[-\pi, \pi]$ with*

$$\boxed{F(z) = B(z) \exp\left(-\frac{1}{2\pi} \int_{-\pi}^{\pi} \frac{e^{it} + z}{e^{it} - z}\, d\sigma(t)\right) e^{ic} \cdot \exp\left(\frac{1}{2\pi} \int_{-\pi}^{\pi} \frac{e^{it} + z}{e^{it} - z} \log|F(e^{it})|\, dt\right)}$$

for $|z| < 1$, where c is a real constant.

Proof By Subsection B.2, we can write $F(z) = B(z)G(z)$ where $G \in H_p$ and $G(z)$ *has no zeros* in $\{|z| < 1\}$. Thus, we can define $\phi(z) = \log G(z)$ so as to be *analytic* in $\{|z| < 1\}$. Now for $0 \leq r < 1$, by the inequality between arithmetic and geometric means,

$$\frac{1}{2\pi} \int_{-\pi}^{\pi} p \log^+ |G(re^{i\theta})|\, d\theta \leq \frac{1}{2\pi} \int_{-\pi}^{\pi} \log\left[|G(re^{i\theta})|^p + 1\right] d\theta \leq \log \frac{1}{2\pi} \int_{-\pi}^{\pi} (|G(re^{i\theta})|^p + 1)\, d\theta \leq C$$

because $G \in H_p$, so

$$\int_{-\pi}^{\pi} \log^+ |G(re^{i\theta})|\, d\theta$$

is bounded above for $0 \leq r < 1$. Therefore by Subsection 2 (*here* $|G(0)| > 0$!),

$$\int_{-\pi}^{\pi} |\log|G(re^{i\theta})||\, d\theta$$

is bounded above for $0 \leq r < 1$.

Notation Write $\log^- |G| = |\log|G|| - \log^+ |G|$. (Therefore $\log^- |G| \geq 0$!)

Therefore, we have a sequence of $r_\nu \to 1$, $r_\nu < 1$, and measures μ_+, μ_- on $[-\pi, \pi]$, $d\mu_+ \geq 0$, $d\mu_- \geq 0$, with

$$\log^+ |G(r_\nu e^{i\theta})|\, d\theta \underset{\nu}{\longrightarrow} d\mu_+(\theta) \quad w^*$$
$$\log^- |G(r_\nu e^{i\theta})|\, d\theta \underset{\nu}{\longrightarrow} d\mu_-(\theta) \quad w^*$$

Now by the formulas in Subsection 3 with a change of variable, if z, $|z| < 1$, is *fixed*,

$$\phi(r_\nu z) = i\Im\phi(0) + \frac{1}{2\pi}\int_{-\pi}^{\pi} \frac{e^{it}+z}{e^{it}-z}\Re\phi(r_\nu e^{it})\,dt$$

$$= i\Im\phi(0) + \frac{1}{2\pi}\int_{-\pi}^{\pi} \frac{e^{it}+z}{e^{it}-z}\log|G(r_\nu e^{it})|\,dt.$$

By the aforementioned w* convergence, the right side tends to

$$i\Im\phi(0) + \frac{1}{2\pi}\int_{-\pi}^{\pi} \frac{e^{it}+z}{e^{it}-z}[\,d\mu_+(t) - d\mu_-(t)],$$

and the left side tends to $\phi(z)$. So

$$\phi(z) = i\Im\phi(0) + \frac{1}{2\pi}\int_{-\pi}^{\pi} \frac{e^{it}+z}{e^{it}-z}[\,d\mu_+(t) - d\mu_-(t)].$$

Here, μ_+ *is absolutely continuous.* Indeed, since $G \in H_p$, we can apply the lemma of Subsection 1 — we have

$$\log^+|G(r_\nu e^{i\theta})|\,d\theta \underset{\nu}{\longrightarrow} d\mu_+(\theta) \quad \text{w}^*$$

whilst

$$\int_{-\pi}^{\pi} |G(r_\nu e^{i\theta})|^p\,d\theta \le c,$$

so surely,

$$\int_{-\pi}^{\pi} \exp\left(p\log^+|G(r_\nu e^{i\theta})|\right)\,d\theta \le c + 2\pi,$$

and $(\exp x)/x \to \infty$ for $x \to \infty$. We therefore have

$$d\mu_+(\theta) = h_+(\theta)\,d\theta, \quad h_+ \in L_1.$$

However, all we can say about μ_- *is that*

$$d\mu_-(\theta) = h_-(\theta)\,d\theta + d\sigma(\theta)$$

where $h_- \in L_1$ and $\sigma \ge 0$ is *singular, not necessarily zero.*

Therefore, with $h = h_+ - h_- \in L_1$,

$$\phi(z) = i\Im\phi(0) + \frac{1}{2\pi}\int_{-\pi}^{\pi} \frac{e^{it}+z}{e^{it}-z}(h(t)\,dt - d\sigma(t)).$$

This yields, in particular (taking real parts – recall Subsection 3!)

$$\log|G(re^{i\theta})| = \Re\phi(re^{i\theta}) = \frac{1}{2\pi}\int_{-\pi}^{\pi} \frac{1-r^2}{1+r^2-2r\cos(\theta-t)}[h(t)\,dt - d\sigma(t)].$$

Therefore, by Chapter I, Subsection D.3, at any θ_0 for which $\sigma'(\theta_0)$ exists and the derivative of the indefinite integral of $h(t)$ equals $h(\theta_0)$,

$$\log|G(z)| \to h(e^{i\theta_0}) + \sigma'(\theta_0) \quad \text{as } z \underset{\angle}{\longrightarrow} e^{i\theta_0}.$$

But, σ *being singular*, $\sigma'(\theta) = 0$ a.e. Also, G being $\in H_p$, the $\angle$ limit $G(e^{i\theta})$ exists a.e. and is $\neq 0$ a.e. (Subsections A.2, B.2, and Chapter III, Section B). So it makes sense to talk about $\log|G(e^{i\theta})|$ a.e., and *since* $|G(e^{i\theta})| = |F(e^{i\theta})|$ a.e., we have the identification for the values of the *real* L_1 function h:

$$h(e^{i\theta}) = \log|F(e^{i\theta})| \quad \text{a.e.}\,.$$

We've thus proven

$$\phi(z) = i\Im\phi(0) + \frac{1}{2\pi}\int_{-\pi}^{\pi}\frac{e^{it}+z}{e^{it}-z}[\log|F(e^{it})|\,dt - d\sigma(t)].$$

Finally,

$$F(z) = B(z)G(z) = B(z)e^{\phi(z)},$$

and our desired representation is established. Q.E.D.

Scholium As a by-product of the above proof we have $\log|F(e^{it})| \in L_1(-\pi,\pi)$ for non-zero $F \in H_p$. This is a *quantitative* form of the theorem in Section B of Chapter III. The result also follows more quickly from Subsection 2 and Fatou's lemma.

Definition The formula

$$F(z) = B(z)\exp\left(-\frac{1}{2\pi}\int_{-\pi}^{\pi}\frac{e^{it}+z}{e^{it}-z}\,d\sigma(t)\right)e^{ic}\,\exp\left(\frac{1}{2\pi}\int_{-\pi}^{\pi}\frac{e^{it}+z}{e^{it}-z}\log|F(e^{it})|\,dt\right)$$

is called the *canonical representation* of the function $F \not\equiv 0$ in H_p. We call (after Beurling)

$$I_F(z) = e^{ic}B(z)\exp\left\{-\frac{1}{2\pi}\int_{-\pi}^{\pi}\frac{e^{it}+z}{e^{it}-z}\,d\sigma(t)\right\}$$

the *inner factor* of $F(z)$, and

$$O_F(z) = \exp\left\{\frac{1}{2\pi}\int_{-\pi}^{\pi}\frac{e^{it}+z}{e^{it}-z}\log|F(e^{it})|\,dt\right\}$$

the *outer factor* of $F(z)$.

Note that $|I_F(z)| \leq 1$ for $|z| < 1$ (since $d\sigma \geq 0$!), and by the above discussion, $|I_F(\zeta)| \equiv 1$ a.e., $|\zeta| = 1$. (Cf. argument showing that $\Re\phi(z) \to h(e^{i\theta})$ a.e. for $z \underset{\angle}{\longrightarrow} e^{i\theta}$.)

E. Beurling's Theorem

1. Polynomial approximation

Lemma *Let $p > 0$. Then, if $F \in H_p$, there exist polynomials $P(z)$ with $\|F - P\|_p$ arbitrarily small.*

Proof By Subsection C.2, $\|F(z) - F(rz)\|_p < \epsilon$ if $r < 1$ is sufficiently close to 1. But the Taylor series of $F(rz)$ converges *even uniformly* for $|z| \leq 1$; cutting it off far enough out, we get a polynomial $P(z)$ with $\|F - P\|_p < 2\epsilon$.

2. General Smirnov theorem

Lemma (Generalization of Smirnov's theorem) *Let F and G (both $\not\equiv 0$) belong to (perhaps different) H_p spaces. If the ratio of their outer factors $k(z) = O_F(z)/O_G(z)$ has $k(e^{i\theta}) \in L_{p'}$, say, then $O_F(z)/O_G(z) \in H_{p'}$.*

Proof For $r < 1$, by the inequality between arithmetic and geometric means

$$\left(\text{N.B.}\quad \frac{1}{2\pi}\int_{-\pi}^{\pi} \frac{1-r^2}{1+r^2-2r\cos(\theta-t)}\, dt = 1\,!\right),$$

$$p\log|k(re^{i\theta})| = \frac{p}{2\pi}\int_{-\pi}^{\pi} \frac{1-r^2}{1+r^2-2r\cos(\theta-t)} \log\left|\frac{F(e^{it})}{G(e^{it})}\right| dt$$

$$\leq \log \frac{1}{2\pi}\int_{-\pi}^{\pi} \frac{1-r^2}{1+r^2-2r\cos(\theta-t)} \left|\frac{F(e^{it})}{G(e^{it})}\right|^p dt,$$

so

$$\int_{-\pi}^{\pi} |k(re^{i\theta})|^p\, d\theta \leq \frac{1}{2\pi}\int_{-\pi}^{\pi}\int_{-\pi}^{\pi} \frac{1-r^2}{1+r^2-2r\cos(\theta-t)} \left|\frac{F(e^{it})}{G(e^{it})}\right|^p dt\, d\theta = \int_{-\pi}^{\pi} \left|\frac{F(e^{it})}{G(e^{it})}\right|^p dt,$$

which is enough.

3. Beurling's theorem in H_p

Theorem (General case due to Srinivasan and Wang; for $p = 2$ this is due to Beurling) *Let $F = I_F O_F \in H_p$, $p > 0$. The closure of $F(z)\cdot\{$polynomials in $z\}$ in H_p is precisely $I_F \cdot H_p$.*

Proof (a) The closure is *not more* than $I_F \cdot H_p$. Indeed, let $G \in H_p$, and let the $P_n(z)$ be polynomials with $\|FP_n - G\|_p \xrightarrow[n]{} 0$. We may, without loss of generality, assume that $\|FP_n - FP_{n+1}\|_p < 2^{-n}$. Then, since $|I_F(e^{i\theta})| = 1$, a.e., by Subsection C.1, if $0 < p < 1$,

$$\int_{-\pi}^{\pi} |O_F(e^{i\theta})P_n(e^{i\theta}) - O_F(e^{i\theta})P_{n+1}(e^{i\theta})|^p\, d\theta < 2^{-n},$$

and if $p \geq 1$, a similar inequality holds with the integral affected by a pth root. Since $O_F \in H_p$ if F is (cf. proof in Subsection 2 above!), $O_F P_n \in H_p$ and by Subsection C.1, $\|O_F P_n - O_F P_{n+1}\|_p \leq 2^{-n}$. By inequalities like those used in Subsection 2, it is easy to see that

$$O_F(z)P_1(z) + \sum_{n=1}^{\infty} [O_F(z)P_{n+1}(z) - O_F(z)P_n(z)]$$

converges uniformly in the interior of $\{|z| < 1\}$ to an analytic function $R(z)$, say, and by the triangle inequality for $\|\ \|_p$ (N.B. valid also for $0 < p < 1$!) we have, for $r < 1$, (assuming $p < 1$; if $p \geq 1$, the integral has a pth root):

$$\int_{-\pi}^{\pi} |R(re^{i\theta})|^p\, d\theta \leq \|O_F P_1\|_p + \sum_{n=1}^{\infty} \|O_F P_{n+1} - O_F P_n\|_p < \infty,$$

so $R \in H_p$.

Now

$$FP_n = I_F O_F P_n = I_F \left(O_F P_1 + \sum_{m=1}^{n-1} (O_F P_{m+1} - O_F P_m) \right) \underset{n}{\longrightarrow} I_F R.$$

So, since $\|FP_n - G\|_p \underset{n}{\longrightarrow} 0$, we have $G = I_F R$, a member of $I_F H_p$.

(b) The closure is *at least* $I_F \cdot H_p$. Since $|I_F(z)| \le 1$ for $|z| < 1$, it is enough to show that $O_F \cdot \{\text{polynomials}\}$ is *dense* in H_p.

First, suppose $p \ge 1$. Then we can argue *by duality*. Let

$$\frac{1}{q} + \frac{1}{p} = 1,$$

assume that $k \in L_q$ is not zero a.e., and that

$$\int_{-\pi}^{\pi} e^{in\theta} O_F(e^{i\theta}) k(\theta)\, d\theta = 0 \quad \text{for } n = 0, 1, 2, \ldots .$$

It is enough to show that then

$$0 = \int_{-\pi}^{\pi} G(e^{i\theta}) k(\theta)\, d\theta$$

for any $G \in H_p$. Since $O_F(e^{i\theta})k(\theta) \in L_1$,

$$\int_{-\pi}^{\pi} e^{in\theta} O_F(e^{i\theta}) k(\theta)\, d\theta = 0 \quad \text{for } n = 0, 1, 2, \ldots$$

makes $O_F(e^{i\theta})k(\theta) = e^{i\theta} H(e^{i\theta})$ where $H \in H_1$ is (here) $\not\equiv 0$. Therefore

$$k(\theta) = e^{i\theta} I_H(e^{i\theta}) \frac{O_H(e^{i\theta})}{O_F(e^{i\theta})},$$

and

$$\left| \frac{O_H(e^{i\theta})}{O_F(e^{i\theta})} \right| = |k(\theta)| \in L_q.$$

So by Subsection 2, $O_H/O_F = R$ is in H_q, and $k(\theta) = e^{i\theta} I_H(e^{i\theta}) R(e^{i\theta})$. Therefore, if $G \in H_p$, $G(e^{i\theta})k(\theta) = e^{i\theta} I_H(e^{i\theta}) G(e^{i\theta}) R(e^{i\theta})$ with $I_H G R \in H_1$, so

$$\int_{-\pi}^{\pi} e^{in\theta} k(\theta) G(e^{i\theta})\, d\theta = 0 \quad \text{for } n = 0, 1, 2, \ldots .$$

We're done in case $p \ge 1$.

Now, using Zygmund's idea, suppose that $F \in H_p$ with $p \ge 1/2$. *Clearly* we can take $F_1(z) = \sqrt{O_F(z)}$ and then $F_1(z) \in H_{2p}$ with $O_{F_1} = F_1$. Therefore, if G is any element of H_{2p} we can find a *polynomial* P with $\|F_1 P - G\|_{2p} < \epsilon$ *by the above*. Using Schwarz, we then get

$$\|O_F P - F_1 G\|_p \le \sqrt{\|O_F\|_p} \sqrt{\epsilon}$$

(or another variant, depending on whether or not $p > 1$).

In particular, if Q is a *polynomial* we can find another polynomial P such that $\|O_F P - F_1 Q\|_p \le \delta$, say, $\delta > 0$ being arbitrary. *Now*, given $H \in H_p$, use first the result of Subsection 1 to find a polynomial R with $\|H - R\|_p \le \delta$. Then, as remarked above, we can find another polynomial Q making $||F_1Q - R||_{2p}$ as small as we like – using Schwarz, we see that we can choose Q with $||F_1Q - R||_p < \delta$. Finally, we have the polynomial P with $||O_FP - F_1Q||_p < \delta$, whence

$$||H - O_FP||_p \leq ||H - R||_p + ||R - F_1Q||_p + ||F_1Q - O_FP||_p \leq 3\delta.$$

This extends the result we are proving to all values of $p \geq 1/2$.

The same argument may now be used to extend it to values of $p \geq 1/4$, then to values $\geq 1/8$, etc.

We're done.

F. Invariant subspaces

Let $\mathscr{H}$ be a Hilbert space with orthonormal basis

$$\{e_n;\ n = 0, \pm 1, \pm 2, \pm 3, \ldots\},$$

and consider the unitary transformation V defined on $\mathscr{H}$ by putting

$$Ve_n = e_{n+1}.$$

The problem is to study the invariant subspaces of V.

We map $\mathscr{H}$ isometrically onto $L_2(-\pi, \pi)$ by making e_n correspond to $e^{in\theta}/\sqrt{2\pi}$; then V corresponds to multiplication by $e^{i\theta}$.

Theorem *Let a closed subspace E of $L_2(-\pi, \pi)$ be such that $e^{i\theta}E = E$. Then $E = \chi_A L_2$ where χ_A is the characteristic function of some measurable set $A \subset (-\pi, \pi)$.*

Proof If $E \subset L_2(-\pi, \pi)$ *properly*, then $1 \notin E$. Otherwise $e^{i\theta}$ and therefore $e^{2i\theta}$, $e^{3i\theta}$, etc. would be in E. Also, from $e^{i\theta}E = E$ we get $e^{-i\theta}E = E$, so from $1 \in E$ we would get $e^{-i\theta} \in E$, $e^{-2i\theta} \in E$, etc., and finally E would $= L_2$.

So let ϕ be the *closest* element of E to 1. Then, arguing as above,

$$e^{in\theta}\phi \in E, \quad n = 0, \pm 1, \pm 2, \ldots ,$$

so

$$1 - \phi \perp e^{in\theta}\phi, \quad n \in \mathbb{Z}.$$

Therefore

$$\int_{-\pi}^{\pi} e^{in\theta}(\phi(\theta) - |\phi(\theta)|^2)\, d\theta = 0, \quad n \in \mathbb{Z},$$

so $\phi = |\phi|^2$ a.e., and, finally, $\phi(\theta) = 1$ or 0 a.e., so $\phi = \chi_A$ for some measurable A. Therefore $\chi_A \in E$ so $\chi_A L_2 \subseteq E$. $\chi_A L_2$ is clearly a closed subspace of L_2. If $\chi_A L_2$ is not *all* of E, there is a $\psi \in E$ which is *orthogonal* to $\chi_A L_2$.

In particular,

$$\int_{-\pi}^{\pi} \chi_A(\theta)\psi(\theta)e^{in\theta}\, d\theta = 0, \quad n \in \mathbb{Z},$$

so $\psi\chi_A \equiv 0$, so $\psi \equiv 0$ a.e. on A. But at the same time $e^{in\theta}\psi \in E$ for $n \in \mathbb{Z}$, so since $1 - \chi_A = 1 - \phi \perp E$,

$$\int_{-\pi}^{\pi} (1 - \chi_A(\theta))\psi(\theta)e^{in\theta}\, d\theta = 0, \quad n \in \mathbb{Z},$$

i.e.,

$$\int_{-\pi}^{\pi} \chi_{\sim A}(\theta)\psi(\theta)e^{in\theta}\,d\theta = 0, \quad n \in \mathbb{Z}.$$

So $\psi \equiv 0$ a.e. on $\sim A$. So $\psi \equiv 0$ a.e., a contradiction, and $\chi_A L_2 = E$. Q.E.D.

Theorem *Let E be a closed subspace of L_2 and suppose that $e^{i\theta}E \subset E$ properly. Then $E = \omega H_2$, where $|\omega(\theta)| \equiv 1$ a.e.*

Proof $e^{i\theta}E$ is also closed. Let $\omega \in E$ be $\neq 0$, $\omega \perp e^{i\theta}E$. Then in particular,

$$\omega \perp e^{in\theta}\omega, \quad n = 1, 2, \ldots,$$

so

$$\int_{-\pi}^{\pi} |\omega(\theta)|^2 e^{in\theta}\,d\theta = 0, \quad n = 1, 2, \ldots,$$

and by conjugation

$$\int_{-\pi}^{\pi} |\omega(\theta)|^2 e^{-in\theta}\,d\theta = 0, \quad n = 1, 2, \ldots,$$

so *finally* $|\omega(\theta)| =$ const. a.e. Without loss of generality, $|\omega(\theta)| = 1$ a.e.

Now ω, $e^{i\theta}\omega$, $e^{2i\theta}\omega, \ldots$ etc. are all in E so $\omega H_2 \subseteq E$. I claim $\omega H_2 = E$. ωH_2 is closed in L_2, because $|\omega(\theta)| = 1$, and the orthogonal complement of ωH_2 in L_2 is $e^{-i\theta}\omega(\theta)\overline{H}_2$, as is easily seen by direct calculation.

But if $F \in E$ and g is a *polynomial* in $e^{i\theta}$, $gF \in E$ so $e^{i\theta}gF \in e^{i\theta}E$, and then $\omega \perp e^{i\theta}gF$ by choice of ω. That is,

$$\int_{-\pi}^{\pi} e^{-i\theta}\overline{g(\theta)}\omega(\theta)\overline{F(\theta)}\,d\theta = 0.$$

This in fact must hold for *all* $g \in H_2$ because polynomials in $e^{i\theta}$ are $\|\ \|_2$-dense in H_2 (Subsection E.1) and $|\omega(\theta)| = 1$ a.e. Thus, $e^{-i\theta}\omega\overline{H}_2 \perp F$ and, $F \in E$ being arbitrary,

$$e^{-i\theta}\omega(\theta)\overline{H}_2 \perp E.$$

In other words, $P \in L_2$ and $P \perp \omega H_2$ imply $P \perp E$. Therefore $E \subseteq \omega H_2$. So $E = \omega H_2$.

Corollary *Let E be a subspace of H_2 and satisfy $e^{i\theta}E \subseteq E$, $E \neq \{0\}$. Then $E = \omega H_2$ where ω is an* inner function*.

Proof If $e^{i\theta}E = E$, then $e^{in\theta}E = E$ for $n = 1, 2, \ldots$, so $E \subseteq \bigcap_1^\infty e^{in\theta}H_2$. But this intersection consists only of 0. So $e^{i\theta}E \subset E$ properly. Therefore $E = \omega H_2$ where $|\omega(\theta)| \equiv 1$ a.e. $\omega \in E \subseteq H_2$. So ω is an inner function. Q.E.D.

Corollary *Let $f_\alpha = I_\alpha O_\alpha$, with each I_α* inner *and each O_α* outer *in H_2, and suppose that*

$$I_\alpha(z) = B_\alpha(z)\exp\left\{-\frac{1}{2\pi}\int_{-\pi}^{\pi}\frac{e^{it}+z}{e^{it}-z}\,d\sigma_\alpha(t)\right\},$$

---

* i.e., one whose *outer factor* is identically 1.

with Blaschke products B_α *and* singular $d\sigma_\alpha \geq 0$. *Let E be the invariant subspace of* H_2 generated *by the* f_α*; in other words, let E be the* smallest *closed subspace of* H_2 *containing* all *the* f_α, *such that* $e^{i\theta}E \subset E$.

Then $E = \omega H_2$ *where*

$$\omega(z) = B(z)\exp\left\{-\frac{1}{2\pi}\int_{-\pi}^{\pi}\frac{e^{it}+z}{e^{it}-z}\,d\sigma(t)\right\}.$$

Here, $B(z)$ *is the greatest common divisor of the* $B_\alpha(z)$, *and* $d\sigma \geq 0$ *is the* largest measure $\leq$ all *of the* $d\sigma_\alpha$.

Proof E is, by the preceding corollary, of the form ωH_2, where ω is an inner function whose constant factor e^{ic} we may just as well take equal to 1.

Each $f_\alpha \in E$, so $f_\alpha = I_\alpha O_\alpha$ is of the form ωg, $g \in H_2$. Therefore if we write

$$\omega(z) = B(z)\exp\left(-\frac{1}{2\pi}\int_{-\pi}^{\pi}\frac{e^{it}+z}{e^{it}-z}\,d\sigma(t)\right),$$

we certainly see that $B|B_\alpha$ and $d\sigma \leq d\sigma_\alpha$ by taking into account the inner factor of g.

Now suppose that there is a Blaschke product Б and a singular measure σ' with Б| every B_α and $d\sigma' \leq$ every $d\sigma_\alpha$. Then each f_α is in ΩH_2, where

$$\Omega(z) = Б(z)\exp\left(-\frac{1}{2\pi}\int_{-\pi}^{\pi}\frac{e^{it}+z}{e^{it}-z}\,d\sigma'(t)\right).$$

Therefore $E \subseteq \Omega H_2$, i.e., $\omega H_2 \subseteq \Omega H_2$. This makes Б$|B$ and $d\sigma' \leq d\sigma$ by the preceding reasoning, and ω is as specified in the statement of the corollary. Q.E.D.

Problem 5

(1) Let the $f_n(z)$ be *outer* functions in some H_p, $p > 0$, and suppose that $|f_1(z)| \geq |f_2(z)| \geq |f_3(z)| \geq \ldots$ for $|z| < 1$. Suppose that

$$f_n(z) \xrightarrow[n]{} f(z)$$

uniformly in the interior of $\{|z| < 1\}$. Prove that if $f(z)$ is not identically zero, it is *outer*.

(2) If $f(z) \in H_1$ and $\Re f(z) > 0$ for $|z| < 1$, $f(z)$ is *outer*. (Hint: consider the functions $(1/n) + f(z)$.)

(3) Let $\omega(z)$ be a non-constant *inner function*. For which complex α is $\omega(z) - \alpha$ an *outer function*? (Hint: In case $|\alpha| < 1$ look at $(\omega(z) - \alpha)/(1 - \bar{\alpha}\omega(z))$.)

G. Approximation of inner functions by Blaschke products

Definition H_∞ denotes the set of functions *analytic* and *bounded* in $\{|z| < 1\}$. If $F \in H_\infty$ we put

$$\|f\|_\infty = \sup\{|F(z)|;\ |z| < 1\}.$$

Most of the results of Sections A–D hold for H_∞ if $\|\ \|_p$ is replaced by $\|\ \|_\infty$.

> A *notable exception* is the theorem of Subsection C.2 saying that $\|F(re^{i\theta}) - F(e^{i\theta})\|_p \to 0$ as $r \to 1$ if $F \in H_p$. This is in general *false* for $F \in H_\infty$ if we use the norm $\|\ \|_\infty$!

Indeed, if $F \in H_\infty$ and we *have* $\|F(e^{i\theta}) - F(re^{i\theta})\|_\infty \to 0$ as $r \to \infty$, then $F(e^{i\theta})$ is *continuous*. And there are plenty of $F \in H_\infty$ with $F(e^{i\theta})$ *not continuous*!

However, we *do* have, for $F \in H_\infty$,

$$\|F\|_\infty = \operatorname{ess\,sup}_\theta |F(e^{i\theta})|.$$

(If $F \in H_\infty$, F is in every H_p, so the $\angle$ boundary value $F(e^{i\theta})$ exists a.e.!) The material in Section D also applies.

Proofs that all this work carries over to H_∞ are left to the reader – they follow easily from the results presented thus far in this book.

A special class of functions in H_∞ is constituted by the *inner functions*, i.e., those whose *outer factor* is equal to 1. An *inner function* $\omega(z)$ *has the property that* $|\omega(e^{i\theta})| \equiv 1$ a.e. The importance of these functions is already seen in Sections E, F above.

As we see from Subsection D.4, an inner function $\omega(z)$ has the representation

$$\omega(z) = e^{ic}B(z)\exp\left\{-\frac{1}{2\pi}\int_{-\pi}^{\pi}\frac{e^{it}+z}{e^{it}-z}\,d\sigma(t)\right\}$$

where $c \in \mathbb{R}$, $B(z)$ is a *Blaschke product*, and $d\sigma \geq 0$ is a *singular measure* on $[-\pi, \pi]$.

Lemma *An inner function* $\omega(z)$ *is a constant multiple of a Blaschke product iff*

$$\int_{-\pi}^{\pi} \log|\omega(re^{i\theta})|\,d\theta \to 0 \quad \text{as } r \to 1.$$

Proof By the work of Subsection A.2,

$$\int_{-\pi}^{\pi} \log|B(re^{i\theta})|\,d\theta \to 0 \quad \text{as } r \to 1$$

for the Blaschke product factor $B(z)$ of $\omega(z)$. As for the 'singular' factor,

$$S(z) = \exp\left\{-\frac{1}{2\pi}\int_{-\pi}^{\pi}\frac{e^{it}+z}{e^{it}-z}\,d\sigma(t)\right\},$$

$$\log|S(re^{i\theta})| = -\frac{1}{2\pi}\int_{-\pi}^{\pi}\frac{1-r^2}{1+r^2-2r\cos(\theta-t)}\,d\sigma(t),$$

so

$$\int_{-\pi}^{\pi} \log|S(re^{i\theta})|\,d\theta = -\int_{-\pi}^{\pi} d\sigma(t)$$

as $r \to 1$, and since $d\sigma \geq 0$, this last expression is zero iff $\sigma \equiv 0$, i.e., iff $S(z) \equiv 1$.

Now there is the remarkable

Theorem (Frostman, rediscovered years later by D. J. Newman) *Let $\omega(z)$ be any inner function. Then, given any $\epsilon > 0$ there is a Blaschke product $B(z)$ and a real c with*

$$\left\|\omega - e^{ic}B\right\|_\infty < \epsilon.$$

Remark Thus, the *measurable* function $e^{ic}B(e^{i\theta})$ is (a.e.) *uniformly within* ϵ of $\omega(e^{i\theta})$.

Proof Let the *number* ω satisfy $|\omega| \le 1$, and let $0 < \rho < 1$.

Then we have the elementary formula

$$\frac{1}{2\pi}\int_{-\pi}^{\pi} \log\left|\frac{\omega - \rho e^{i\phi}}{1 - \rho e^{-i\phi}\omega}\right| d\phi = \max(\log\rho, \log|\omega|).$$

Putting $\omega = \omega(re^{it})$ and integrating again,

$$\frac{1}{2\pi}\int_{-\pi}^{\pi}\int_{-\pi}^{\pi} \log\left|\frac{\omega(re^{it}) - \rho e^{i\phi}}{1 - \rho e^{-i\phi}\omega(re^{it})}\right| d\phi\, dt = \int_{-\pi}^{\pi} \max(\log\rho, \log|\omega(re^{it})|)\, dt.$$

But $|\omega(re^{it})| \le 1$ and $\log|\omega(re^{it})| \to \log|\omega(e^{it})| = 0$ a.e. as $r \to 1$. So ($\log\rho > -\infty$!),

$$\int_{-\pi}^{\pi} \max(\log\rho, \log|\omega(re^{it})|)\, dt \;\to\; 0$$

as $r \to 1$ *by bounded convergence*.

Therefore, by Fubini's theorem,

$$\frac{1}{2\pi}\int_{-\pi}^{\pi}\left\{\int_{-\pi}^{\pi} \log\left|\frac{\omega(re^{it}) - \rho e^{i\phi}}{1 - \rho e^{-i\phi}\omega(re^{it})}\right| dt\right\} d\phi \;\to\; 0 \quad \text{as } r \to 1.$$

Since

$$\log\left|\frac{\omega(z) - \rho e^{i\phi}}{1 - \rho e^{-i\phi}\omega(z)}\right| \le 0,$$

the *inner integral* in the above expression is ≤ 0, and, by *Fatou's lemma applied to* $(1/2\pi)\int_{-\pi}^{\pi}(\ \ )\,d\phi$,

$$\frac{1}{2\pi}\int_{-\pi}^{\pi}\left\{\limsup_{r\to 1}\int_{-\pi}^{\pi} \log\left|\frac{\omega(re^{it}) - \rho e^{i\phi}}{1 - \rho e^{-i\phi}\omega(re^{it})}\right| dt\right\} d\phi = 0.$$

Now, since $0 < \rho < 1$, for each ϕ,

$$\omega_\phi(z) = \frac{\omega(z) - \rho e^{i\phi}}{1 - \rho e^{-i\phi}\omega(z)}$$

is regular and bounded in $\{|z| < 1\}$, ($|\omega(z)| \le 1$ there) and is *indeed an inner function, because* $|\omega_\phi(e^{it})| \equiv 1$ a.e., *due to the fact that* $|\omega(e^{it})| \equiv 1$ a.e.

The lemma (and its proof) shows that

$$\lim_{r\to 1}\int_{-\pi}^{\pi} \log|\omega_\phi(re^{it})|\, dt$$

exists, and that if it's *zero*, $\omega_\phi(z)$ is a *constant multiple of a Blaschke product*. By the calculation made above, this limit *must* be zero for *almost every* ϕ. So *almost every* ω_ϕ *is a constant multiple of a Blaschke product*.

Finally, if $|z| < 1$,

$$|\omega(z) - \omega_\phi(z)| = \left| \frac{\rho e^{i\phi} - \rho e^{-i\phi}(\omega(z))^2}{1 - \rho e^{-i\phi}\omega(z)} \right| \leq \frac{2\rho}{1-\rho},$$

and this is $< \epsilon$ (for all z, $|z| < 1$) if $\rho > 0$ is sufficiently small.

We are done.

Scholium This result has several applications in the deeper study of H_∞ (e.g., Carleson's proof of the corona conjecture, not given here, but to be found in Duren's book and in Garnett's more recent one).

V

Norm Inequalities for Harmonic Conjugation

A. Review: Hilbert transforms of L_2 functions

Recall, from Chapter I, Subsection E.4, the

Theorem *Let $f(\theta) \in L_2(-\pi, \pi)$ and write, for $0 \le r < 1$,*

$$U(re^{i\theta}) = \frac{1}{2\pi}\int_{-\pi}^{\pi} \frac{1-r^2}{1+r^2-2r\cos(\theta-t)} f(t)\,dt,$$

$$\tilde{U}(re^{i\theta}) = \frac{1}{2\pi}\int_{-\pi}^{\pi} \frac{2r\sin(\theta-t)}{1+r^2-2r\cos(\theta-t)} f(t)\,dt,$$

so that $\tilde{U}(z)$ is a harmonic conjugate of $U(z)$, and $U(0) = 0$.

Then (taking $f(\theta+2\pi)$ equal to $f(\theta)$)

$$\tilde{f}(\theta) = \lim_{\epsilon\to 0} \frac{1}{\pi} \int_{\epsilon\le|t|\le\pi} \left\{ f(\theta-t) \Big/ 2\tan\frac{t}{2} \right\} dt$$

exists a.e., $\|\tilde{f}\|_2 \le \|f\|_2$, and

$$\tilde{U}(re^{i\theta}) = \frac{1}{2\pi}\int_{-\pi}^{\pi} \frac{1-r^2}{1+r^2-2r\cos(\theta-t)} \tilde{f}(t)\,dt,$$

so that $\tilde{U}(z) \to \tilde{f}(\theta)$ a.e. for $z \underset{\angle}{\longrightarrow} e^{i\theta}$.

The Cauchy principal value defining $\tilde{f}(\theta)$ in this theorem is usually written

$$\frac{1}{\pi} \fint_{-\pi}^{\pi} \frac{f(\theta-t)}{2\tan(t/2)}\,dt.$$

Given $f(\theta) \in L_2(-\pi, \pi)$, it follows immediately from the theorem that $F(z) = U(z)+i\tilde{U}(z)$

is in H_2. The boundary values $F(e^{i\theta})$ satisfy $F(e^{i\theta}) = f(\theta) + i\tilde{f}(\theta)$ a.e., so if $f \in L_2$ is *real*, *we have a way of constructing* $F \in H_2$ with $\Re F(e^{i\theta}) = f(\theta)$ a.e.

Now we saw in Chapter III, Subsection C.2 that as long as $f \in L_1(-\pi, \pi)$ and $f(\theta + 2\pi) = f(\theta)$,

$$\tilde{f}(\theta) = \frac{1}{\pi} \int_{-\pi}^{\pi} \frac{f(\theta - t)}{2\tan(t/2)} \, dt$$

exists and is finite a.e.

Definition $\tilde{f}(\theta)$ is called the *Hilbert transform* of $f(\theta)$. Sometimes, by abuse of language, we call it the *harmonic conjugate* of $f(\theta)$, on account of the theorem quoted at the beginning of this section. It would be better to call it the *function conjugate* to $f(\theta)$.

The inequality $\|\tilde{f}\|_2 \leq \|f\|_2$ has a valid analogue for other L_p spaces.

B. Hilbert transforms of L_p functions, $1 < p < \infty$

1. An identity

Lemma *Let* $U(z)$, $V(z)$ *be harmonic in* $\{|z| < 1\}$, *and let* $\tilde{U}(z)$, $\tilde{V}(z)$ *be their harmonic conjugates with* $\tilde{U}(0) = 0$, $\tilde{V}(0) = 0$. *Then, if* $0 \leq r < 1$,

$$\int_{-\pi}^{\pi} U(re^{i\theta})\tilde{V}(re^{i\theta}) \, d\theta = -\int_{-\pi}^{\pi} \tilde{U}(re^{i\theta})V(re^{i\theta}) \, d\theta.$$

Proof If

$$U(re^{i\theta}) = \sum_{-\infty}^{\infty} A_n r^{|n|} e^{in\theta},$$

$$V(re^{i\theta}) = \sum_{-\infty}^{\infty} B_n r^{|n|} e^{in\theta},$$

we have

$$\tilde{U}(re^{i\theta}) = \sum_{-\infty}^{\infty} (-i \operatorname{sgn} n) A_n r^{|n|} e^{in\theta},$$

$$\tilde{V}(re^{i\theta}) = \sum_{-\infty}^{\infty} (-i \operatorname{sgn} n) B_n r^{|n|} e^{in\theta}.$$

All the series here are absolutely convergent if $r < 1$, so that the desired relation may be verified directly by termwise integration.

2. M. Riesz' theorem

Theorem (M. Riesz) *Let* $f(\theta) \in L_p$, $1 < p < \infty$, (sic!) *and let*

$$U(re^{i\theta}) = \frac{1}{2\pi} \int_{-\pi}^{\pi} \frac{1 - r^2}{1 + r^2 - 2r\cos(\theta - t)} f(t) \, dt,$$

with $\tilde{U}$ the harmonic conjugate of U satisfying $\tilde{U}(0) = 0$. Then (taking always f as 2π-periodic)

$$\tilde{f}(\theta) = \frac{1}{\pi} \lim_{\epsilon \to 0} \int_{\epsilon}^{\pi} \frac{f(\theta - t) - f(\theta + t)}{2\tan(t/2)} \, dt$$

exists a.e., $\tilde{f} \in L_p$, and

$$\tilde{U}(re^{i\theta}) = \frac{1}{2\pi} \int_{-\pi}^{\pi} \frac{1 - r^2}{1 + r^2 - 2r\cos(\theta - t)} \tilde{f}(t) \, dt$$

for $r < 1$. There is a constant K_p depending only on p with $\|\tilde{f}\|_p \le K_p \|f\|_p$.

Proof For $p = 2$, this is the result quoted in Section A, so we may suppose that $1 < p < 2$ or $2 < p < \infty$.

As remarked in Section A, *existence* of $\tilde{f}(\theta)$ a.e. was already proved in Chapter III, Subsection C.2 under the assumption that $f \in L_1$; this is certainly fulfilled for $f \in L_p$, $p > 1$. *It is thus enough to show that*

$$\int_{-\pi}^{\pi} |\tilde{U}(re^{i\theta})|^p \, d\theta \le (K_p \|f\|_p)^p$$

for $0 \le r < 1$, for then, by Chapter I, Section C there is a $g \in L_p$, $\|g\|_p \le K_p \|f\|_p$, with

$$\tilde{U}(re^{i\theta}) = \frac{1}{2\pi} \int_{-\pi}^{\pi} \frac{1 - r^2}{1 + r^2 - 2r\cos(\theta - t)} g(t) \, dt,$$

and we can argue as in Chapter I, Subsection E.4 to show that $g(\theta) = \tilde{f}(\theta)$ a.e. (The argument of Chapter I, Subsection E.4 also furnishes a proof, valid under the present circumstances, that $\tilde{f}(\theta)$ *exists* a.e. – and is equal to $g(\theta)$. So in fact, we do not really *require* the theorem of Chapter III, Subsection C.2 at this point!)

We first prove that

$$\int_{-\pi}^{\pi} |\tilde{U}(re^{i\theta})|^p \, d\theta \le (K_p \|f\|_p)^p, \quad 0 \le r < 1,$$

in the case that $1 < p < 2$. Afterwards, we'll use the lemma of Subsection 1 and a duality argument to obtain the same result for $2 < p < \infty$.

The following argument is due to Katznelson. (It is, however, essentially an elegant manner of presenting reasoning already found in the second edition of Zygmund's book.) *It is enough* to prove the desired inequality for the case $f(\theta) \ge 0$, for in the general case we can break up a real $f \in L_p$ into the difference of two positive ones, and use the triangle inequality for L_p norms, together with the fact that

$$\|f\|_p^p = \|f_+\|_p^p + \|f_-\|_p^p \ge C_p \left(\|f_+\|_p + \|f_-\|_p \right)^p$$

for $f = f_+ - f_-$, if $f_+ \cdot f_- \equiv 0$ a.e.

Assuming $f(\theta) \ge 0$, and $f \not\equiv 0$ (!), write $F(z) = U(z) + i\tilde{U}(z)$ for $|z| < 1$. $\Re F(z) = U(z) > 0$ for $|z| < 1$, so $F(z)$ is free of zeros there, and we can take the analytic function $G(z) = (F(z))^p$.

Since $F(0) = U(0) \geq 0$, we have, for $0 \leq r < 1$, by Cauchy's theorem

$$\int_{-\pi}^{\pi} \Re G(re^{i\theta})\,d\theta = 2\pi\Re G(0) = 2\pi(U(0))^p \geq 0.$$

Now, for given $r < 1$, we break up $[-\pi, \pi]$ into two complementary sets, E_1 and E_2. Take a γ, $0 < \gamma < \pi/2$, such that $\pi/2 < p\gamma < p\pi/2 < \pi$. *There is such a* γ, *because* $1 < p < 2$. *Then*, let

$$E_1 = \{\theta;\ -\gamma \leq \arg F(re^{i\theta}) \leq \gamma\}$$

and

$$E_2 = \{\theta;\ \gamma < |\arg F(re^{i\theta})| \leq \pi/2\}.$$

Since $\Re F(z) > 0$, $|\arg F(z)| < \pi/2$, so $E_1 \cup E_2 = [-\pi, \pi]$.

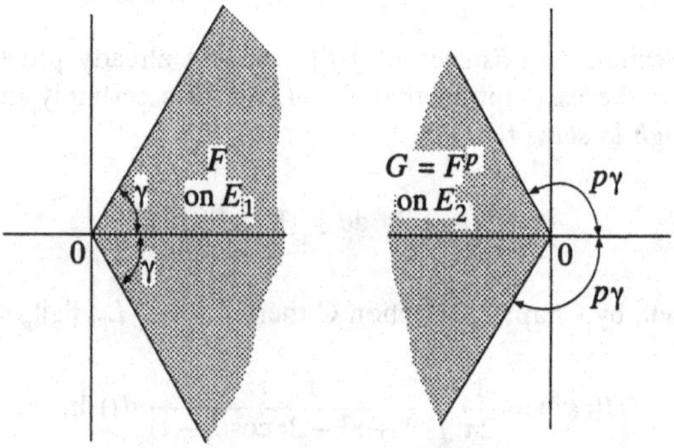

We now have

$$(*) \qquad \int_{E_1} \Re G(re^{i\theta})\,d\theta + \int_{E_2} \Re G(re^{i\theta})\,d\theta = \int_{-\pi}^{\pi} \Re G(re^{i\theta})\,d\theta \geq 0.$$

For $\theta \in E_2$, $\Re G(re^{i\theta})$ is *negative* and $\leq -|G(re^{i\theta})||\cos p\gamma|$ – look at the *second* of the above two diagrams. So by $(*)$,

$$|\cos p\gamma| \int_{E_2} |G(re^{i\theta})|\,d\theta \leq \int_{E_1} \Re G(re^{i\theta})\,d\theta.$$

For $\theta \in E_1$ (*first* of the above two diagrams!)

$$|F(re^{i\theta})| \leq |U(re^{i\theta})| / \cos\gamma,$$

so

$$|G(re^{i\theta})| \leq \cos^{-p}\gamma\, |U(re^{i\theta})|^p,$$

which, substituted in the previous, yields

$$\int_{E_2} |G(re^{i\theta})|\,d\theta \leq |\sec p\gamma| \int_{E_1} |G(re^{i\theta})|\,d\theta \leq (\cos^{-p}\gamma)|\sec p\gamma| \int_{E_1} |U(re^{i\theta})|^p\,d\theta.$$

We also see that

$$\int_{E_1} |G(re^{i\theta})|\,d\theta \leq \cos^{-p}\gamma \int_{E_1} |U(re^{i\theta})|^p\,d\theta.$$

Since $E_1 \cup E_2 = [-\pi, \pi]$ and $E_1 \cap E_2 = \emptyset$, we have, adding the above two inequalities,

$$\int_{-\pi}^{\pi} |G(re^{i\theta})|\, d\theta \leq \frac{1 + |\sec p\gamma|}{\cos^p \gamma} \int_{E_1} |U(re^{i\theta})|^p\, d\theta,$$

whence, a fortiori,

$$\int_{-\pi}^{\pi} |\tilde{U}(re^{i\theta})|^p\, d\theta \leq \frac{1 + |\sec p\gamma|}{\cos^p \gamma} \int_{-\pi}^{\pi} |U(re^{i\theta})|^p\, d\theta,$$

since $G(z) = (U(z) + i\tilde{U}(z))^p$. Finally,

$$\int_{-\pi}^{\pi} |\tilde{U}(re^{i\theta})|^p\, d\theta \leq \frac{1 + |\sec p\gamma|}{\cos^p \gamma} \, ||f||_p^p,$$

by the properties of $P_r(\theta)$ (Chapter I, Subsection D.1), and we are done in case $1 < p < 2$.

Remark This very clever idea is like the one used in Chapter IV, Subsection D.2 to show that $\int_{-\pi}^{\pi} |\log |F(re^{i\theta})|| r\, d\theta$ is bounded if $\int_{-\pi}^{\pi} \log^+ |F(re^{i\theta})|\, d\theta$ is.

In case $2 < p < \infty$, we prove

$$\int_{-\pi}^{\pi} |\tilde{U}(re^{i\theta})|^p\, d\theta \leq C_p \int_{-\pi}^{\pi} |U(re^{i\theta})|^p\, d\theta$$

in the following manner. Let $1/q = 1 - (1/p)$, then $1 < q < 2$, and by Hölder,

$$||\tilde{U}(re^{i\theta})||_p = \left(\int_{-\pi}^{\pi} |\tilde{U}(re^{i\theta})|^p\, d\theta \right)^{1/p}$$

is the *supremum* of

$$\left| \int_{-\pi}^{\pi} \tilde{U}(re^{i\theta}) T(re^{i\theta})\, d\theta \right|$$

taken over all *finite sums*

$$T(re^{i\theta}) = \sum_n B_n r^{|n|} e^{in\theta}$$

with

$$||T(re^{i\theta})||_q = \left(\int_{-\pi}^{\pi} |T(re^{i\theta})|^q\, d\theta \right)^{1/q} \leq 1.$$

However, since $1 < q < 2$, for any such T, by what has just been proven,

$$||\tilde{T}(re^{i\theta})||_q \leq K_q \, ||T(re^{i\theta})||_q \, ;$$

therefore, by Subsection 1,

$$\left| \int_{-\pi}^{\pi} \tilde{U}(re^{i\theta}) T(re^{i\theta})\, d\theta \right| = \left| \int_{-\pi}^{\pi} U(re^{i\theta}) \tilde{T}(re^{i\theta})\, d\theta \right|$$

$$\leq ||\tilde{T}(re^{i\theta})||_q \, ||U(re^{i\theta})||_p \leq K_q \, ||T(re^{i\theta})||_q \, ||U(re^{i\theta})||_p \leq K_q \, ||U(re^{i\theta})||_p,$$

and

$$\|\tilde{U}(re^{i\theta})\|_p \le K_q \|U(re^{i\theta})\|_p$$

for $2 < p < \infty$, as required. We are done.

C. $\tilde{f}(\theta)$ for f in L_1

Let $f(\theta + 2\pi) = f(\theta)$, $f \in L_1(-\pi, \pi)$. By Chapter III, Subsection C.2,

$$\tilde{f}(\theta) = \frac{1}{2\pi} \fint_{-\pi}^{\pi} \frac{f(\theta - t)}{\tan(t/2)} \, dt$$

exists and is finite a.e.

1. Definition of $m_h(\lambda)$; Kolmogorov's theorem

Definition If h is a measurable function defined on $[-\pi, \pi]$, we write*, for $\lambda \ge 0$,

$$m_h(\lambda) = |\{\theta \in [-\pi, \pi];\ |h(\theta)| \ge \lambda\}|.$$

Clearly $m_h(0) = 2\pi$, $m_h(\lambda)$ decreases, and $m_h(\lambda) \to 0$ as $\lambda \to \infty$ if $h(\theta)$ is finite a.e. m_h is called the *distribution function* of h.

Theorem (Kolmogorov) *If $f \in L_1$ and $\lambda > 0$,*

$$m_{\tilde{f}}(\lambda) \le \frac{K}{\lambda} \|f\|_1,$$

where K is a constant independent of f and λ.

Proof (Carleson, via Katznelson) Suppose $\tilde{f}(\theta) = g(\theta) + h(\theta)$. Then, if $|\tilde{f}(\theta)| \ge \lambda$ we *must have* $|g(\theta)| \ge \lambda/2$ or $|h(\theta)| \ge \lambda/2$, so $m_{\tilde{f}}(\lambda) \le m_g(\lambda/2) + m_h(\lambda/2)$. Therefore it is enough to prove the theorem for the case where $f(\theta) \ge 0$, since, by our observation, we can extend it therefrom to general f by taking a larger constant K.

So assume $f(\theta) \ge 0$; since harmonic conjugation is *linear*, we may, without loss of generality, take $\|f\|_1$ to be 2π, then we're done if we show that

$$m_{\tilde{f}}(\lambda) \le \frac{2\pi}{\lambda}.$$

For $|z| < 1$, write

$$F(z) = \frac{1}{2\pi} \int_{-\pi}^{\pi} \frac{e^{it} + z}{e^{it} - z} f(t) \, dt;$$

then $\Re F(z) > 0$, and, from Chapter III, Subsection C.2, for almost all θ,

$$F(z) \to f(\theta) + i\tilde{f}(\theta) \quad \text{as } z \underset{\angle}{\longrightarrow} e^{i\theta}.$$

Given $\lambda > 0$, let

$$\phi(z) = 1 + \frac{F(z) - \lambda}{F(z) + \lambda}.$$

---

* $|E|$ denotes the *Lebesgue measure* of the set E

Observe that $w \mapsto 1 + (w - \lambda)/(w + \lambda)$ is the following conformal mapping:

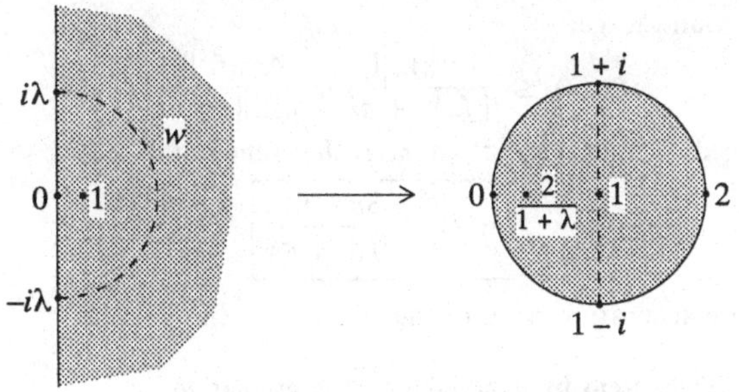

In this mapping, $0 \mapsto 0$, $1 \mapsto 2/(1+\lambda)$, and the semi-circle joining $i\lambda$ to $-i\lambda$ goes onto the diameter from $1+i$ to $1-i$. We surely have $|\phi(z)| \leq 2$, $|z| < 1$, so by Chapter II, Subsection B.1,

$$2\pi\phi(0) = \int_{-\pi}^{\pi} \phi(e^{i\theta})\,d\theta,$$

i.e.,

$$2\pi\Re\phi(0) = \int_{-\pi}^{\pi} \Re\phi(e^{i\theta})\,d\theta.$$

Now

$$F(0) = \frac{1}{2\pi}\int_{-\pi}^{\pi} f(t)\,dt = \frac{\|f\|_1}{2\pi} = 1$$

by assumption, since $f(t) \geq 0$, therefore, $\phi(0) = 2/(1+\lambda)$. Also $\Re\phi(e^{i\theta}) \geq 0$. Finally, if $|\tilde{f}(\theta)| \geq \lambda$ we have $|F(e^{i\theta})| \geq \lambda$ so $\Re\phi(e^{i\theta}) \geq 1$. Thus, from

$$\int_{-\pi}^{\pi} \Re\phi(e^{i\theta})\,d\theta = \frac{4\pi}{1+\lambda}$$

we get

$$|\{\theta;\ \Re\phi(e^{i\theta}) \geq 1\}| \leq \frac{4\pi}{1+\lambda},$$

i.e.

$$m_{\tilde{f}}(\lambda) \leq \frac{4\pi}{1+\lambda},$$

more than what was needed.

We are done.

Scholium If $f \geq 0$ but $\|f\|_1 \neq 2\pi$, we can make a change of variable in the more precise result found at the end of the above proof to obtain

$$m_{\tilde{f}}(\lambda) \leq \frac{4\pi\,\|f\|_1}{\|f\|_1 + 2\pi\lambda}.$$

Finally, if f is merely *real*, $f = f_+ - f_-$ with $f_+, f_- \geq 0$, $\|f\|_1 = \|f_+\|_1 + \|f_-\|_1$, we use

$$m_{\tilde{f}}(\lambda) \leq m_{\tilde{f}_+}(\lambda/2) + m_{\tilde{f}_-}(\lambda/2)$$

and the result just found to get

$$m_{\tilde{f}}(\lambda) \leq \frac{4\pi\,\|f_+\|_1}{\|f_+\|_1 + \pi\lambda} + \frac{4\pi\,\|f_-\|_1}{\|f_-\|_1 + \pi\lambda}.$$

Observe that for fixed λ, $x/(x + \pi\lambda)$ is an *increasing* function of $x \geq 0$, so for f *real* in L_1,

$$\boxed{m_{\tilde{f}}(\lambda) \leq \frac{8\pi\,\|f\|_1}{\|f\|_1 + \pi\lambda}}.$$

Sometimes this more accurate version of the above theorem is useful.

2. Proof of M. Riesz' theorem by Marcinkiewicz interpolation

We use the result in Subsection 1 together with that quoted in Section A to give a new proof of M. Riesz' theorem (Subsection B.2) in order to show an example of *Marcinkiewicz interpolation* for linear operations. As the duality argument at the end of Subsection B.2 shows, the version proved below *actually covers all values of* p, $1 < p < \infty$, if we throw in the easy case $p = 2$, given in Section A.

Theorem (M. Riesz) *If* $1 < p < 2$ *there is a constant* K_p *depending on* p *with*

$$\|\tilde{U}(re^{i\theta})\|_p \leq K_p \|U(re^{i\theta})\|_p$$

if $r < 1$ *and* $U(z)$ *is harmonic in* $\{|z| < 1\}$.

Proof Write $f(\theta) = U(re^{i\theta})$; then $\tilde{f}(\theta) = \tilde{U}(re^{i\theta})$. Here, f and $\tilde{f}$ are both *continuous*, so by the *definition* of integrals,

$$\int_{-\pi}^{\pi} |f(\theta)|^p \, d\theta = -\int_0^\infty \lambda^p \, dm_f(\lambda),$$

$$\int_{-\pi}^{\pi} |\tilde{f}(\theta)|^p \, d\theta = -\int_0^\infty \lambda^p \, dm_{\tilde{f}}(\lambda).$$

Our job is to estimate the *second* integral in terms of the *first*. Since $\tilde{f}(\theta)$ is here continuous, $m_{\tilde{f}}(\lambda) = 0$ for λ large enough, and integration by parts gives

$$\int_{-\pi}^{\pi} |\tilde{f}(\theta)|^p \, d\theta = p\int_0^\infty \lambda^{p-1} m_{\tilde{f}}(\lambda) \, d\lambda.$$

Suppose $f = g + h$. Then, as we argued in Subsection 1, $\tilde{f} = \tilde{g} + \tilde{h}$ makes

$$m_{\tilde{f}}(\lambda) \leq m_{\tilde{g}}(\lambda/2) + m_{\tilde{h}}(\lambda/2).$$

Marcinkiewicz had the crazy idea of using this inequality to estimate $m_{\tilde{f}}(\lambda)$ *by breaking up* f *as* $g + h$ *in a way depending on* λ !

Given $\lambda > 0$, *take*:

$$g(\theta) = \begin{cases} f(\theta) & \text{if } |f(\theta)| < \lambda \\ 0 & \text{otherwise.} \end{cases}$$

$$h(\theta) = \begin{cases} 0 & \text{if } |f(\theta)| < \lambda \\ f(\theta) & \text{otherwise.} \end{cases}$$

The idea here is that g is *not large*, so $g \in L_2$.

Indeed, by Section A,

$$\|\tilde{g}\|_2^2 \leq \|g\|_2^2 = \int_{|f(\theta)|<\lambda} |f(\theta)|^2 \, d\theta = -\int_0^\lambda s^2 \, dm_f(s),$$

so

$$m_{\tilde{g}}(\lambda/2) \leq \frac{4}{\lambda^2}\|\tilde{g}\|_2^2 \leq -\frac{4}{\lambda^2}\int_0^\lambda s^2 \, dm_f(s).$$

For h, we use the result of Subsection 1, which says that

$$m_{\tilde{h}}(\lambda/2) \leq \frac{2K}{\lambda}\|h\|_1 = \frac{2K}{\lambda}\int_{|f(\theta)|\geq\lambda} |f(\theta)| \, d\theta = -\frac{2K}{\lambda}\int_\lambda^\infty s \, dm_f(s).$$

Therefore, finally,

$$\boxed{m_{\tilde{f}}(\lambda) \leq -\frac{4}{\lambda^2}\int_0^\lambda s^2 \, dm_f(s) - \frac{2K}{\lambda}\int_\lambda^\infty s \, dm_f(s).}$$

Now we have:

$$\int_{-\pi}^{\pi} |\tilde{f}(\theta)|^p \, d\theta = p\int_0^\infty \lambda^{p-1} m_{\tilde{f}}(\lambda) \, d\lambda$$

$$\leq -4p\int_0^\infty \lambda^{p-3}\int_0^\lambda s^2 \, dm_f(s) \, d\lambda - 2Kp\int_0^\infty \lambda^{p-2}\int_\lambda^\infty s \, dm_f(s) \, d\lambda.$$

By Fubini's theorem:

$$-\int_0^\infty \lambda^{p-3}\int_0^\lambda s^2 \, dm_f(s) \, d\lambda = -\int_0^\infty\int_s^\infty \lambda^{p-3}s^2 \, d\lambda \, dm_f(s)$$

$$= -\frac{1}{2-p}\int_0^\infty s^{p-2}s^2 \, dm_f(s) = -\frac{1}{2-p}\int_0^\infty s^p \, dm_f(s)$$

$$= \frac{\|f\|_p^p}{2-p},$$

because $-2 < p-3 < -1$.

Similarly,

$$-\int_0^\infty \lambda^{p-2}\int_\lambda^\infty s \, dm_f(s) \, d\lambda = -\int_0^\infty\int_0^s \lambda^{p-2}s \, d\lambda \, dm_f(s) = -\frac{1}{p-1}\int_0^\infty s^p \, dm_f(s)$$

$$= \frac{\|f\|_p^p}{p-1},$$

because $-1 < p-2 < 0$.

So finally,

$$\|\tilde{f}\|_p^p \le 2p\left[\frac{2}{2-p} + \frac{K}{p-1}\right]\|f\|_p^p.$$

Q.E.D.

3. Zygmund's $L \log L$ theorem

It turns out that $f \in L_1$ is *not enough* to guarantee $\tilde{f} \in L_1$. The theorem which applies here is due to Zygmund:

Theorem (Zygmund) *There is a constant c with*

$$\int_{-\pi}^{\pi} |\tilde{f}(\theta)|\,d\theta \le c\int_{-\pi}^{\pi} |f(\theta)|(1+\log^+|f(\theta)|)\,d\theta + 2\pi,$$

and a similar inequality holds with $f(\theta)$ replaced by $U(re^{i\theta})$ and $\tilde{f}(\theta)$ replaced by $\tilde{U}(re^{i\theta})$.

Remark $x(1+\log^+ x)$ is *convex* for $x > 0$ so, since $P_r \ge 0$ and

$$\frac{1}{2\pi}\int_{-\pi}^{\pi} P_r(t)\,dt = 1,$$

if

$$U(re^{i\theta}) = \frac{1}{2\pi}\int_{-\pi}^{\pi} P_r(\theta - t)f(t)\,dt,$$

we have, for $r < 1$, by Jensen's inequality,

$$\int_{-\pi}^{\pi} |U(re^{i\theta})|(1+\log^+|U(re^{i\theta})|)\,d\theta \le \int_{-\pi}^{\pi} |f(\theta)|(1+\log^+|f(\theta)|)\,d\theta.$$

Therefore $|f(\theta)|\log^+|f(\theta)| \in L_1(-\pi,\pi)$ *guarantees that*

$$F(z) = \frac{1}{2\pi}\int_{-\pi}^{\pi} \frac{e^{i\theta}+z}{e^{i\theta}-z}f(\theta)\,d\theta \in H_1.$$

Proof of theorem We use the estimate for $m_{\tilde{f}}(\lambda)$ obtained in Subsection 2 (boxed formula) to get a bound on

$$\int_{-\pi}^{\pi} |\tilde{f}(\theta)|\,d\theta = -\int_0^{\infty} \lambda\,dm_{\tilde{f}}(\lambda) = \int_0^{\infty} m_{\tilde{f}}(\lambda)\,d\lambda \le 2\pi + \int_1^{\infty} m_{\tilde{f}}(\lambda)\,d\lambda$$

($m_{\tilde{f}}(\lambda) \le 2\pi$ for all $\lambda \ge 0$!). We have, by the formula in Subsection 2,

$$\int_1^{\infty} m_{\tilde{f}}(\lambda)\,d\lambda \le -4\int_1^{\infty} \lambda^{-2}\int_0^{\lambda} s^2\,dm_f(s)\,d\lambda - 2K\int_1^{\infty} \lambda^{-1}\int_{\lambda}^{\infty} s\,dm_f(s)\,d\lambda.$$

Since $dm_f(s) \le 0$, this is

$$\le -4\int_0^\infty\int_0^\lambda \frac{s^2}{\lambda^2}\, dm_f(s)\, d\lambda - 2K\int_1^\infty\int_1^s \frac{s}{\lambda}\, d\lambda\, dm_f(s)$$

$$= -4\int_0^\infty\int_s^\infty \frac{s^2}{\lambda^2}\, d\lambda\, dm_f(s) - 2K\int_1^\infty s\log s\, dm_f(s)$$

$$= -\int_0^\infty (4s + 2Ks\log^+ s)\, dm_f(s)$$

$$= -\int_{-\pi}^{\pi} |f(\theta)|(4 + 2K\log^+|f(\theta)|)\, d\theta.$$

This does it.

4. Converse of Zygmund's theorem for positive f

In case $f(\theta) \ge 0$, the above theorem has a converse!

Theorem (M. Riesz) *Let $f(\theta) \ge 0$, let*

$$F(z) = \frac{1}{2\pi}\int_{-\pi}^{\pi} \frac{e^{it}+z}{e^{it}-z} f(t)\, dt,$$

and suppose $F \in H_1$. Then

$$|f(\theta)|\log^+|f(\theta)| \in L_1(-\pi,\pi).$$

Proof Let $G(z) = 1 + F(z)$; then $\Re G(z) > 1$ for $|z| < 1$ and $G(0) = 1 + F(0)$ is *real*. So we can define a regular branch of $\log G(z)$ in $\{|z| < 1\}$ with $\log G(0)$ *real and* > 0.

By Cauchy's theorem, for $r < 1$,

$$G(0)\log G(0) = \frac{1}{2\pi}\int_{-\pi}^{\pi} G(re^{i\theta})\log G(re^{i\theta})\, d\theta,$$

i.e., the left side being real,

$$G(0)\log G(0) = \frac{1}{2\pi}\int_{-\pi}^{\pi} \left[\Re G(re^{i\theta})\log|G(re^{i\theta})| - \Im G(re^{i\theta})\arg G(re^{i\theta})\right] d\theta.$$

In the present case, $\Re G(z) > 0$, so $-\pi/2 < \arg G(z) < \pi/2$, and

$$\int_{-\pi}^{\pi} \Re G(re^{i\theta})\log|G(re^{i\theta})|\, d\theta \le 2\pi G(0)\log G(0) + \frac{\pi}{2}\int_{-\pi}^{\pi} |\Im G(re^{i\theta})|\, d\theta.$$

As $r \to 1$,

$$\Re G(re^{i\theta})\log|G(re^{i\theta})| \to (1 + f(\theta))\log|1 + f(\theta) + i\tilde{f}(\theta)| \quad \text{a.e.},$$

whilst

$$\int_{-\pi}^{\pi} |\Im G(re^{i\theta})|\, d\theta \le \|F\|_1.$$

So by Fatou's lemma, we get

$$\int_{-\pi}^{\pi} (1 + f(\theta)) \log |1 + f(\theta) + i\tilde{f}(\theta)| \, d\theta$$

$$\leq 2\pi \left(1 + ||F||_1\right) \log \left(1 + ||F||_1\right) + (\pi/2)\, ||F||_1 ,$$

using a trivial estimate for $G(0) = 1 + F(0)$. Since $f(\theta) > 0$, this is enough.

5. Another theorem of Kolmogorov

Using the result of Subsection 4, it is easy to construct examples of $f \in L_1$ *with* $\tilde{f}(\theta)$ *not in* L_1 *on any interval*. However,

Theorem (Kolmogorov) *If* $f \in L_1$,

$$\int_{-\pi}^{\pi} |\tilde{f}(\theta)|^p \, d\theta < \infty$$

whenever $0 < p < 1$, *so*

$$F(z) = \frac{1}{2\pi} \int_{-\pi}^{\pi} \frac{e^{it} + z}{e^{it} - z} f(t) \, dt$$

belongs to H_p *for* $0 < p < 1$.

Proof By the result in Subsection 1, if $0 < p < 1$,

$$\int_{-\pi}^{\pi} |\tilde{f}(\theta)|^p \, d\theta = -\int_0^\infty \lambda^p \, dm_{\tilde{f}}(\lambda) = p \int_0^\infty \lambda^{p-1} m_{\tilde{f}}(\lambda) \, d\lambda$$

$$\leq 2\pi + p \int_1^\infty \lambda^{p-1} m_{\tilde{f}}(\lambda) \, d\lambda \leq 2\pi + Kp\, ||f||_1 \int_1^\infty \lambda^{p-2} \, d\lambda$$

$$= 2\pi + Kp\, ||f||_1 \cdot \frac{1}{1-p}.$$

Q.E.D.

Remark Using the inequality *actually proved* in Subsection 1, and given in the scholium there,

$$m_{\tilde{f}}(\lambda) \leq \frac{8\pi \, ||f||_1}{||f||_1 + \pi\lambda}$$

(valid for real f), we obtain a better inequality of the form

$$\int_{-\pi}^{\pi} |\tilde{f}(\theta)|^p \, d\theta \leq C_p \, ||f||_1^p, \quad 0 < p < 1.$$

Indeed,

$$\int_{-\pi}^{\pi} |\tilde{f}(\theta)|^p \, d\theta = p \int_0^\infty \lambda^{p-1} m_{\tilde{f}}(\lambda) \, d\lambda \le 8\pi p \int_0^\infty \frac{\lambda^{p-1} \|f\|_1 \, d\lambda}{\|f\|_1 + \pi\lambda}$$

$$= 8\pi p \|f\|_1^p \int_0^\infty \frac{s^{p-1} \, ds}{1 + \pi s},$$

the integral being finite for $0 < p < 1$.

There is, however, a much *easier* way to establish the same result. Taking, without loss of generality and as usual, $f(\theta) \ge 0$, form the analytic function

$$F(z) = \frac{1}{2\pi} \int_{-\pi}^{\pi} \frac{e^{it} + z}{e^{it} - z} f(t) \, dt,$$

then apply Cauchy's theorem to $(F(re^{i\theta}))^p$, $0 < p < 1$, observing that $|\arg F(z)| \le \pi/2$. The desired inequality comes out quite easily.

D. $\tilde{f}(\theta)$ for bounded f and for continuous periodic $f(\theta)$

1. Integrability of $\exp \lambda |\tilde{f}|$

Theorem *If* $f(\theta)$ *is* real *and*

$$\|f\|_\infty = \operatorname*{ess\,sup}_{-\pi \le \theta \le \pi} |f(\theta)| \le \frac{\pi}{2},$$

then

$$\int_{-\pi}^{\pi} e^{\lambda |\tilde{f}(\theta)|} \, d\theta \le C_\lambda,$$

a constant depending on λ *alone, for each* $\lambda < 1$.

Remark Thus, although $f \in L_\infty$ *does not* imply $\tilde{f} \in L_\infty$, $\tilde{f}$ *still* has *very strong* boundedness properties.

Remark The theorem *cannot be improved* to cover the case $\lambda = 1$, as is seen by considering the example where

$$f(\theta) = \frac{\pi}{2} \times \text{the characteristic function of an } \textit{interval}.$$

Proof Write

$$F(z) = \frac{1}{2\pi} \int_{-\pi}^{\pi} \frac{e^{it} + z}{e^{it} - z} f(t) \, dt,$$

then, if f is *real* and $|f(t)| \le \pi/2$, $|\Re F(z)| \le \pi/2$ and $\Im F(0) = 0$, hence, if $0 \le \lambda < 1$, by Cauchy,

$$\int_{-\pi}^{\pi} \exp\left(-i\lambda F(re^{i\theta})\right) d\theta = 2\pi e^{-i\lambda \Re F(0)}.$$

Taking real parts,

$$\int_{-\pi}^{\pi} \Re \exp\left(-i\lambda F(re^{i\theta})\right)\, d\theta \le 2\pi |e^{-i\lambda \Re F(0)}| = 2\pi.$$

But

$$|\arg \exp\left(-i\lambda F(re^{i\theta})\right)| \le \frac{\pi}{2}\lambda,$$

so

$$\int_{-\pi}^{\pi} |\exp\left(-i\lambda F(re^{i\theta})\right)| \cos\frac{\pi}{2}\lambda\, d\theta = \int_{-\pi}^{\pi} \exp\left(\lambda \Im F(re^{i\theta})\right) \cos\frac{\pi}{2}\lambda\, d\theta$$

$$\le \int_{-\pi}^{\pi} \Re \exp\left(-i\lambda F(re^{i\theta})\right)\, d\theta \le 2\pi,$$

and finally,

$$\int_{-\pi}^{\pi} \exp\left(\lambda \Im F(re^{i\theta})\right)\, d\theta \le 2\pi \Big/ \cos\frac{\pi}{2}\lambda.$$

Similarly, using $\exp(i\lambda F(z))$;

$$\int_{-\pi}^{\pi} \exp\left(-\lambda \Im F(re^{i\theta})\right)\, d\theta \le 2\pi \Big/ \cos\frac{\pi}{2}\lambda.$$

Since $\Im F(re^{i\theta}) \to \tilde{f}(\theta)$ a.e. as $r \to 1$, the theorem follows by Fatou's lemma.

2. Case of continuous periodic f

Corollary *If $f(\theta)$ is* continuous *of period 2π, and real,* $\exp \lambda|\tilde{f}(\theta)|$ *is in L_1 for* all λ.

Proof Given $\epsilon > 0$, we can find a *finite sum*

$$S(\theta) = \sum_n A_n e^{in\theta}$$

with $g(\theta) = f(\theta) - S(\theta)$ satisfying $\|g(\theta)\|_\infty \le \epsilon$. Then $\exp \lambda|\tilde{g}(\theta)|$ is in L_1 for $\lambda < \pi/2\epsilon$. But $\tilde{f}(\theta) = \tilde{g}(\theta) + \tilde{S}(\theta)$, and $\tilde{S}(\theta)$, being a finite sum of the same form as $S(\theta)$, is *bounded*.

Q.E.D.

E. Lipschitz classes

It is useful to know:

Theorem (Privalov) *Let $f(\theta)$ be 2π-periodic and* Lip α, $0 < \alpha < 1$. *Then $\tilde{f}(\theta)$ is* Lip α.

Proof First of all, if $h > 0$,

$$\left| \tilde{f}(\theta) - \frac{1}{\pi}\int_h^{\pi} \frac{f(\theta - t) - f(\theta + t)}{2\tan(t/2)}\, dt \right| \le \frac{1}{\pi}\int_0^h \frac{|f(\theta - t) - f(\theta + t)|}{t}\, dt$$

$$\le \frac{1}{\pi}\int_0^h \frac{ct^\alpha}{t}\, dt = O(h^\alpha)$$

for $0 < \alpha < 1$. So we may work with

$$\frac{1}{\pi}\int_{2h}^{\pi} \frac{f(\theta - t) - f(\theta + t)}{2\tan(t/2)}\,dt,$$

say, instead of $\tilde{f}(\theta)$, if we wish to estimate $\tilde{f}(\theta + h) - \tilde{f}(\theta)$.

We may also assume $f(\theta) = 0$, since $\tilde{f}(\phi)$ and

$$\frac{1}{2\pi}\int_{2h}^{\pi} [f(\phi - t) - f(\phi + t)] \cot\frac{t}{2}\,dt$$

don't change if $f(s)$ is replaced by $f(s) - f(\theta)$. *Thus, we may assume* $|f(\theta - t)| \le C|t|^{\alpha}$. Now

$$\int_{2h}^{\pi} \frac{f(\theta + h - t) - f(\theta + h + t)}{\tan(t/2)}\,dt - \int_{2h}^{\pi} \frac{f(\theta - t) - f(\theta + t)}{\tan(t/2)}\,dt$$

$$= \int_{2h}^{\pi} \left[\frac{f(\theta + h - t)}{\tan(t/2)} - \frac{f(\theta - t)}{\tan(t/2)}\right] dt + \int_{2h}^{\pi} \frac{f(\theta + t) - f(\theta + h + t)}{\tan(t/2)}\,dt.$$

We evaluate the *first* integral on the right, the second being handled similarly. The first integral is

$$\int_{h}^{2h} \frac{f(\theta - s)}{\tan\left((s + h)/2\right)}\,ds + \int_{2h}^{\pi - h} \frac{f(\theta - s)}{\tan\left((s + h)/2\right)}\,ds - \int_{2h}^{\pi - h} \frac{f(\theta - t)}{\tan(t/2)}\,dt - \int_{\pi - h}^{\pi} \frac{f(\theta - t)}{\tan(t/2)}\,dt$$

$$= O(h^{\alpha}) + \int_{2h}^{\pi - h} f(\theta - t)\left(\cot\frac{t + h}{2} - \cot\frac{t}{2}\right) dt + o(h)$$

since $|f(\theta - t)| \le O(t^{\alpha})$. The integral is in absolute value

$$\le \text{const}\cdot\int_{2h}^{\pi - h} t^{\alpha}\cdot\frac{h}{\sin^2(t/2)}\,dt \le \text{const}\cdot\int_{2h}^{\infty}\frac{ht^{\alpha}\,dt}{t^2} = \text{const}\cdot h^{\alpha}$$

since $0 < \alpha < 1$.

We are done.

F. Return to conformal mapping

We consider again, as in Chapter II, Subsection C.3, a conformal mapping $f(z)$ of the unit disc onto the domain bounded by a Jordan curve Γ with continuously turning tangent.

Theorem *If $f(z)$ maps $\{|z| < 1\}$ conformally onto the region bounded by a Jordan curve Γ with continuously turning tangent, $f'(z)$ and $1/f'(z)$ belong to all the spaces H_p, $0 < p < \infty$.*

Proof By the theorem of Chapter II, Section C.3, each harmonic determination of $\arg f'(z)$ in $\{|z| < 1\}$ (these are *available* since $f'(z)$ never vanishes there) has a *continuous* extension up to $\{|z| = 1\}$. Any such determination comes from an *analytic* branch of $\log f'(z)$ defined in the open unit disc, so $\log|f'(z)| = \Re \log f'(z)$ is a *harmonic conjugate* there of $-\Im \log f'(z) = -\arg f'(z)$, equal to $\log|f'(0)|$ plus the *usual* harmonic conjugate studied in this chapter.

The corollary in Subsection D.2 (or rather a combination of its *proof* with the one of the theorem in Subsection D.1) now tells us that the *integrals*

$$\int_0^{2\pi} \exp(\pm\lambda \log|f'(re^{i\theta})|)\, d\theta$$

are, for *any* $\lambda > 0$, *bounded* for $0 \le r < 1$. But this means that $f'(z)$ and $1/f'(z)$ are both in H_λ for any $\lambda > 0$. We are done.

Corollary *Under the hypothesis of the theorem, Γ is rectifiable.*

Proof By the fundamental theorem of Chapter II, Subsection C.1, $f(e^{i\theta})$ is a parametrization of Γ. Let us check that for any partition $0 = \theta_0 < \theta_1 < \ldots < \theta_n = 2\pi$ of $[0, 2\pi]$, we have

$$\sum_{k=1}^{n} |f(e^{i\theta_k}) - f(e^{i\theta_{k-1}})| \le \|f'\|_1 .$$

According to the result referred to, the sum on the left is equal to

$$\lim_{r\to 1} \sum_{k=1}^{n} |f(re^{i\theta_k}) - f(re^{i\theta_{k-1}})|,$$

and for each $r < 1$, the present sum is

$$\sum_{k=1}^{n} \left| \int_{\theta_{k-1}}^{\theta_k} f'(re^{i\theta}) \cdot ire^{i\theta}\, d\theta \right| \le \int_0^{2\pi} |f'(re^{i\theta})|\, d\theta \le \|f'\|_1 .$$

Since $\|f'\|_1 < \infty$ by the above theorem, our corollary follows.

Under the conditions of the theorem, $f'(z)$, as a function in every space H_p, $0 < p < \infty$, has finite $\angle$ boundary values a.e. on the unit circumference.

Notation This boundary value at $z = e^{i\theta}$ is denoted by $f'(e^{i\theta})$.

Corollary *Under the conditions of the theorem, if J is an arc of the unit circumference and $f(J)$ the corresponding arc of Γ we have, for any p, $1 < p < \infty$,*

$$A_p|J|^p \le \text{length}\, f(J) \le B_p|J|^{1/p}$$

where A_p and B_p are constants depending on Γ (and the particular conformal mapping f) with $A_p > 0$, $B_p < \infty$.

Proof By the preceding corollary and the first theorem in Chapter II, Subsection D.2,

$$\text{length}\, f(J) = \int_J |f'(e^{i\theta})|\, d\theta.$$

Taking $1/q = 1 - (1/p)$, this relation yields, by Hölder,

$$\text{length}\, f(J) \le \|f'\|_q |J|^{1/p}$$

with $\|f'\|_q < \infty$ by the above theorem.

Again, still by Hölder,

$$|J| = \int_J |f'(e^{i\theta})|^{1/p}|f'(e^{i\theta})|^{-1/p}\,d\theta$$

$$\leq \left(\int_J |f'(e^{i\theta})|\,d\theta\right)^{1/p} \cdot \left(\int_0^{2\pi} |f'(e^{i\theta})|^{-q/p}\,d\theta\right)^{1/q}.$$

Raising to the pth power and using the previous relation, we get from this

$$\operatorname{length} f(J) \geq A_p|J|^p$$

with

$$A_p = \left(\int_0^{2\pi} |f'(e^{i\theta})|^{-q/p}\,d\theta\right)^{-p/q} > 0$$

according to the above theorem.

This does it.

The observation following the first theorem of Chapter II, Subsection D.2 enables us to extend this corollary from *arcs* J of the unit circumference to *Borel sets* E thereon. If, for such E, we denote by $|f(E)|$ the *linear measure* of $f(E)$ on Γ, we still have

$$|f(E)| = \int_E |f'(e^{i\theta})|\,d\theta,$$

and thence, by the reasoning used to prove the corollary,

$$A_p|E|^p \leq |f(E)| \leq B_p|E|^{1/p}.$$

It is *important* to note that one *cannot* conclude from the hypothesis of the above theorem that $f'(z) \in H_\infty$, nor can one take $p = 1$ in the last corollary. Possession of a continuously turning tangent by Γ is thus *not enough* to guarantee that $f'(z)$ extends *continuously* up to $\{|z| = 1\}$.

For *that*, still *more* regularity is required of Γ. One of the best known results concerning this is due to Kellogg.

Because Γ is rectifiable (by the first of the above corollaries), its arcs have finite length. For w and w' on Γ, we denote by

$$\operatorname{length}_\Gamma(w, w')$$

the length of the *shorter* of the two arcs on Γ joining w to w'. We also denote the *line* tangent to Γ at w thereon by T_w, as in Chapter II, Subsection C.3. Then we have

Kellogg's Theorem *Let $0 < \alpha < 1$ (N.B.!), and suppose that for Γ, with a continuously turning tangent, the acute angle between T_w and $T_{w'}$ is $\leq \text{const} \cdot \{\operatorname{length}_\Gamma(w, w')\}^\alpha$ whenever w and w', on Γ, are sufficiently close together. Then, if $f(z)$ is a conformal mapping of $\{|z| < 1\}$ onto the domain bounded by Γ, $f'(z)$ extends continuously up to the unit circumference, and*

$$|f'(e^{i\theta'}) - f'(e^{i\theta})| \leq \text{const} \cdot |\theta' - \theta|^\alpha.$$

Proof Fix any harmonic determination of $\arg f'(z)$ in $\{|z| < 1\}$. According to the theorem of Chapter II, Subsection C.3, that function extends continuously up to $\{|z| = 1\}$; we write

$$u(\theta) = \lim_{z \to e^{i\theta}} \arg f'(z),$$

making $u(\theta)$ continuous and periodic, of period 2π. In proving the theorem just referred to, it was seen that

$$u(\theta) = \arg\left(\frac{p(e^{i\theta})}{e^{i\theta}}\right) - \frac{\pi}{2}$$

(for suitable determination of the argument), where $p(e^{i\theta})$ denotes a unit vector tangent to Γ at $f(e^{i\theta})$, pointing in the direction 'of increasing θ', and hence *parallel* to $T_{f(e^{i\theta})}$. For small values of $|\theta' - \theta|$, the acute angle between $T_{f(e^{i\theta'})}$ and $T_{f(e^{i\theta})}$ is therefore equal to the one between $p(e^{i\theta'})$ and $p(e^{i\theta})$. However, by the *continuity* of $\arg(p(e^{i\theta})/e^{i\theta})$, that acute angle must be

$$\left|\arg\left(p(e^{i\theta'})/e^{i\theta'}\right) - \arg\left(p(e^{i\theta})/e^{i\theta}\right) + \theta' - \theta\right|;$$

the hypothesis hence makes this last expression

$$\leq \text{const}\cdot\left\{\text{length}_\Gamma\,(f(e^{i\theta'}), f(e^{i\theta}))\right\}^\alpha$$

when $|\theta' - \theta|$ is small. Since, by the second of the above corollaries, we have

$$\text{length}_\Gamma\,(f(e^{i\theta'}), f(e^{i\theta})) \leq \text{const}\cdot|\theta' - \theta|^{1/2}$$

(say), the observation just made implies that

$$|u(\theta') - u(\theta)| \leq \text{const}\cdot|\theta' - \theta|^{\alpha/2},$$

first for $|\theta' - \theta|$ *small*, but then (by cooking the constant) for *all* values of θ and θ'.

We proceed to run through the same argument *twice*. The last relation and the theorem of Section E yield

$$|\tilde{u}(\theta') - \tilde{u}(\theta)| \leq \text{const}\cdot|\theta' - \theta|^{\alpha/2},$$

with the Hilbert transform $\tilde{u}(\theta)$ of $u(\theta)$ now evidently continuous, and periodic as well, of period 2π. But $\log|f'(z)|$ is, on $\{|z| < 1\}$, a harmonic conjugate of the *bounded* harmonic function $-\arg f'(z)$ having, at $z = e^{i\theta}$, the (continuous) boundary value $-u(\theta)$, so

$$\log|f'(z)| = \log|f'(0)| - \frac{1}{2\pi}\int_0^{2\pi}\frac{1-|z|^2}{|z - e^{i\theta}|^2}\tilde{u}(\theta)\,d\theta$$

for $|z| < 1$. Thence, by the continuity and periodicity of $\tilde{u}(\theta)$, we find that $\log|f'(z)|$ has a *continuous* extension up to $\{|z| = 1\}$. The boundary value $|f'(e^{i\theta})|$ is in particular *bounded* (and also bounded away from 0).

Denoting by M an *upper bound* on $|f'(e^{i\theta})|$ we *now* get, as in the proof of the second of the above corollaries,

$$\text{length}_\Gamma\,(f(e^{i\theta'}), f(e^{i\theta})) \leq M|\theta' - \theta|.$$

From this, we see that *in fact*

$$|u(\theta') - u(\theta)| \leq \text{const}\cdot|\theta' - \theta|^\alpha$$

(with exponent α on the right, and not just $\alpha/2$). Appealing *again* to the theorem of Section E, we find, since $0 < \alpha < 1$, that

$$|\tilde{u}(\theta') - \tilde{u}(\theta)| \leq \text{const}\cdot|\theta' - \theta|^\alpha.$$

From the above Poisson representation of $\log|f'(z)|$ we obtain, for that function's continuous extension to the unit circumference, the value $\log|f'(0)| - \tilde{u}(\theta)$ at $z = e^{i\theta}$. As we know, $\arg f'(z)$ has at the same point the continuous boundary value $u(\theta)$. Therefore

$\log f'(z) = \log|f'(z)| + i\arg f'(z)$ extends continuously up to $\{|z| = 1\}$ and takes, at $z = e^{i\theta}$, the boundary value

$$\log|f'(0)| - \tilde{u}(\theta) + iu(\theta).$$

Writing $\log f'(e^{i\theta})$ for that boundary value, we get, by the last two relations,

$$|\log f'(e^{i\theta'}) - \log f'(e^{i\theta})| \leq \text{const} \cdot |\theta' - \theta|^\alpha.$$

By exponentiation, we see from this result that the boundary value $f'(e^{i\theta})$ of $f'(z)$ exists everywhere and is continuous, and that

$$|f'(e^{i\theta'}) - f'(e^{i\theta})| \leq \text{const} \cdot |\theta' - \theta|^\alpha,$$

e^W being uniformly Lip 1 in W on any bounded region of the W-plane.

The theorem is proved.

Remark In the course of the proof it was seen that $|f'(e^{i\theta})|$ is *bounded away from zero*, as well as being bounded. Therefore, under the conditions of the theorem, we have, with a constant M depending on Γ (and on the particular choice of the mapping $f(z)$),

$$M^{-1}|E| \leq |f(E)| \leq M|E|$$

for Borel subsets E of $\{|z| = 1\}$ – cf. the observation following the second corollary above. This result is useful in the estimation of harmonic measures.

The last theorem is no longer valid for $\alpha = 1$. The reader is invited to obtain a substitute for that case after perusal of the proof in Section E. He or she is also invited to look for a substitute in the case $\alpha = 0$, i.e., for a condition on the variation of T_w as w moves along Γ which is *weaker* than that in the theorem's hypothesis (for *every* $\alpha > 0$), but still strong enough to guarantee *continuity* of $f'(e^{i\theta})$ and the relation stated in the last remark.

VI

H_p Spaces for the Upper Half Plane

In order to study certain classes of functions analytic or harmonic in $\Im z > 0$ we make the conformal mapping

$$z \mapsto w = \frac{i-z}{i+z}$$

onto the unit circle $\{|w| < 1\}$ and apply the results already obtained in previous chapters.

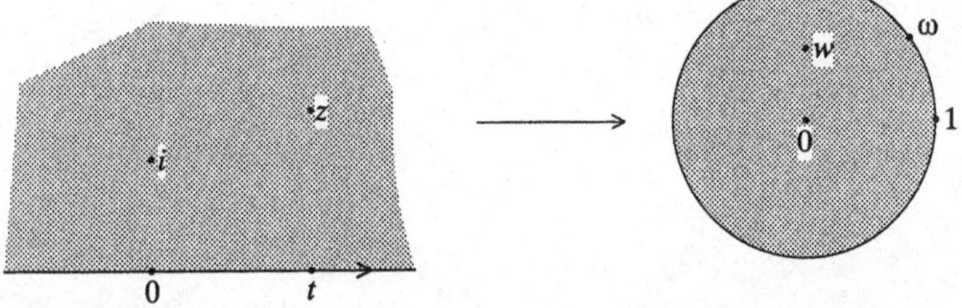

A. Poisson's formula for the half plane

If $\Im z > 0$ and t is real, z corresponds in the above conformal mapping to

$$w = \frac{i-z}{i+z} = re^{i\theta},$$

say, and t corresponds to

$$\omega = \frac{i-t}{i+t} = e^{i\tau},$$

say.

We want to work out the Poisson kernel

$$\frac{1-r^2}{1+r^2-2r\cos(\theta-\tau)}\,d\tau \;=\; \frac{1-|w|^2}{|w-\omega|^2}\frac{d\omega}{i\omega}.$$

Now

$$\omega = \frac{2i}{i+t} - 1,$$

so

$$d\omega = -\frac{2i\,dt}{(i+t)^2},$$

and

$$\frac{d\omega}{i\omega} = -\frac{2\,dt}{(i+t)(i-t)} = \frac{2\,dt}{t^2+1}.$$

Thus,

$$\frac{1-|w|^2}{|w-\omega|^2}\cdot\frac{d\omega}{i\omega} = \frac{1-\left|\frac{i-z}{i+z}\right|^2}{\left|\frac{i-z}{i+z}-\frac{i-t}{i+t}\right|^2}\cdot\frac{2\,dt}{t^2+1}$$

$$= \frac{\dfrac{|z+i|^2-|z-i|^2}{\cancel{|z+i|^2}}}{\dfrac{|(i+t)(i-z)-(i-t)(i+z)|^2}{\cancel{|z+i|^2\,|t+i|^2}}}\cdot\frac{2\,dt}{\cancel{t^2+1}}$$

$$= \frac{4y}{|2i(z-t)|^2}\cdot 2\,dt = \frac{2y\,dt}{|z-t|^2}$$

Then by theorems of Chapter I:

Theorem *Let $V(z)$ be harmonic and bounded for $\Im z > 0$. Then the limit $V(t) = \lim V(z)$ for $z \underset{L}{\longrightarrow} t$ exists a.e. for t on $\mathbb{R}$, and for $\Im z > 0$*

$$V(z) = \frac{1}{\pi}\int_{-\infty}^{\infty}\frac{y}{(x-t)^2+y^2}V(t)\,dt.$$

Theorem *Let $V(z)$ be harmonic and ≥ 0 for $\Im z > 0$. Then there is a constant $\alpha \geq 0$ and a* measure *$\mu \geq 0$ on $\mathbb{R}$ with*

$$\int_{-\infty}^{\infty}\frac{d\mu(t)}{1+t^2} < \infty,$$

such that, for $\Im z > 0$,

$$V(z) = \alpha y + \frac{1}{\pi}\int_{-\infty}^{\infty}\frac{y\,d\mu(t)}{(x-t)^2+y^2}.$$

Proof If, in the correspondence $z \mapsto w = (i-z)/(i+z)$, we let $U(w) = V(z)$, we have

$$U(re^{i\theta}) = \frac{1}{2\pi}\int_{-\pi}^{\pi}\frac{1-r^2}{1+r^2-2r\cos(\theta-\tau)}\,d\nu(\tau)$$

where ν is a (finite) positive measure on $(-\pi, \pi)$. For

$$e^{i\tau} = \frac{i-t}{i+t},$$

$d\mu(t)$ is given by

$$d\mu(t) = \tfrac{1}{2}(1+t^2)\,d\nu(\tau).$$

α is simply the point mass $(1/2\pi)\{\nu(\{-\pi\}) + \nu(\{\pi\})\}$, if there be any.

Remark It is common to think of $\pi\alpha$ as 'the mass at ∞'.

What is the harmonic conjugate of

$$V(z) = \frac{1}{2\pi}\int_{-\infty}^{\infty} \frac{y}{(x-t)^2+y^2}\,d\mu(t)\ ?$$

Observe that

$$\frac{y}{(x-t)^2+y^2} + \frac{i(x-t)}{(x-t)^2+y^2} = \frac{i}{z-t}$$

is, for each $t \in \mathbb{R}$, *analytic* in $\Im z > 0$.

> So one choice for the harmonic conjugate of $V(z)$ would be
>
> $$\frac{1}{\pi}\int_{-\infty}^{\infty} \frac{x-t}{(x-t)^2+y^2}\,d\mu(t).$$
>
> This particular choice is frequently used.

The boxed formula is applicable as long as

$$\int_{-\infty}^{\infty} \frac{|d\mu(t)|}{1+|t|} < \infty.$$

But in the most general situations, *all we know is that*

$$\int_{-\infty}^{\infty} \frac{|d\mu(t)|}{1+t^2} < \infty$$

(cf. above theorem for the Poisson representation of *positive harmonic functions*).

> In that case we use the harmonic conjugate
>
> $$\frac{1}{\pi}\int_{-\infty}^{\infty} \left(\frac{x-t}{(x-t)^2+y^2} + \frac{t}{t^2+1}\right) d\mu(t).$$

Here, the integral is absolutely convergent as long as

$$\int_{-\infty}^{\infty} \frac{|d\mu(t)|}{1+t^2} < \infty.$$

So, for $V(z) = \frac{1}{\pi}\int_{-\pi}^{\pi} \frac{y}{(x-t)^2+y^2}\,d\mu(t)$ we have *the following two choices* for the harmonic conjugate:

$$\frac{1}{\pi}\int_{-\infty}^{\infty} \frac{x-t}{(x-t)^2+y^2}\,d\mu(t) \quad \textit{and}$$

$$\frac{1}{\pi}\int_{-\infty}^{\infty} \left(\frac{x-t}{(x-t)^2+y^2} + \frac{t}{t^2+1}\right) d\mu(t).$$

B. Boundary behaviour

Theorem *Let*

$$V(z) = \frac{1}{\pi}\int_{-\infty}^{\infty} \frac{y}{(x-t)^2+y^2}\,d\mu(t),$$

where μ is a signed measure on $\mathbb{R}$ with

$$\int_{-\infty}^{\infty} \frac{|\,d\mu(t)|}{1+t^2} < \infty.$$

Then, at any t_0 where $\mu'(t_0)$ exists and is finite, $V(z) \to \mu'(t_0)$ as $z \underset{L}{\longrightarrow} t_0$.

Proof If

$$\frac{i-z}{i+z} = re^{i\theta}, \quad \frac{i-t}{i+t} = e^{i\tau},$$

then

$$V(z) = u(re^{i\theta}) = \frac{1}{2\pi}\int_{-\pi}^{\pi} \frac{1-r^2}{1+r^2-2r\cos(\theta-\tau)}\,d\nu(\tau),$$

where the relation between ν and μ is given by

$$d\nu(\tau) = \frac{2\,d\mu(t)}{1+t^2}.$$

We also have $d\tau = 2\,dt/(1+t^2)$, so

$$\frac{d\nu(\tau)}{d\tau} = \frac{d\mu(t)}{dt}.$$

By Chapter I, wherever $\nu'(\tau_0)$ exists and is finite,

$$u(w) \to \nu'(\tau_0) \quad \text{for} \quad w \underset{L}{\longrightarrow} e^{i\tau_0}.$$

So our result follows by direct carrying-over.

Remark It is also instructive to go through a direct proof of this statement, keeping the kernel $y/[(x-t)^2+y^2]$ but following the procedure of Chapter I (integration by parts). It will be found that the computational details are *simpler* here!

> *In fact, this is generally the case* – so much so that frequently results for $\{|w| < 1\}$ are proved *by first going over to the half plane* $\Im z > 0$.
>
> $$\frac{y}{(x-t)^2+y^2} \quad \text{and} \quad \frac{x-t}{(x-t)^2+y^2}$$
>
> are just *easier to handle* than
>
> $$\frac{1-r^2}{1+r^2-2r\cos(\theta-\tau)} \quad \text{and} \quad \frac{2r\sin(\theta-\tau)}{1+r^2-2r\cos(\theta-\tau)}$$
>
> respectively!

Similarly, by translation from Chapter I:

Theorem *Let*

$$V(z) = \frac{1}{\pi}\int_{-\infty}^{\infty} \frac{y}{(x-t)^2+y^2}\,d\mu(t),$$

where

$$\int_{-\infty}^{\infty} \frac{|\,d\mu(t)|}{1+t^2} < \infty.$$

If $\mu'(t_0)$ exists *and* is infinite,

$$V(t_0+iy) \to \mu'(t_0) \quad \text{as } y \to 0+.$$

Theorem *Let*

$$\int_{-\infty}^{\infty} \frac{|F(t)|\,dt}{1+t^2} < \infty$$

and let

$$\tilde{V}(z) = \frac{1}{\pi}\int_{-\infty}^{\infty}\left(\frac{x-t}{(x-t)^2+y^2} + \frac{t}{t^2+1}\right)F(t)\,dt.$$

Then, for almost all $x \in \mathbb{R}$,

$$\frac{1}{\pi}\int_{|t-x|\geq y'}\left(\frac{1}{x-t} + \frac{t}{t^2+1}\right)F(t)\,dt \;-\; \tilde{V}(x+iy) \;\to\; 0$$

as $y \to 0$, *where* $y' = \tan(y/(1+y))$.*

Remark In case

$$\int_{-\infty}^{\infty} \frac{|F(t)|}{|t|+1}\,dt < \infty$$

---

* Here, one can just as well replace y' by y. That is most easily seen when the direct procedure suggested in the preceding remark is followed.

we have simply

$$\frac{1}{\pi}\int_{-\infty}^{\infty}\frac{x-t}{(x-t)^2+y^2}F(t)\,\mathrm{d}t-\frac{1}{\pi}\int_{|t-x|\geq y'}\frac{F(t)}{x-t}\,\mathrm{d}t\ \to\ 0\quad \text{a.e.}$$

as $y \to 0$, with y' as above.

Thus, existence of the principal value

$$\frac{1}{\pi}\;⨍_{-\infty}^{\infty}\frac{F(t)}{x-t}\,\mathrm{d}t=\lim_{y\to 0}\frac{1}{\pi}\int_{|t-x|\geq y}\frac{F(t)}{x-t}\,\mathrm{d}t$$

is here *equivalent almost everywhere to existence of*

$$\lim_{y\to 0}\frac{1}{\pi}\int_{-\infty}^{\infty}\frac{x-t}{(x-t)^2+y^2}F(t)\,\mathrm{d}t.$$

There is a coarser kind of boundary behaviour which is easy to verify. *Note that*

$$\frac{1}{\pi}\int_{-\infty}^{\infty}\frac{y}{(x-t)^2+y^2}\,\mathrm{d}t=1,$$

that

$$\frac{y}{(x-t)^2+y^2}\geq 0,$$

and that

$$\frac{1}{\pi}\int_{|t-x|\geq\delta}\frac{y}{(x-t)^2+y^2}\,\mathrm{d}t\to 0\quad\text{as } y\to 0$$

for *any* $\delta > 0$.

From this we get easily, by the usual arguments:

Theorem *If $\phi(t)$ is bounded on $\mathbb{R}$,*

$$\frac{1}{\pi}\int_{-\infty}^{\infty}\frac{y}{(x-t)^2+y^2}\phi(t)\,\mathrm{d}t\ \to\ \phi(x_0)$$

for $z \to x_0$, x_0 being any point of continuity of ϕ. If $\phi(t)$ is uniformly continuous, the convergence is uniform.

Theorem *Let μ be a totally finite measure on $\mathbb{R}$ and write*

$$\mathrm{d}\mu_y(x)=\left\{\frac{1}{\pi}\int_{-\infty}^{\infty}\frac{y}{(x-t)^2+y^2}\,\mathrm{d}\mu(t)\right\}\mathrm{d}x$$

for $y > 0$. Then $\mathrm{d}\mu_y(x)\to\mathrm{d}\mu(x)$ w as $y \to 0$.*

Proof By the preceding and duality!

Theorem *Let $F \in L_p(-\infty,\infty)$, $1 \leq p < \infty$. Let*

$$F_y(x)=\frac{1}{\pi}\int_{-\infty}^{\infty}\frac{y}{(x-t)^2+y^2}F(t)\,\mathrm{d}t.$$

Then $\|F_y\|_p \le \|F\|_p$ *for* $y > 0$ *and* $\|F_y - F\|_p \to 0$ *as* $y \to 0$.

C. The H_p spaces for $\Im z > 0$

Definition $F(z)$, analytic for $\Im z > 0$, is said to belong to $H_p(\Im z > 0)$ or, just H_p if we *know* we are dealing with the upper half plane, provided that there is a constant $C < \infty$ with

$$\int_{-\infty}^{\infty} |F(x+iy)|^p \, dx \le C$$

for all $y > 0$. We use this definition for all $p > 0$.

Remark $H_p(\Im z > 0)$ is *not* just obtained from H_p for $\{|w| < 1\}$ by conformal mapping! *There is a factor* (depending on p) which *also* comes in!

Lemma *If* $F \in H_p$, $|F(z)| \le k/y^{1/p}$ *with a constant k depending on F.*

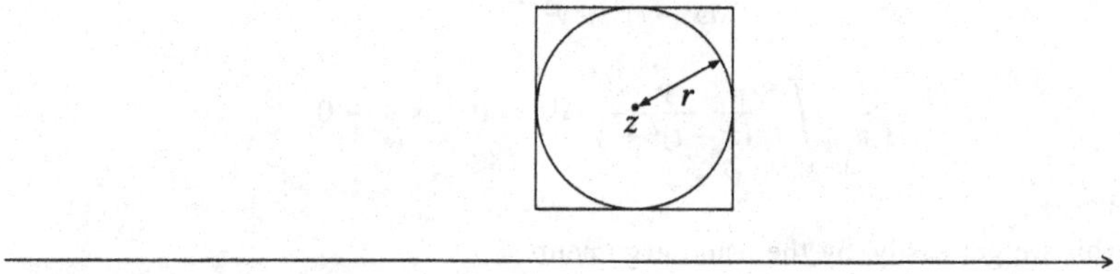

Proof If $\Im z > 0$, by *subharmonicity* of $|F(x)|^p$ for $p > 0$,

$$|F(z)|^p \le \frac{1}{2\pi} \int_0^{2\pi} |F(z + \rho e^{i\phi})|^p \, d\phi$$

for $0 < \rho < y$. Therefore, integrating ρ from 0 to $r < y$,

$$\frac{r^2}{2} |F(z)|^p \le \frac{1}{2\pi} \int_0^r \int_0^{2\pi} |F(z + \rho e^{i\phi})|^p \rho \, d\phi \, d\rho$$

$$\le \frac{1}{2\pi} \int_{x-r}^{x+r} \int_{y-r}^{y+r} |F(\xi + i\eta)|^p \, d\eta \, d\xi \quad \text{(see picture)},$$

which, by hypothesis, is (change the order of integration)

$$\le \frac{1}{2\pi} \cdot C \cdot 2r = \frac{Cr}{\pi},$$

so $|F(z)|^p \le 2C/\pi r$. Make $r \to y$. Done.

Lemma *Let $F \in H_p$, $p \geq 1$, and let $h > 0$. Then*

$$F(z+ih) = \frac{1}{\pi}\int_{-\infty}^{\infty} \frac{yF(t+ih)}{(x-t)^2+y^2}\,dt.$$

Proof By Cauchy's theorem.

Let Γ_R be the contour shown below, with R very large, and let $R\sin\theta_0 = h$. *Then*

$$F(z+ih) = \frac{1}{2\pi i}\int_{\Gamma_R} \frac{F(\zeta)}{\zeta-(z+ih)}\,d\zeta$$

$$= \frac{1}{2\pi i}\int_{\Gamma_R} \frac{F(\zeta)}{(\zeta-ih)-z}\,d\zeta = \frac{1}{2\pi i}\int_{-R\cos\theta_0}^{R\cos\theta_0} \frac{F(t+ih)\,dt}{t-z} + \frac{1}{2\pi i}\int_{\theta_0}^{\pi-\theta_0} \frac{F(Re^{i\theta})}{Re^{i\theta}-ih-z}iRe^{i\theta}\,d\theta.$$

For *very large R*, the *second* integral is bounded by

$$J = \frac{1}{2\pi}\int_{\theta_0}^{\pi-\theta_0} |F(Re^{i\theta})|(1+O(1/R))\,d\theta.$$

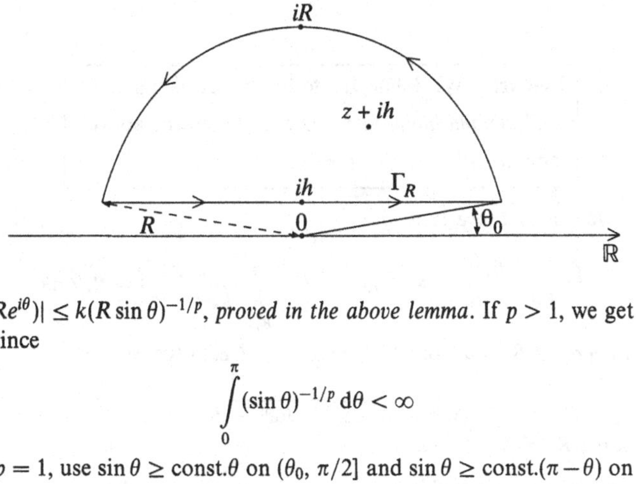

Now use $|F(Re^{i\theta})| \leq k(R\sin\theta)^{-1/p}$, proved in the above lemma. If $p > 1$, we get immediately $J \leq k'R^{-1/p}$ since

$$\int_0^{\pi} (\sin\theta)^{-1/p}\,d\theta < \infty$$

for $p > 1$. If $p = 1$, use $\sin\theta \geq \text{const.}\theta$ on $(\theta_0, \pi/2]$ and $\sin\theta \geq \text{const.}(\pi-\theta)$ on $[\pi/2, \pi-\theta_0)$ to get

$$J \leq \frac{k''}{R}\int_{\theta_0}^{\pi/2}\frac{d\theta}{\theta} = \frac{k''}{R}\log\frac{\pi}{2\theta_0} \leq \frac{k''}{R}\log\frac{\pi}{2\sin\theta_0} = \frac{k''}{R}\log\frac{\pi R}{2h}.$$

This also tends to 0 as $R \to \infty$.

So

$$\boxed{F(z+ih) = \lim_{R'\to\infty}\frac{1}{2\pi i}\int_{-R'}^{R'}\frac{F(t+ih)\,dt}{t-z}.}$$

Now $\bar{z}+ih$ is *outside* Γ_R, so, *again by Cauchy's theorem*,

$$0=\frac{1}{2\pi i}\int_{\Gamma_R}\frac{F(\zeta)\,d\zeta}{\zeta-ih-\bar{z}}=\frac{1}{2\pi i}\int_{-R\cos\theta_0}^{R\cos\theta_0}\frac{F(t+ih)\,dt}{t-\bar{z}}+\frac{1}{2\pi i}\int_{\theta_0}^{\pi-\theta_0}\frac{F(Re^{i\theta})}{Re^{i\theta}-ih-\bar{z}}iRe^{i\theta}\,d\theta.$$

By the argument given above the second integral tends to 0 as $R\to\infty$. So finally

$$\boxed{0=\lim_{R'\to\infty}\frac{1}{2\pi i}\int_{-R'}^{R'}\frac{F(t+ih)\,dt}{t-\bar{z}}.}$$

Subtract this from the previous boxed relation and note the important identity

$$\frac{1}{t-z}-\frac{1}{t-\bar{z}}=\frac{2iy}{|t-z|^2}=\frac{2iy}{(x-t)^2+y^2}.$$

We get

$$F(z+ih)=\frac{1}{\pi}\lim_{R'\to\infty}\int_{-R'}^{R'}\frac{yF(t+ih)\,dt}{(x-t)^2+y^2}.$$

Q.E.D.

> **Remark** We define H_∞ to be the set of functions *analytic and bounded* in $\Im z>0$. Then the above lemma *still holds* for $p=\infty$.

Note that *here* we *cannot prove* that

$$\int_{\theta_0}^{\pi-\theta_0}\frac{F(Re^{i\theta})}{Re^{i\theta}-ih-z}Re^{i\theta}\,d\theta \quad\text{and}\quad \int_{\theta_0}^{\pi-\theta_0}\frac{F(Re^{i\theta})}{Re^{i\theta}-ih-\bar{z}}Re^{i\theta}\,d\theta$$

separately go to zero as $R\to\infty$. *But their difference does*, because

$$\frac{1}{Re^{i\theta}-ih-z}-\frac{1}{Re^{i\theta}-ih-\bar{z}}$$

is $O(R^{-2})$ for large R !

Remark If, in the above proof (for $1\le p<\infty$) we *add* instead of *subtracting*, we get

$$\boxed{F(z+ih)=\frac{i}{\pi}\int_{-\infty}^{\infty}\frac{(x-t)F(t+ih)\,dt}{(x-t)^2+y^2},}$$

the integral being absolutely convergent for $1\le p<\infty$. This *fails* if $p=\infty$.

Theorem *Let* $F(z)\in H_p(\Im z>0)$, $p\ge 1$. *Then, for almost all* $t\in\mathbb{R}$,

$$\lim_{z\,\overrightarrow{\angle}\,t}F(z)=F(t)\quad\textit{exists,}$$

$$F(t)\in L_p(-\infty,\infty),\quad\textit{and}$$

$$\boxed{F(z) = \frac{1}{\pi}\int_{-\infty}^{\infty} \frac{y}{(x-t)^2+y^2} F(t)\,dt, \quad \Im z > 0.}$$

Proof If $p > 1$, we can get a *sequence* of $h > 0$ *tending to zero* such that the

$$F_h(t) = F(t+ih)$$

tend weakly to some $F \in L_p(-\infty, \infty)$ — for $\|F_h\|_p \le C$, $h > 0$, by hypothesis. Substitute in the formula derived in the above lemma: for *fixed* z, $\Im z > 0$, *we have*

$$F(z+ih) = \frac{1}{\pi}\int_{-\infty}^{\infty} \frac{y}{(x-t)^2+y^2} F_h(t)\,dt \longrightarrow \frac{1}{\pi}\int_{-\infty}^{\infty} \frac{y}{(x-t)^2+y^2} F(t)\,dt$$

as $h \to 0$ through the the particular sequence of values given. Since $F(z+ih) \to F(z)$ also, we get the above boxed formula.

Now $F(t) \in L_p(-\infty, \infty)$, $p > 1$, so if

$$\phi(x) = \int_0^x F(t)\,dt,$$

$\phi'(x)$ exists and equals $F(x)$ a.e. Observe that $d\phi(t)/(1+t^2) = F(t)\,dt/(1+t^2)$ is absolutely integrable, so a theorem of Section B teaches us that $F(z) \to F(x_0)$ for $z \underset{\angle}{\longrightarrow} x_0$ a.e. in $x_0 \in \mathbb{R}$. We are done in case $p > 1$. (N.B. Including for the case $p = \infty$, by a preceding remark!)

Now for $p = 1$ all we know is that

$$F_h(t)\,dt \to d\mu(t) \quad w^*$$

as $h \to 0$ through its sequence, where μ is some complex *measure* on $\mathbb{R}$ with

$$\int_{-\infty}^{\infty} |d\mu(t)| < \infty.$$

The above argument then yields

$$F(z) = \frac{1}{\pi}\int_{-\infty}^{\infty} \frac{y}{(x-t)^2+y^2}\,d\mu(t).$$

But now we can apply the theorem of F. and M. Riesz.

If $\Im z > 0$, the *second* boxed relation in the proof of the above lemma reads, for $h > 0$,

$$\int_{-\infty}^{\infty} \frac{F_h(t)\,dt}{t-\bar{z}} = 0.$$

Therefore, since $F_h(t)\,dt \to d\mu(t)$ w^*,

$$\int_{-\infty}^{\infty} \frac{d\mu(t)}{t-\bar{z}} = 0, \quad \Im z > 0.$$

Take $z = ik$, $k > 0$; we see that

$$\int_{-\infty}^{\infty} \frac{d\mu(t)}{t+ik} = 0, \quad k > 0.$$

Differentiating successively with respect to k we get

$$\int_{-\infty}^{\infty} \frac{d\mu(t)}{(t+ik)^n} = 0, \quad n = 1, 2, \ldots .$$

Finally, we have

$$\int_{-\infty}^{\infty} \frac{1}{(i+t)^n}\, d\mu(t) = 0, \quad n = 1, 2, \ldots .$$

Now, in the conformal mapping

$$z \mapsto \frac{i-z}{i+z}, \qquad t \mapsto \frac{i-t}{i+t} = e^{i\tau},$$

define a measure ν on $[-\pi, \pi]$ by $d\nu(\tau) = d\mu(t)/(i-t)$. Then, for $n = 1, 2, \ldots,$

$$\int_{-\pi}^{\pi} e^{in\tau}\, d\nu(\tau) = \int_{-\infty}^{\infty} \frac{(i-t)^{n-1}}{(i+t)^n}\, d\mu(t),$$

which can be rewritten in the form

$$\sum_{k=1}^{n} a_k \int_{-\infty}^{\infty} \frac{d\mu(t)}{(i+t)^k},$$

and *is hence equal to* 0. So

$$\int_{-\pi}^{\pi} e^{in\tau}\, d\nu(\tau) = 0, \quad n = 1, 2, 3,$$

and by the F. and M. Riesz theorem, $d\nu(\tau)$ is *absolutely continuous*. So $d\mu(t)/(i-t)$, *and hence* $d\mu(t)$ *is*; $d\mu(t) = F(t)\,dt$ for some $F \in L_1(-\infty, \infty)$. We now conclude the proof as in the case $p > 1$. Q.E.D.

Theorem *If* $F \in H_p$, $1 \le p < \infty$, *(sic!) and* $\Im z > 0$, *then*

$$\frac{1}{2\pi i} \int_{-\infty}^{\infty} \frac{F(t)}{t-z}\, dt = F(z) \qquad \text{and} \qquad \int_{-\infty}^{\infty} \frac{F(t)\,dt}{t-\bar{z}} = 0.$$

Proof By the two boxed formulas in the proof of a preceding lemma,

$$F(z+ih) = \frac{1}{2\pi i} \int_{-\infty}^{\infty} \frac{F_h(t)\,dt}{t-z},$$

$$0 = \int_{-\infty}^{\infty} \frac{F_h(t)\,dt}{t-\bar{z}},$$

where $F_h(t) = F(t+ih)$. But by the above theorem

$$F_h(x) = \frac{1}{\pi} \int_{-\infty}^{\infty} \frac{hF(t)\,dt}{(x-t)^2 + h^2}$$

where $F \in L_p(-\infty, \infty)$. So by a result of Section B, $\|F_h - F\|_p \to 0$ as $h \to 0$. Substitution into the preceding formulas gives the desired result.

The *first* of the above two theorems has its analogue for *harmonic functions*.

Theorem *Let $U(z)$ be harmonic in $\Im z > 0$ and suppose that, for some $p \geq 1$,*

$$\int_{-\infty}^{\infty} |U(t+iy)|^p \, dt \leq C < \infty$$

independently of y. If $p > 1$, then

$$U(z) = \frac{1}{\pi} \int_{-\infty}^{\infty} \frac{y}{(x-t)^2 + y^2} u(t) \, dt, \quad \Im z > 0,$$

where $u(t) \in L_p(-\infty, \infty)$.
If $p = 1$,

$$U(z) = \frac{1}{\pi} \int_{-\infty}^{\infty} \frac{y}{(x-t)^2 + y^2} \, d\mu(t), \quad \Im z > 0,$$

where

$$\int_{-\infty}^{\infty} |\, d\mu(t)| < \infty.$$

Proof $|U(z)|^p$ is *subharmonic*, as is easily verified from the mean value formula of Gauss and Hölder's inequality — therefore the argument of the *first* lemma of this section can be applied to *it*, and we see that $|U(z)| \leq C/y^{1/p}$ for $\Im z = y > 0$.

In particular each function $U_h(z) = U(z + ih)$, $h > 0$, is *bounded* (and harmonic!) for $\Im z > 0$. Now we can use the *conformal mapping*

$$\frac{i-z}{i+z} = w,$$

put $U_h(z) = \check{U}_h(w)$, and *use* the *known* representation (Poisson's formula) for *functions $\check{U}_h(w)$ harmonic* and *bounded* in $\{|w| < 1\}$. If we go back to the z-plane, the Poisson formula for $\check{U}_h(re^{i\theta})$ in terms of $\check{U}_h(e^{i\tau})$ goes over into one for $U_h(z)$ in terms of $U_h(t)$, $t \in \mathbb{R}$. The calculation has *already been carried out* in Section A, and we find

$$U_h(z) = \frac{1}{\pi} \int_{-\infty}^{\infty} \frac{yU_h(t)}{(x-t)^2 + y^2} \, dt,$$

i.e.,

$$U(z + ih) = \frac{1}{\pi} \int_{-\infty}^{\infty} \frac{yU_h(t)}{(x-t)^2 + y^2} \, dt.$$

The rest of the argument consists in passing to the limit $h \to 0$ and is *exactly* like what was done in proving Poisson's formula for H_p functions above. We're done.

Remark The above proof works also when $p = \infty$ ($U(z)$ *bounded* in $\Im z > 0$).

Theorem *If $F(z) \in H_p(\Im z > 0)$, $p \geq 1$, and, for $w = (i-z)/(i+z)$, $f(w) = F(z)$, then $f(w) \in H_p(|w| < 1)$.*

Proof By the first of the above theorems,

$$F(z) = \frac{1}{\pi} \int_{-\infty}^{\infty} \frac{y}{|z-t|^2} F(t)\, dt,$$

where $F(t) \in L_p(-\infty, \infty)$. The change of variable $z \mapsto w$ was carried out in Section A where we found

$$f(re^{i\phi}) = \frac{1}{2\pi} \int_{-\pi}^{\pi} \frac{1-r^2}{1+r^2-2r\cos(\phi-\tau)} f(e^{i\tau})\, d\tau,$$

putting

$$F(t) = f\left(\frac{i-t}{i+t}\right).$$

Now

$$\int_{-\pi}^{\pi} |f(e^{i\tau})|^p\, d\tau = \int_{-\infty}^{\infty} \frac{2|F(t)|^p\, dt}{1+t^2} \leq 2 \int_{-\infty}^{\infty} |F(t)|^p\, dt < \infty,$$

so, by Chapter I, the means

$$\int_{-\pi}^{\pi} |f(re^{i\phi})|^p\, d\phi$$

are bounded. So $f \in H_p(|w| < 1)$. Q.E.D.

Remark The computation used in proving the above theorem shows that the **converse**

$$f(w) \in H_p(\{|w| < 1\}) \Rightarrow F(z) \in H_p(\Im z > 0)$$

is false!

The correct equivalence is obviously, for

$$f\left(\frac{i-z}{i+z}\right) = F(z),$$

$$f(w) \in H_p(\{|w| < 1\}) \quad \text{iff} \quad \frac{F(z)}{(z+i)^{2/p}} \in H_p(\Im z > 0).$$

Theorem *If $0 < p < 1$ and $F(z) \in H_p$, $f(w) \in H_p(\{|w| < 1\})$ for*

$$f\left(\frac{i-z}{i+z}\right) = F(z).$$

Proof $|F(z)|^p$ is *still subharmonic*, so we *still* have $|F(z)| \leq K/y^{1/p}$, and in particular $|F(z)|$, and *hence* $|F(z)|^p$, is *still bounded* in each half plane $\Im z \geq h > 0$. Therefore

$|F_h(z)|^p = |F(z+ih)|^p$, being *bounded* and *subharmonic* in $\Im z > 0$, is there *majorized* by the *bounded harmonic function taking the same boundary values on* $\mathbb{R}$, i.e.

$$|F(z+ih)|^p \leq \frac{1}{\pi}\int_{-\infty}^{\infty} \frac{y|F_h(t)|^p \, dt}{(x-t)^2+y^2}.$$

Now

$$\int_{-\infty}^{\infty} |F_h(t)|^p \, dt \leq C, \quad h > 0,$$

so, making $h \to 0$ we find, in the usual way, a measure μ,

$$\int_{-\infty}^{\infty} |\, d\mu(t)| < \infty,$$

with

$$|F(z)|^p \leq \frac{1}{\pi}\int_{-\infty}^{\infty} \frac{y}{|z-t|^2} \, d\mu(t).$$

Going over to

$$f\left(\frac{i-z}{i+z}\right) = F(z),$$

we get

$$|f(re^{i\theta})|^p \leq \frac{1}{2\pi}\int_{-\infty}^{\infty} \frac{1-r^2}{1+r^2-2r\cos(\theta-\tau)} \, d\nu(\tau),$$

where, for $(i-t)/(i+t) = e^{i\tau}$,

$$d\nu(\tau) = \frac{2\, d\mu(t)}{1+t^2}.$$

Therefore

$$\int_{-\pi}^{\pi} |f(re^{i\theta})|^p \, d\theta \leq 2\int_{-\infty}^{\infty} \frac{|\, d\mu(t)|}{1+t^2}$$

by Fubini's theorem, for any $r < 1$. But

$$\int_{-\infty}^{\infty} \frac{|\, d\mu(t)|}{1+t^2} < \infty.$$

So done.

Because of these theorems, functions $\not\equiv 0$ *in* $H_p(\Im z > 0)$ *can be decomposed into inner and outer factors.*

Theorem *Let* $F(z) \not\equiv 0$ *belong to* $H_p(\Im z > 0)$. *Then, for* $\Im z > 0$,

$$F(z) = I_F(z) \cdot O_F(z),$$

where

(i) $I_F(z)$, *the* inner factor *of* F, *is*

$$I_F(z) = e^{i\gamma} B(z) \exp\left(\frac{1}{\pi}\int_{-\infty}^{\infty}\left(\frac{i}{z-t}+\frac{it}{t^2+1}\right) d\sigma(t)\right) e^{i\alpha z}$$

with

(a) γ *real*

(b) $B(z)$ *a Blaschke product for* $\Im z > 0$,

$$B(z) = \prod_k \left(e^{i\alpha_k}\cdot\frac{z-z_k}{z-\bar{z}_k}\right),$$

where the z_k *are the* zeros *of* $F(z)$ *in* $\Im z > 0$ *and the real* α_k *are so chosen that*

$$e^{i\alpha_k}\cdot\frac{i-z_k}{i-\bar{z}_k} \geq 0,$$

(c) $d\sigma(t) \leq 0$ *a* singular measure *on* $\mathbb{R}$ *with*

$$\int_{-\infty}^{\infty}\frac{|d\sigma(t)|}{1+t^2} < \infty,$$

(d) $\alpha \geq 0$ *(the 'mass at* ∞*' is* $-\pi\alpha$*);*

(ii) *the* outer factor *of* F, O_F, *is*

$$O_F(z) = \exp\left(\frac{i}{\pi}\int_{-\infty}^{\infty}\left(\frac{1}{z-t}+\frac{t}{t^2+1}\right)\log|F(t)|\, dt\right).$$

Proof Take the corresponding factorization for functions $f(w)$ in $H_p(\{|w|<1\})$ and *change the variables* in it:

$$w = \frac{i-z}{i+z}, \quad f(w) = F(z), \quad e^{i\tau} = \frac{i-t}{i+t}.$$

The calculations have mostly already been made in Section A.

Scholium In terms of the z_k, the *necessary and sufficient condition* for the *convergence* of the Blaschke product $B(z)$ is that only a finite number of z_k equal i and that

$$\prod_{z_k \neq i}\left|\frac{i-z_k}{i-\bar{z}_k}\right| > 0.$$

This can be written more easily as follows: Firstly,

$$1-\left|\frac{i-z_k}{i-\bar{z}_k}\right|^2 = \frac{|z_k+i|^2-|z_k-i|^2}{|z_k+i|^2} = \frac{4\Im z_k}{|z_k+i|^2}.$$

Therefore

$$\prod_k\left|\frac{i-z_k}{i-\bar{z}_k}\right|$$

converges iff

$$\sum_k \frac{\Im z_k}{|z_k+i|^2} < \infty.$$

This is usually stated as follows:

The Blaschke product $\prod_k (e^{i\alpha_k}(z - z_k)/(z - \bar{z}_k))$ is convergent in $\Im z > 0$ iff $\sum_{|z_k|<1} \Im z_k < \infty$ and $\sum_{|z_k|\geq 1} \Im z_k/|z_k|^2 < \infty$.

The following picture shows that it is *much easier* to visualise Blaschke products for the upper half plane than for the circle:

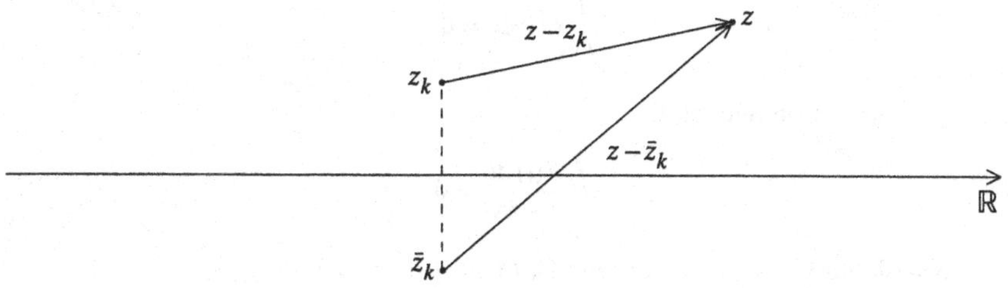

From this it is manifest why the individual factors have modulus < 1 precisely in $\Im z > 0$!

Scholium Of course, if $F(z) \in H_p(\Im z > 0)$,

$$\int_{-\infty}^{\infty} \frac{|\log|F(t)||\, dt}{1 + t^2} < \infty.$$

This follows from the corresponding result for the circle (Chapter IV) by a change of variable.

Remark The above given factorization for the *half plane* is sometimes credited to Nevanlinna. Most of the obvious analogues of Beurling's theorem (Chapter IV, Section E) hold.

D. M. Riesz' theorems for the Hilbert transform

If $u(t) \in L_p(-\infty, \infty)$, $1 \leq p \leq \infty$, we can form the function, harmonic in $\Im z > 0$,

$$U(z) = \frac{1}{\pi} \int_{-\infty}^{\infty} \frac{y u(t)\, dt}{(x - t)^2 + y^2},$$

such that $U(z) \to u(t)$ for $z \xrightarrow[\angle]{} t$ a.e. in t (Section B).

If, also, $p < \infty$ (sic!), we can form the harmonic conjugate

$$\boxed{\tilde{U}(z) = \frac{1}{\pi}\int_{-\infty}^{\infty} \frac{(x-t)u(t)\,dt}{(x-t)^2+y^2}}$$

(Section A).

The function $U(z)+i\tilde{U}(z)$ is *analytic* in $\Im z > 0$.

Lemma *If* $F(z) \in H_1(\Im z > 0)$,

$$\int_{-\infty}^{\infty} F(t)\,dt = 0.$$

Proof By a theorem in Section C,

$$\int_{-\infty}^{\infty} \frac{F(t)\,dt}{t+iN} = 0$$

for each $N > 0$. Since $F \in H_1$, by Section C, $F(t) \in L_1(-\infty,\infty)$, so

$$\int_{-\infty}^{\infty} F(t)\,dt = \lim_{N\to\infty}\int_{-\infty}^{\infty} \frac{iN}{t+iN}F(t)\,dt = 0. \qquad \text{Q.E.D.}$$

Lemma *Let* $1 < p \le 2$, *let* $u \in L_p(-\infty,\infty)$, *and let* $U(z)$ *and* $\tilde{U}(z)$ *be related to* $u(t)$ *by the above boxed formulas. There is a constant* K_p depending only on p *such that, for each* $h > 0$,

$$\int_{-\infty}^{\infty} |\tilde{U}(x+ih)|^p\,dx \le K_p \int_{-\infty}^{\infty} |U(x+ih)|^p\,dx.$$

Proof Clearly there is no loss of generality in assuming $u(t)$ *real*, since the general case follows from this, with perhaps a worse value of K_p.

We can also assume that $u(t)$ *has compact support*. For, in general, we can find $u_n(t)$ of compact support with $\|u_n - u\|_p \xrightarrow[n]{} 0$, so that, if

$$U_n(z) = \frac{1}{\pi}\int_{-\infty}^{\infty} \frac{yu_n(t)\,dt}{|z-t|^2} \quad\text{and}\quad \tilde{U}_n(z) = \frac{1}{\pi}\int_{-\infty}^{\infty} \frac{(x-t)u_n(t)\,dt}{|z-t|^2},$$

we have for each fixed h, $h > 0$,

$$\int_{-\infty}^{\infty} |U_n(x+ih) - U(x+ih)|^p\,dx \le \int_{-\infty}^{\infty} |u_n(t)-u(t)|^p\,dt \xrightarrow[n]{} 0,$$

whilst $\tilde{U}_n(x+ih) \xrightarrow[n]{} \tilde{U}(x+ih)$ u.c.c. in x, whence, by Fatou's lemma,

$$\int_{-\infty}^{\infty} |\tilde{U}(x+ih)|^p\,dx \le \liminf_{n\to\infty}\int_{-\infty}^{\infty} |\tilde{U}_n(x+ih)|^p\,dx,$$

and the validity of the inequality for each n would imply the same for $\tilde{U}$ and U.

So, henceforth, without loss of generality, we assume that $u(t)$ is of compact support.

We now assume also that $u(t) \geq 0$ and $u \not\equiv 0$.

Let $F(z) = U(z) + i\tilde{U}(z)$. Then, for $h > 0$, $U(z) = \Re F(z)$ is > 0 for $\Im z \geq h$, so $F(z)$ has no zeros in any such half plane. Therefore

$$G(z) = [F(z+ih)]^p$$

is *also analytic* for $\Im z \geq 0$.

Now $G(z) \in H_1(\Im z > 0)$. Indeed,

$$\int_{-\infty}^{\infty} [U(x+iy+ih)]^p \, dx \leq C < \infty$$

for each $y > 0$, whilst

$\lvert\tilde{U}(z+ih)\rvert$ is *bounded* for $\Im z \geq 0$ and now *also* $O(1/z)$ *there* because $u(t)$ is of *compact support*.

So, since $p > 1$,

$$\int_{-\infty}^{\infty} \lvert\tilde{U}(x+iy+ih)\rvert^p \, dx \leq \text{const}$$

for $y \geq 0$. (N.B. The constant *depends* upon h, but we *don't care* about that now!)

So

$$\int_{-\infty}^{\infty} \lvert F(x+iy+ih)\rvert^p \, dx \leq c, \quad y > 0,$$

and $G(z) \in H_1$.

Therefore by the previous lemma $$\int_{-\infty}^{\infty} G(x)\,dx = 0.$$

Now we can use the method of Katznelson and Zygmund (Chapter V). We have

$$G(z) = (F_h(z))^p, \quad 1 < p \leq 2,$$

with $\Re F_h = U_h > 0$, where $F_h(z) = F(z+ih)$ and $U_h(z) = U(z+ih)$.

Choose a γ, $0 < \gamma < \pi/2$, such that $\pi/2 < p\gamma < p\pi/2 \leq \pi$ — *because* $1 < p \leq 2$ *we can do that*. Let

$$E = \{x \in \mathbb{R};\ \lvert\arg F_h(x)\rvert \leq \gamma\}.$$

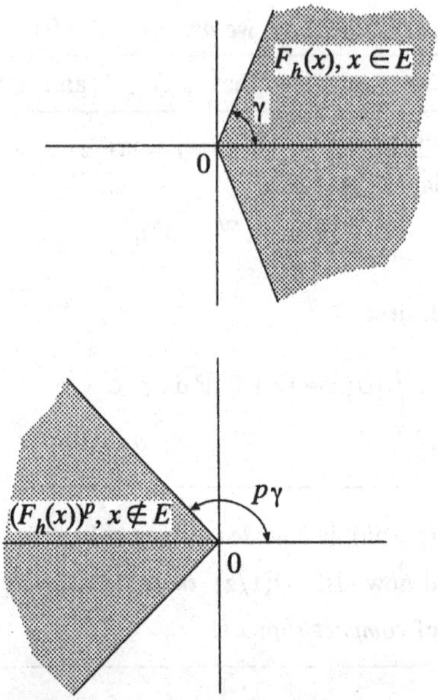

We play these two situations against each other.

From the previous boxed relation,

$$\int_{\sim E} \Re(F_h(x))^p \, dx + \int_E \Re(F_h(x))^p \, dx = 0.$$

But for $x \in \sim E$,

$$\Re(F_h(x))^p \le -|F_h(x)|^p| \cos p\gamma|$$

(see second picture!).

Therefore

$$\int_{\sim E} |F_h(x)|^p| \cos p\gamma| \, dx \le \int_E \Re(F_h(x))^p \, dx \le \int_E |F_h(x)|^p \, dx.$$

But for $x \in E$ (first picture!),

$$|F_h(x)| \le \frac{U_h(x)}{\cos\gamma},$$

so

$$\int_E |F_h(x)|^p \, dx \le \frac{1}{\cos^p\gamma} \int_E [U_h(x)]^p \, dx.$$

Finally we see that

$$\int_{-\infty}^{\infty} |F_h(x)|^p \, dx = \left(\int_{\sim E} + \int_E \right) |F_h(x)|^p \, dx$$

$$\le (\cos\gamma)^{-p} \left(1 + \frac{1}{|\cos p\gamma|} \right) \int_E [U_h(x)]^p \, dx,$$

or, since $|\tilde{U}_h(x)| \leq |F_h(x)|$ for $\tilde{U}_h(x) = \tilde{U}(x+ih)$,

$$\int_{-\infty}^{\infty} |\tilde{U}_h(x)|^p \, dx \leq (\cos\gamma)^{-p}\left(1+\frac{1}{|\cos p\gamma|}\right)\int_{-\infty}^{\infty} [U_h(x)]^p \, dx,$$

as desired, in case $u(t) \geq 0$.

Now we wish to remove that assumption. Observe that we *certainly have*

$$\int_{-\infty}^{\infty} |\tilde{U}_h(x)|^p \, dx \leq \check{K}_p \int_{-\infty}^{\infty} [u(t)]^p \, dt$$

with

$$\check{K}_p = (\cos\gamma)^{-p}\left(1+\frac{1}{|\cos p\gamma|}\right)$$

in case $u \geq 0$. For a *general* real u of compact support, we write $u = u_+ - u_-$, with

$$\int_{-\infty}^{\infty} |u(t)|^p \, dt = \int_{-\infty}^{\infty} (u_+(t))^p \, dt + \int_{-\infty}^{\infty} (u_-(t))^p \, dt,$$

where u_+ and u_- have disjoint bounded measurable supports, and *both* are ≥ 0.

Then, with obvious notation,

$$\tilde{U}_h(x) = \widetilde{U_h^+}(x) - \widetilde{U_h^-}(x),$$

so $\left\|\tilde{U}_h\right\|_p \leq \|\widetilde{U_h^+}\|_p + \|\widetilde{U_h^-}\|_p$, and

$$\int_{-\infty}^{\infty} |\tilde{U}_h(x)|^p \, dx \leq 2^{p-1}\left(\int_{-\infty}^{\infty} |\widetilde{U_h^+}(x)|^p \, dx + \int_{-\infty}^{\infty} |\widetilde{U_h^-}(x)|^p \, dx\right)$$

which, by what has already been done, is

$$\leq \frac{2^{p-1}}{\cos^p\gamma}\left(1+\frac{1}{|\cos p\gamma|}\right)\left[\int_{-\infty}^{\infty} (u_+(t))^p \, dt + \int_{-\infty}^{\infty} (u_-(t))^p \, dt\right]$$

$$= \frac{2^{p-1}}{\cos^p\gamma}\left(1+\frac{1}{|\cos p\gamma|}\right)\int_{-\infty}^{\infty} |u(t)|^p \, dt.$$

We see that

$$\int_{-\infty}^{\infty} |\tilde{U}(x+ih)|^p \, dx \leq K_p \int_{-\infty}^{\infty} |u(t)|^p \, dt$$

is *now proven for general real u of variable sign and compact support* (and hence for real u *without compact support* by the preliminary discussion given above). This *would suffice* for many purposes.

We wish, however, to prove the lemma as stated.

That is based on a trick with the Poisson and conjugate Poisson kernels.

Observe that for fixed real x_0 and $h > 0$,

$$\Re\frac{1}{z - x_0 + ih}$$

is *harmonic and bounded* in $\Im z > 0$, so by Section C,

$$\Re\frac{1}{z - x_0 + ih} = \frac{1}{\pi}\int_{-\infty}^{\infty} \Re\frac{1}{\xi - x_0 + ih}\frac{y}{|z-\xi|^2}\, d\xi,$$

i.e., changing sign,

$$\frac{x_0 - x}{(x_0 - x)^2 + (y+h)^2} = \frac{1}{\pi}\int_{-\infty}^{\infty} \frac{x_0 - \xi}{(x_0-\xi)^2 + h^2}\cdot\frac{y}{(\xi - x)^2 + y^2}\, d\xi.$$

Therefore

$$\tilde{U}(x_0 + iy + ih) = \frac{1}{\pi}\int_{-\infty}^{\infty} \frac{x_0 - t}{(x_0 - t)^2 + (y+h)^2} u(t)\, dt =$$

$$= \left(\frac{1}{\pi}\right)^2 \int_{-\infty}^{\infty}\int_{-\infty}^{\infty} \frac{x_0 - \xi}{(x_0-\xi)^2 + h^2}\cdot\frac{y}{(\xi - t)^2 + y^2} u(t)\, d\xi\, dt.$$

By Fubini's theorem ($u(t)$ can be assumed of compact support!), this works out to

$$\frac{1}{\pi}\int_{-\infty}^{\infty} \frac{x_0 - t}{(\xi - x_0)^2 + h^2} U(\xi + iy)\, d\xi.$$

Thus,

$$\tilde{U}(x + iy + ih) = \frac{1}{\pi}\int_{-\infty}^{\infty} \frac{x - t}{(x-t)^2 + h^2} U(t + iy)\, dt.$$

By what has just been proven,

$$\int_{-\infty}^{\infty} |\tilde{U}(x + iy + ih)|^p\, dx \le K_p \int_{-\infty}^{\infty} |U(t + iy)|^p\, dt$$

for *all* $h > 0$ with a K_p depending *only* on p. Now, *keeping* $y > 0$ *fixed*, make $h \to 0$ so that

$$\tilde{U}(x + iy + ih) \to \tilde{U}(x + iy)$$

u.c.c. in x and use Fatou's lemma. We find in the limit that

$$\int_{-\infty}^{\infty} |\tilde{U}(x + iy)|^p\, dx \le K_p \int_{-\infty}^{\infty} |U(t + iy)|^p\, dt,$$

the *full strength* of what was required. Q.E.D.

Scholium In case $p = 2$, we can say more. For then $(F(z))^2$ is regular in $\Im z > 0$ *whether* $U(z)$ *is* ≥ 0 *or not*, so

$$\int_{-\infty}^{\infty} (F(x + ih))^2\, dx = 0$$

for each $h > 0$, F_h^2 being in H_1. Taking real parts, we have (in the case of *real* U)

$$\int_{-\infty}^{\infty} [(U(x+ih))^2 - (\tilde{U}(x+ih))^2]\,dx = 0,$$

i.e.,

$$\boxed{\int_{-\infty}^{\infty} [\tilde{U}(x+ih)]^2\,dx = \int_{-\infty}^{\infty} [U(x+ih)]^2\,dx.}$$

For $p = 2$ *we have equality*!

Now we remove the restriction $p \le 2$ by a duality argument.

Theorem (M. Riesz) *Let* $1 < p < \infty$ *and let* $u(t) \in L_p(-\infty, \infty)$. *Write, for* $\Im z > 0$,

$$\tilde{U}(z) = \frac{1}{\pi}\int_{-\infty}^{\infty} \frac{x-t}{|z-t|^2}u(t)\,dt.$$

Then, for each $h > 0$,

$$\int_{-\infty}^{\infty} \left|\tilde{U}(x+ih)\right|^p dx \le K_p \int_{-\infty}^{\infty} |u(t)|^p\,dt.$$

Proof If $1 < p \le 2$, the result is contained in the above lemma, so assume $2 < p < \infty$ and let

$$\frac{1}{p} + \frac{1}{q} = 1;$$

then $1 < q \le 2$. By real variable theory, given any $\epsilon > 0$ we can, corresponding to $h > 0$, find a $v(t) \in L_q(-\infty, \infty)$ *of compact support* with

$$\int_{-\infty}^{\infty} |v(t)|^q\,dt = 1$$

such that, for $\tilde{U}_h(x) = \tilde{U}(x+ih)$,

$$\left\|\tilde{U}_h\right\|_p - \epsilon \le \left|\int_{-\infty}^{\infty} \tilde{U}_h(x)v(x)\,dx\right|.$$

But now, by Fubini,

$$\int_{-\infty}^{\infty} \tilde{U}_h(x)v(x)\,dx = \frac{1}{\pi}\int_{-\infty}^{\infty}\int_{-\infty}^{\infty} \frac{(x-t)u(t)v(x)}{(x-t)^2+h^2}\,dt\,dx$$

$$= -\frac{1}{\pi}\int_{-\infty}^{\infty}\left(\int_{-\infty}^{\infty} \frac{t-x}{(t-x)^2+h^2}v(x)\,dx\right)u(t)\,dt = -\int_{-\infty}^{\infty} u(t)\tilde{V}_h(t)\,dt,$$

where $\tilde{V}_h(t) = \tilde{V}(t+ih)$ with

$$\tilde{V}(z) = \frac{1}{\pi}\int_{-\infty}^{\infty} \frac{x-s}{|z-s|^2} v(s)\,\mathrm{d}s.$$

By the lemma, since $1 < q \le 2$,

$$\left\|\tilde{V}_h\right\|_q \le K_q \|v\|_q = K_q,$$

so by Hölder again,

$$\left|\int_{-\infty}^{\infty} u(t)\tilde{V}_h(t)\,\mathrm{d}t\right| \le K_q \|u\|_p.$$

Thus, $\left\|\tilde{U}_h\right\|_p - \epsilon \;\le\; K_q \|u\|_p$. Squeeze ϵ. The theorem is proved. Q.E.D.

This important theorem has many corollaries.

Corollary *If $u(t) \in L_p(-\infty,\infty)$ is real valued and $1 < p < \infty$, then*

$$F(z) = \frac{i}{\pi}\int_{-\infty}^{\infty} \frac{u(t)\,\mathrm{d}t}{z-t} \quad \in H_p(\Im z > 0)$$

and $\Re F(z) \to u(t)$ as $z \underset{L}{\longrightarrow} t$ a.e. in t.

Corollary *If $u(t) \in L_p(-\infty,\infty)$ and $1 < p < \infty$, then*

$$\tilde{u}(x) = \lim_{\epsilon\to 0}\frac{1}{\pi}\int_{|t-x|\ge\epsilon} \frac{u(t)}{x-t}\,\mathrm{d}t$$

exists a.e., $\|\tilde{u}\|_p \le K_p \|u\|_p$, and

$$\tilde{U}(z) = \frac{1}{\pi}\int_{-\infty}^{\infty} \frac{x-t}{|z-t|^2} u(t)\,\mathrm{d}t = \frac{1}{\pi}\int_{-\infty}^{\infty} \frac{y}{|z-t|^2}\tilde{u}(t)\,\mathrm{d}t.$$

Remark Thus, when u is *real*, the H_p function

$$F(z) = \frac{i}{\pi}\int_{-\infty}^{\infty} \frac{u(t)\,\mathrm{d}t}{z-t}$$

of the previous corollary has $\Im F(t) = \tilde{u}(t)$ a.e., whilst $\Re F(t) = u(t)$.

Proof of second corollary Take $\tilde{U}(z)$ – it is harmonic in $\Im z > 0$. By M. Riesz' theorem, if $u \in L_p$,

$$\int_{-\infty}^{\infty} |\tilde{U}(x+iy)|^p\,\mathrm{d}x \le C, \quad y > 0.$$

Therefore, since $1 < p < \infty$, by a theorem of Section C, there is a function $v(t) \in L_p(-\infty,\infty)$, such that

$$\tilde{U}(z) = \frac{1}{\pi}\int_{-\infty}^{\infty} \frac{yv(t)\,\mathrm{d}t}{|z-t|^2}.$$

By Section B, $\tilde{U}(z) \to v(t)$ as $z \underset{L}{\longrightarrow} t$ for almost all t. In particular, $\lim_{y\to 0} \tilde{U}(t+iy)$ exists and is finite for almost all $t \in \mathbb{R}$. So, by another result of Section B, for almost all x,

$$\lim_{\epsilon\to 0} \frac{1}{\pi} \int_{|t-x|\geq\epsilon} \frac{u(t)\,dt}{x-t}$$

exists and has the same value as $\lim_{y\to 0} \tilde{U}(x+iy) = v(x)$. So, with the notation for principal values which is now familiar,

$$\frac{1}{\pi} \fint_{-\infty}^{\infty} \frac{u(t)\,dt}{x-t} = v(x) \quad \text{a.e.}$$

This is enough.

Definition

$$\tilde{u}(x) = \frac{1}{\pi} \fint_{-\infty}^{\infty} \frac{u(t)\,dt}{x-t}$$

is called the *Hilbert transform* of $u \in L_p(-\infty,\infty)$, $1 < p < \infty$.

Corollary *If* $1 < p < \infty$ *and* $u \in L_p(-\infty,\infty)$, $\tilde{\tilde{u}} = -u$.

Proof Without loss of generality, take u to be *real*, and write

$$F(z) = \frac{i}{\pi} \int_{-\infty}^{\infty} \frac{u(t)\,dt}{z-t};$$

then $F \in H_p$, $\Re F(t) = u(t)$ and $\Im F(t) = \tilde{u}(t)$. Let

$$G(z) = \frac{i}{\pi} \int_{-\infty}^{\infty} \frac{\tilde{u}(t)\,dt}{z-t}.$$

Then

$$-iF(z) + G(z) = \frac{1}{\pi} \int_{-\infty}^{\infty} \frac{[u(t) + i\tilde{u}(t)]\,dt}{z-t} = \frac{1}{\pi} \int_{-\infty}^{\infty} \frac{F(t)}{z-t}\,dt = -2iF(z)$$

by Subsection 3. So $G(z) = -iF(z)$. Therefore for almost all x,

$$\tilde{\tilde{u}}(x) = \lim_{y\to 0} \Im G(x+iy) = -\lim_{y\to 0} \Re F(x+iy) = -u(x).$$

Q.E.D.

Corollary *If* $1 < p < \infty$, *there are two constants* C_p *and* D_p *with*

$$\boxed{C_p \|u\|_p \leq \|\tilde{u}\|_p \leq D_p \|u\|_p,}$$

and the Hilbert transform is an isomorphic mapping of $L_p(-\infty,\infty)$ onto itself.

Proof By the above two corollaries.

Remark In case $p = 2$, we have an *isometry*,

$$\boxed{\int_{-\infty}^{\infty} |\tilde{u}(x)|^2\,dx = \int_{-\infty}^{\infty} |u(t)|^2\,dt.}$$

It suffices to verify this for real u. Then, $u(t) + i\tilde{u}(t) = F(t)$ with $F \in H_2$, so $F^2 \in H_1$, so

$$\int_{-\infty}^{\infty} (f(t))^2\,dt = 0.$$

Take real parts.

Corollary *If* $(1/p) + (1/q) = 1$, $1 < p < \infty$, $u \in L_p$ *and* $v \in L_q$, *we have*

$$\boxed{\int_{-\infty}^{\infty} u(t)v(t)\,dt = \int_{-\infty}^{\infty} \tilde{u}(x)\tilde{v}(x)\,dx}$$

and

$$\boxed{\int_{-\infty}^{\infty} u(t)\tilde{v}(t)\,dt = -\int_{-\infty}^{\infty} \tilde{u}(t)v(t)\,dt,}$$

all the integrals being absolutely convergent.

Proof When u and v are *real*, we can put

$$F(x) = \frac{i}{\pi}\int_{-\infty}^{\infty} \frac{u(t)\,dt}{z-t}, \quad G(z) = \frac{i}{\pi}\int_{-\infty}^{\infty} \frac{v(t)\,dt}{z-t};$$

then $F \in H_p$, $G \in H_q$, $F(t) = u(t) + i\tilde{u}(t)$ a.e. and $G(t) = v(t) + i\tilde{v}(t)$ a.e. We have $F(z)G(z) \in H_1$, so by the first lemma of this section,

$$\int_{-\infty}^{\infty} F(t)G(t)\,dt = 0.$$

Taking real parts gives us the *first* boxed relation. Taking imaginary parts gives us the *second*. Absolute convergence of the integrals in question follows by Hölder's inequality, and the result for complex u, v is finally deduced from the real case, using bilinearity.

E. Fourier transforms. The Paley-Wiener theorem

Let $F(t) \in L_p(-\infty, \infty)$. The Hausdorff-Young theorem says that if $1 \le p \le 2$ (and in general for *no larger* values of p !), and if we write

$$\hat{F}_N(\lambda) = \int_{-N}^{N} e^{i\lambda t} F(t)\,dt,$$

then, as $N \to \infty$, the $\hat{F}_N(\lambda)$ tend in $L_q(-\infty, \infty)$ to a function $\hat{F}(\lambda)$, called the *Fourier transform* of F, where $1/q = 1-(1/p)$. The Fourier transform $\hat{F}$ satisfies

$$\|\hat{F}\|_q \le K_q \|F\|_p.$$

If $p = 1$, these statements are obvious; in that case $\hat{F}(\lambda)$ is even *continuous* and *zero* at $\pm\infty$. For $p = 2$, they are part of *Plancherel's theorem*. These two cases ($p = 1$ and $p = 2$) suffice for many applications.

Theorem *Let $F(t) \in L_p(-\infty,\infty)$ with $1 \le p \le 2$ be the boundary-value function of a function $F \in H_p$. Then $\hat{F}(\lambda) = 0$ a.e. for $\lambda \ge 0$.*

Proof First of all, if $p = 1$, i.e., $F \in H_1$, then $e^{i\lambda z}F(z)$ is also in H_1 for $\lambda \ge 0$ ($|e^{i\lambda z}| \le 1$ for $\Im z \ge 0$ then!). So then

$$\int_{-\infty}^{\infty} e^{i\lambda t}F(t)\,dt = 0,$$

i.e., $\hat{F}(\lambda) = 0$, by a lemma of Section D.

If we just have $F \in H_p$ for $1 < p \le 2$, we use $F_{(k)}(z) = (ik/(z+ik))F(z)$ with $k > 0$. Clearly $F_{(k)} \in H_1$. So $\hat{F}_{(k)}(\lambda) \equiv 0$, $\lambda > 0$. But now $\|F_{(k)} - F\|_p \to 0$ as $k \to \infty$ so, by the Hausdorff–Young theorem (or just plain old Plancherel in case $p = 2$), $\|\hat{F}_{(k)}(\lambda) - \hat{F}(\lambda)\|_q \to 0$ as $k \to \infty$, where $1/q = 1-(1/p)$. Since each $\hat{F}_{(k)}(\lambda)$ vanishes identically on $[0,\infty)$, $\hat{F}(\lambda) \equiv 0$ a.e. there. Q.E.D.

Remark In case $p = 2$, *the converse of the above theorem is true.* Namely, if $\Phi(\lambda) \in L_2(-\infty,\infty)$ and $\Phi(\lambda) \equiv 0$ a.e. for $\lambda \ge 0$, there *is* an $F \in H_2$ with $\hat{F}(\lambda) = \Phi(\lambda)$. It suffices to take

$$F(t) = \frac{1}{2\pi}\operatorname*{l.i.m.}_{N\to\infty}\int_{-N}^{N} e^{-i\lambda t}\Phi(\lambda)\,d\lambda.$$

(Here, l.i.m. = 'limit in the mean' denotes a limit in L_2-norm.)

Indeed, if $\Phi(\lambda) \equiv 0$ on $(0,\infty)$ we have

$$F(t) = \frac{1}{2\pi}\operatorname*{l.i.m.}_{N\to\infty}\int_{0}^{N} e^{i\lambda t}\Phi(-\lambda)\,d\lambda.$$

For each $y > 0$, $e^{-y\lambda}\Phi(-\lambda) \in L_1(-\infty,\infty)$, so

$$\int_0^\infty e^{i\lambda z}\Phi(-\lambda)\,d\lambda = \int_0^\infty e^{i\lambda x}e^{-y\lambda}\Phi(-\lambda)\,d\lambda$$

converges absolutely and defines a function $2\pi F_1(z)$ regular in $\Im z > 0$. By Plancherel's theorem,

$$2\pi\int_{-\infty}^{\infty} |F_1(x+iy)|^2\,dx = \int_0^\infty e^{-2y\lambda}|\Phi(-\lambda)|^2\,d\lambda \le \|\Phi\|_2^2,$$

so $F_1(z) \in H_2$.

Finally, $\left\|e^{-y\lambda}\Phi(-\lambda) - \Phi(-\lambda)\right\|_2 \to 0$ as $y \to 0$, so by Plancherel again,

$$\int_{-\infty}^{\infty} |F_1(t+iy) - F(t)|^2 \, dt \to 0 \quad \text{as } y \to 0.$$

Therefore $F(t)$ is the boundary-value function $F_1(t)$ of a function $F_1 \in H_2$. *The Fourier integral inversion formula for* L_2 *now shows that* $\hat{F}(\lambda) = \Phi(\lambda)$.

Theorem (Phragmén–Lindelöf) *Let $f(z)$ be analytic in $\Im z > 0$ and continuous in $\Im z \geq 0$. Suppose that*

(i) $|f(z)| \leq \text{const.} \exp(\text{const.}|z|)$ *in* $\Im z \geq 0$

(ii) $|f(z)| \leq M$, $x \in \mathbb{R}$

(iii) $\limsup_{y \to \infty} \left[\log|f(iy)|/y\right] = a$.

Then $|f(z)| \leq Me^{ay}$, $\Im z \geq 0$.

Proof Pick any $\epsilon > 0$. In quadrants I and II, $g(z) = e^{i(a+\epsilon)z} f(z)$ satisfies $\log|g(z)| \leq O(|z|)$ and $|g(z)|$ is bounded on the boundary of each of these quadrants.

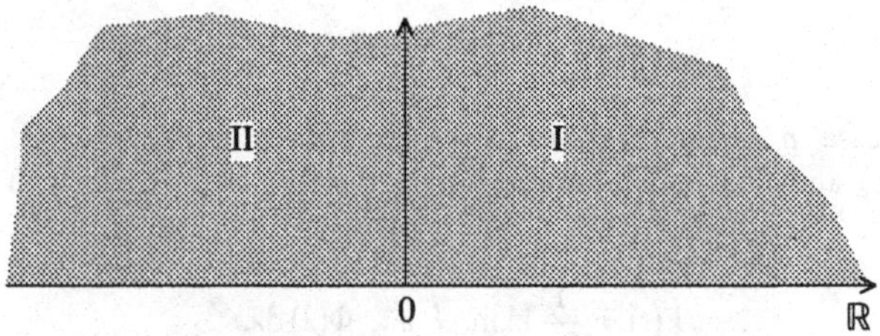

Each quadrant has an opening of $90° < 180°$, so, by the *ordinary* Phragmén–Lindelöf theorem, $g(z)$ is *bounded* in quadrants I and II. *Therefore $g(z)$ is bounded* in $\Im z \geq 0$.

From this, by another elementary Phragmén–Lindelöf theorem,

$$|g(z)| \leq \sup_{t \in \mathbb{R}} |g(t)| \leq M \quad \text{in } \Im z \geq 0,$$

so $|f(z)| \leq Me^{(a+\epsilon)y}$ there. Squeeze ϵ.

Definition An entire function $f(z)$ is said to be of *exponential type* if there are constants A and B with $|f(z)| \leq Ae^{B|z|}$.

Theorem (Paley and Wiener) *Let $f(z)$ be entire of exponential type, and suppose that* $\int_{-\infty}^{\infty} |f(x)|^2 \, dx < \infty$.

Let

$$a = \limsup_{y \to -\infty} \frac{\log|f(iy)|}{|y|}, \qquad b = \limsup_{y \to \infty} \frac{\log|f(iy)|}{y}.$$

Then

$$f(z) = \int_{-a}^{b} e^{-iz\lambda}\varphi(\lambda)\,d\lambda,$$

where

$$\int_{-a}^{b} |\varphi(\lambda)|^2\,d\lambda < \infty.$$

Proof

Step 1. Assume first that we *also know* f is *bounded* on $\mathbb{R}$, say $|f(t)| \le M$ for t real. Then, by the above Phragmén–Lindelöf theorem, $|e^{ibz}f(z)| \le M$ for $\Im z \ge 0$, i.e., $e^{ibz}f(z) \in H_\infty(\Im z > 0)$. Therefore, for $\Im z > 0$, by Section C,

$$e^{ibz}f(z) = \frac{1}{\pi}\int_{-\infty}^{\infty} \frac{y}{|z-t|^2}e^{ibt}f(t)\,dt.$$

So, since $f(t)$, and therefore $e^{ibt}f(t)$, belongs to $L_2(-\infty,\infty)$, we have, by Section B,

$$\int_{-\infty}^{\infty} |e^{ib(t+iy)}f(t+iy)|^2\,dt \;\le\; c, \quad y > 0,$$

i.e., *in fact*, $e^{ibz}f(z) \in H_2(\Im z > 0)$.

So, by the first theorem of this section,

$$\hat{f}(\lambda) = \operatorname*{l.i.m.}_{N\to\infty} \int_{-N}^{N} e^{i(\lambda-b)t}[e^{ibt}f(t)]\,dt$$

is *zero* a.e. for $\lambda - b \ge 0$, i.e., $\hat{f}(\lambda) \equiv 0$ a.e., $\lambda \ge b$.

Working in $\Im z < 0$ instead of in $\Im z > 0$ we see in the same way that $\hat{f}(\lambda) \equiv 0$ a.e., $\lambda \le -a$.

The L_2 inversion formula for Fourier transforms now yields, for $x \in \mathbb{R}$,

$$f(x) = \operatorname*{l.i.m.}_{N\to\infty} \frac{1}{2\pi}\int_{-N}^{N} e^{-ix\lambda}\hat{f}(\lambda)\,d\lambda.$$

In the present case, this simply reduces to

$$f(x) = \frac{1}{2\pi}\int_{-a}^{b} e^{-ix\lambda}\hat{f}(\lambda)\,d\lambda \quad \text{a.e.}$$

Because $\hat{f}(\lambda) \in L_2$, the integral

$$\frac{1}{2\pi}\int_{-a}^{b} e^{-iz\lambda}\hat{f}(\lambda)\,d\lambda$$

actually converges absolutely for all complex z and represents an entire function of z. This entire function coincides a.e. with another, $f(z)$, when z is *real*. So in fact it is *identically equal* to $f(z)$. We are done if f is bounded on $\mathbb{R}$.

Step 2. *But $f(x)$ is indeed bounded on* $\mathbb{R}$! We are given the inequality

$$|f(z)| \leq Ae^{B|z|}$$

with two constants A and B. For each $h > 0$, put

$$f_h(z) = \frac{1}{h}\int_0^h f(z+t)\,dt,$$

then $f_h(z)$ is also entire and satisfies $|f_h(z)| \leq A_h e^{B|z|}$, *with the same B as before, irrespective of the value of h.*

Since $\|f(x)\|_2 < \infty$, Schwarz' inequality now gives

$$|f_h(x)| \leq \frac{\|f\|_2}{\sqrt{h}},$$

i.e., f_h *is bounded on* $\mathbb{R}$, and we also have $\|f_h\|_2 \leq \|f\|_2$. Therefore the result already proved in *Step* 1 guarantees that each Fourier transform $\hat{f}_h(\lambda)$ *vanishes* a.e. for $\lambda < -B$ or for $\lambda > B$. So, by the inversion formula already used,

$$f_h(x) = \frac{1}{2\pi}\int_{-B}^{B} e^{-ix\lambda}\hat{f}_h(\lambda)\,d\lambda,$$

whence, by Schwarz,

$$|f_h(x)|^2 \leq \frac{B}{2\pi^2}\int_{-B}^{B} |\hat{f}_h(\lambda)|^2\,d\lambda.$$

By Plancherel, the right side equals $(B/\pi)\|f_h(x)\|_2^2$ which, as we have observed, is $\leq (B/\pi)\|f\|_2^2$. Thus, for real x,

$$|f_h(x)| \leq \sqrt{\frac{B}{\pi}}\,\|f\|_2$$

for all $h > 0$. Since clearly $f_h(x) \to f(x)$ as $h \to 0$, we finally get $|f(x)| \leq \sqrt{B/\pi}\,\|f\|_2$, and f is bounded on $\mathbb{R}$.

The theorem is completely proved.

Remark Clearly, *if* $\varphi \in L_2$ and

$$f(z) = \int_{-a}^{b} e^{-iz\lambda}\varphi(\lambda)\,d\lambda,$$

$f(z)$ *is* entire of exponential type,

$$\int_{-\infty}^{\infty} |f(x)|^2\,dx < \infty,$$

and

$$|f(iy)| \leq \text{Const.}e^{by} \quad \text{for } y \to \infty,$$

$$|f(iy)| \leq \text{Const.}e^{a|y|} \quad \text{for } y \to -\infty,$$

(assuming, of course, that $-a < b$!). *The Paley–Wiener theorem is a precise converse to this elementary observation.*

Remark If, in the hypothesis of the Paley–Wiener theorem, we suppose

$$\int_{-\infty}^{\infty} |f(x)|\,\mathrm{d}x < \infty$$

instead of

$$\int_{-\infty}^{\infty} |f(x)|^2\,\mathrm{d}x < \infty,$$

but keep the *rest* of the *assumptions* the *same*, we *still* have

$$f(z) = \frac{1}{2\pi}\int_{-a}^{b} e^{-iz\lambda}\hat{f}(\lambda)\,\mathrm{d}\lambda$$

where $\hat{f}(\lambda)$ *is now even continuous.*

The proof of this variant is that same as that of the Paley–Wiener theorem.

F. Titchmarsh's convolution theorem

Definition If $\phi(\lambda),\ \psi(\lambda) \in L_1(-\infty,\infty)$, the *convolution* $\phi * \psi$ is defined by

$$(\phi * \psi)(\lambda) = \int_{-\infty}^{\infty} \phi(\lambda - \tau)\psi(\tau)\,\mathrm{d}\tau.$$

One proves (by use of Fubini's theorem) that the integral on the right *converges absolutely* for almost all $\lambda \in \mathbb{R}$, and that $\phi * \psi \in L_1(-\infty,\infty)$. By a change of variable, one shows that $\phi * \psi = \psi * \phi$. (I am sorry for the heavy use of Greek letters here. The way this chapter has gone, Latin letters serve almost exclusively for *Fourier transforms* of functions in L_1, especially if these transforms are *analytic* in some half plane.) As is well known, *taking Fourier transforms converts convolution to multiplication*, i.e.,

$$(\widehat{\phi * \psi})(t) = \hat{\phi}(t)\hat{\psi}(t).$$

Definition If $\phi \in L_1(-\infty,\infty)$, by the *supporting interval* of ϕ is meant the *minimal closed interval* (*perhaps infinite*!) containing the *support* of ϕ, i.e. *outside* of which $\phi \equiv 0$ a.e.

Let a be the *upper endpoint* of the *supporting interval* of ϕ, and b that of the supporting interval of ψ. Then direct inspection of the convolution integral shows that $(\phi * \psi)(\lambda) \equiv 0$ for $\lambda > a + b$. What is *remarkable* is that if a and $b < \infty$, *no cancellation can take place*, and a *converse* to the remark just made holds:

Theorem *If a is the upper endpoint of the supporting interval of ϕ, b the upper endpoint of the supporting interval of ψ, and if a and b both are* finite *(sic!), then the upper endpoint of the supporting interval of $\phi * \psi$ is* precisely *$a + b$.*

Proof Calling c the upper endpoint of the supporting interval of $\phi * \psi$, we already *know* that $c \le a + b$. We *have to prove that* $a + b \le c$.

To this end, use the inverse Fourier transforms

$$f(z) = \hat{\phi}(-z) = \int_{-\infty}^{a} e^{-iz\lambda}\phi(\lambda)\,d\lambda$$

$$g(z) = \hat{\psi}(-z) = \int_{-\infty}^{b} e^{-iz\lambda}\psi(\lambda)\,d\lambda$$

$$S(z) = (\widehat{\phi * \psi})(-z) = \int_{-\infty}^{c} e^{-iz\lambda}(\phi * \psi)(\lambda)\,d\lambda.$$

We have $S(z) = f(z)g(z)$.

Because a, b and c are *finite*, $f(z)$, $g(z)$ and $S(z)$ are *analytic* for $\Im z > 0$, and because ϕ, ψ and $\phi * \psi \in L_1$, f, g and S are *continuous up to* $\mathbb{R}$ and *bounded* on $\mathbb{R}$. In fact, we clearly have, in $\Im z \geq 0$,

$$|f(z)| \leq Ke^{ay}, \quad |g(z)| \leq Le^{by}, \quad |S(z)| \leq Me^{cy}$$

with certain constants K, L and M.

I say that

$$\limsup_{y\to\infty} \frac{\log|g(iy)|}{y}$$

is *precisely* b. Call the lim sup in question b'; we certainly have $b' \leq b$. Suppose that $b' < b$. For $h > 0$, call

$$\psi_h(\lambda) = \frac{1}{2h}\int_{-h}^{h} \psi(\lambda + \tau)\,d\tau;$$

then $\|\psi_h - \psi\|_1 \to 0$ as $h \to 0$, and, since Fourier transformation takes convolution to *multiplication*,

$$\hat{\psi}_h(-z) = \frac{\sin hz}{hz}\hat{\psi}(-z) = \frac{\sin hz}{hz}g(z).$$

But now $\psi_h(\lambda) \in L_2(-\infty, \infty)$, so the Fourier inversion formula for L_2 gives

$$\psi_h(\lambda) = \frac{1}{2\pi}\operatorname*{l.i.m.}_{N\to\infty}\int_{-N}^{N} e^{i\lambda t}\hat{\psi}_h(-t)\,dt.$$

Observe, however, that for sufficiently large l,

$$e^{ilz}\hat{\psi}_h(-z) = e^{ilz}\frac{\sin hz}{hz}g(z)$$

belongs to $H_2(\Im z > 0)$. Indeed,

$$\limsup_{y\to\infty} \frac{\log|g(iy)|}{y} = b'$$

makes $|g(z)| \leq Le^{b'y}$ by the Phragmén–Lindelöf theorem of Section E and we see by direct inspection that $e^{ilz}\hat{\psi}_h(-z)$ is in H_2 for $l = b' + h$. *Therefore*, by the first theorem of Section E, $\psi_h(\lambda) \equiv 0$ a.e. for $\lambda \geq b' + h$. Making $h \to 0$, we see that $\psi(\lambda) \equiv 0$ a.e. for $\lambda \geq b'$, and *this would contradict our choice of b as the upper limit of the supporting interval* of ψ if $b' < b$. So $b' = b$.

In like manner, we prove that

$$\limsup_{y\to\infty} \frac{\log|f(iy)|}{y}$$

and, indeed,

$$\limsup_{y\to\infty} \frac{\log|f(x_0+iy)|}{y}$$

for any $x_0 \in \mathbb{R}$, are precisely equal to a.

Now, for $k = 1, 2, 3, \ldots$ let $M(k) = \sup_{x\in\mathbb{R}} |g(x+ki)|$. Given any $\epsilon > 0$, we must have $M(k+1)/M(k) > e^{b-\epsilon}$ for *some* k, *otherwise*, we would have

$$\limsup_{y\to\infty} \frac{\log|g(iy)|}{y} \leq b-\epsilon$$

as is easily seen.* So, fixing $\epsilon > 0$, choose a k for which the aforementioned inequality holds, and take $x_0 \in \mathbb{R}$ such that

$$\frac{|g(x_0+(k+1)i)|}{M(k)} \geq e^{b-\epsilon}.$$

Without loss of generality, and to simplify the notation, *we now assume* that $x_0 = 0$.

We work in the half plane $\Im z \geq k$. By the Phragmén–Lindelöf theorem of Section E, $|g(z)| \leq M(k)e^{b(y-k)}$ for $\Im z \geq k$, so $e^{ibz}g(z)$ is *bounded* in that half plane and by Section C we can find a Blaschke product $Б(z)$ for the upper half plane with $g(z)/Б(z-ki)$ *free of zeros* in $\Im z > k$. We see that there is an $h \in H_\infty(\Im z > 0)$ with $\|h\|_\infty \leq 1$, h *free of zeros* in $\Im z > 0$, such that

$$g(z) = e^{-ib(z-ki)}M(k)Б(z-ki)h(z-ki) \quad \text{for } \Im z > k.$$

In the same manner we find a Blaschke product $B(z)$ for the upper half plane and a function $L(z) \in H_\infty(\Im z > 0)$, *free of zeros* in $\Im z > 0$, with

$$S(z) = e^{-ic(z-ki)}B(z-ki)L(z-ki) \quad \text{for } \Im z > k.$$

The ratio $S(z)/g(z) = f(z)$ is regular for $\Im z > k$. *Therefore* $B(z)$ *must have enough zeros to cancel those of* $Б(z)$! So $B(z)/Б(z) = B_1(z)$, *another Blaschke product*, and

$$f(z) = \frac{S(z)}{g(z)} = B_1(z-ki)e^{i(b-c)(z-ki)}\,\frac{L(z-ki)}{M(k)}\cdot\frac{1}{h(z-ki)} \quad \text{for } \Im z > k,$$

whence

$$|f(z)| \leq \text{const.}\frac{e^{(c-b)y}}{|h(z-ki)|} \quad \text{for } \Im z > k.$$

Now $h(z)$ is *free of zeros* in $\Im z > 0$ and in modulus ≤ 1 there, so $\log|1/h(z)|$ is *positive and harmonic* in $\Im z > 0$. Thence, by the Poisson representation given near the beginning of

---

* Then, indeed, the functions $g(z+ik)/M(k)$ would be uniformly bounded for $0 \leq \Im z \leq 1$ by the extended principle of maximum.

this chapter,

$$\log\left|\frac{1}{h(z)}\right| = \frac{1}{\pi}\int_{-\infty}^{\infty}\frac{y}{|z-t|^2}\,d\mu(t) + \beta y \quad \text{for } \Im z > 0,$$

with a *positive* measure μ and $\beta \geq 0$. By our choice of k,

$$e^{b-\epsilon} \leq \frac{|g((k+1)i)|}{M(k)} \leq e^b|h(i)|$$

(since $x_0 = 0$ and $|B(z)| \leq 1$), whence $\log|1/h(i)| \leq \epsilon$. That is,

$$\frac{1}{\pi}\int_{-\infty}^{\infty}\frac{d\mu(t)}{t^2+1} + \beta \;\leq\; \epsilon.$$

But from this we see that for $y \geq 1$

$$\log\left|\frac{1}{h(iy)}\right| = \frac{1}{\pi}\int_{-\infty}^{\infty}\frac{y\,d\mu(t)}{t^2+y^2} + \beta y \;\leq\; \epsilon y$$

since $d\mu(t) \geq 0$ and $\beta \geq 0$. Substituted into a previous relation, this makes

$$|f(iy)| \leq \text{const.}\, e^{(c-b)y} e^{\epsilon(y-k)}$$

for $y \geq k+1$, i.e.,

$$\limsup_{y\to\infty}\frac{\log|f(iy)|}{y} \leq c - b + \epsilon.$$

However, as we saw above, the lim sup on the left is *precisely* a. Therefore $a \leq c - b + \epsilon$. Squeezing ϵ, we get $a + b \leq c$, as required.

We are done.

Corollary (Titchmarsh's convolution theorem) *Let $\phi \in L_1$ have the* finite *supporting interval $[a_1, a_2]$ and $\psi \in L_1$ the* finite *supporting interval $[b_1, b_2]$. Then the supporting interval of $\phi * \psi$ is* precisely *$[a_1 + b_1,\ a_2 + b_2]$.*

Proof The above theorem shows that the *upper endpoint* of $\phi * \psi$'s supporting interval *is* $a_2 + b_2$. The *lower endpoint* is similarly seen to be $a_1 + b_1$ – one way of doing this is to *first* make the *change of variable* $\lambda \to -\lambda$ and *then apply the above theorem again.*

Remark The *finiteness* of a_1 and b_1 in order to get $a_1 + b_1$ (and of a_2 and b_2 in order to get $a_2 + b_2$) is *essential* here. This is easily seen by examples.

Remark Titchmarsh's convolution theorem is a celebrated example of a *purely 'real variable' result proved by complex variable methods.* Mikusinski and others have given *real variable* proofs. Most of them are *harder* than the one given above.

Problem 6

(a) Compute

$$\tilde{k}_\epsilon(x) = \frac{1}{\pi}\int_{-\infty}^{\infty}\frac{k_\epsilon(t)\,dt}{x-t}$$

for the function

$$k_\epsilon(t) = \begin{cases} 0, & |t| < \epsilon \\ 1/t, & |t| \geq \epsilon. \end{cases}$$

(b) If $1 < p < \infty$ and $u(t) \in L_p(-\infty, \infty)$ and $\epsilon > 0$, put

$$(T_\epsilon u)(x) = \frac{1}{\pi} \int_{|t-x| \geq \epsilon} \frac{u(t)\,dt}{x-t}.$$

Using the function $\tilde{k}_\epsilon$ found in (a), show that there is a C_p *independent* of ϵ such that

$$\|T_\epsilon u\|_p \leq C_p \|u\|_p .$$

VII

Duality for H_p spaces

A. H_p spaces and their duals. Sarason's theorem

1. Various spaces and their duals. Tables

We consider mainly the *unit circle*; similar (and in a sense, *more symmetric*) results holding for the *upper half plane* are established by analogous methods, and will be tabulated at the end of this subsection.

By looking just at the boundary values $f(e^{i\theta})$ of $f \in H_p$, we see that H_p can be considered as a $\| \ \|_p$-*closed subspace* of $L_p(-\pi, \pi)$. *This we do* from now on.

We define some more spaces:

$$\mathscr{C} = \{f \text{ continuous on } [-\pi, \pi];\ f(-\pi) = f(\pi)\};$$

$\mathscr{A} = \mathscr{C} \cap H_\infty$ is the set of functions in $\mathscr{C}$ which have an *analytic extension* to $\{|z| < 1\}$, yielding continuous functions on the *closed unit disc*.

We equip $\mathscr{C}$ and $\mathscr{A}$ with the sup-norm $\| \ \|_\infty$.

$$\mathscr{M} = \text{set of finite complex-valued Radon measures on } \{|\zeta| = 1\},$$

equipped with the measure norm

$$\|\mu\| = \int_{-\pi}^{\pi} |\,d\mu(e^{i\theta})|.$$

Note that $\mathscr{M}$ is the dual of $\mathscr{C}$, by a classical theorem of F. Riesz.

Notation

$$H_p(0) = \{f \in H_p;\ \int_{-\pi}^{\pi} f(e^{i\theta})\, d\theta = 0\} = zH_p.$$

Theorem *If $1 < p < \infty$ and $(1/p) + (1/q) = 1$, L_q/H_q has dual $H_p(0)$ and H_p has dual $L_q/H_q(0)$.*

Remark This slight but troublesome *asymmetry* is one of the results of working in the unit circle. It *disappears* if we work in the upper half plane.

Proof of theorem

(a) Let Λ be a bounded linear functional on L_q/H_q. Then Λ certainly gives a linear functional on L_q with the same norm as on L_q/H_q, so for $f \in L_q$, we have

$$\Lambda(f) = \Lambda(f + H_q) = \int_{-\pi}^{\pi} f(e^{i\theta})L(e^{i\theta})\,d\theta,$$

where L is some *function* in L_p with $\|L\|_p$ equal to the norm of Λ (Riesz representation theorem). We have $\Lambda(g) = 0$ for $g \in H_q$, in particular,

$$\int_{-\pi}^{\pi} e^{in\theta}L(e^{i\theta})\,d\theta = 0, \quad n = 0, 1, 2, \ldots,$$

so $L(e^{i\theta})$ has a Fourier series of the form

$$\sum_{1}^{\infty} A_n e^{in\theta},$$

and, since $L(e^{i\theta}) \in L_p$, $L(e^{i\theta}) \in e^{i\theta}H_p = H_p(0)$. *Conversely*, any $L(e^{i\theta})$ of this form *does give* a linear functional Λ on L_q/H_q by the above formula.

(b) If $L(e^{i\theta}) \in L_q$ and $f \in H_q(0)$ is arbitrary, the linear form

$$\Lambda g = \int_{-\pi}^{\pi} \{L(e^{i\theta}) + f(e^{i\theta})\}g(e^{i\theta})\,d\theta,$$

defined for all $g \in H_p$, *does not depend on* $f \in H_q(0)$. So $L + H_q(0)$ *is* a bounded linear functional on H_p.

Conversely, take any linear functional Λ on H_p. By Hahn–Banach, we can first extend Λ to *all* of L_p, thus getting an $L \in L_q$ with

$$\Lambda g = \int_{-\pi}^{\pi} L(e^{i\theta})g(e^{i\theta})\,d\theta, \quad g \in L_p.$$

Restricted back to H_p, Λ is given by the coset $L + H_q(0)$ in the above way. Q.E.D.

Theorem *The dual of*

$$L_1/H_1 \quad \text{is} \quad H_\infty(0).$$

The dual of

$$H_1 \quad \text{is} \quad L_\infty/H_\infty(0).$$

Proof By the same argument used above.

Now, however, comes the

Theorem *The dual of*

$$\mathscr{C}/\mathscr{A} \quad \text{is} \quad H_1(0).$$

Proof Since the dual of $\mathscr{C}$ is $\mathscr{M}$, the dual of $\mathscr{C}/\mathscr{A}$ is

$$\{\mu \in \mathscr{M};\ \int_{-\pi}^{\pi} f(e^{i\theta})\,d\mu(\theta) = 0 \quad \text{for } f \in \mathscr{A}\}.$$

In particular, for μ to be in the dual of $\mathscr{C}/\mathscr{A}$, we must have

$$\int_{-\pi}^{\pi} e^{in\theta}\,d\mu(\theta) = 0, \quad n = 0, 1, 2, \ldots$$

Therefore, by the theorem of the brothers Riesz, $d\mu(\theta) = g(e^{i\theta})\,d\theta$ with a $g \in L_1(-\pi, \pi)$, and we easily check that $g \in H_1(0)$.

Conversely, any $g \in H_1(0)$ *clearly defines* a functional Λ on $\mathscr{C}/\mathscr{A}$ by putting, for $\phi \in \mathscr{C}$,

$$\Lambda(\phi + \mathscr{A}) = \int_{-\pi}^{\pi} g(e^{i\theta})\phi(e^{i\theta})\,d\theta.$$

We are done.

Remark Using an obvious extension of the above notation, write

$$\mathscr{A}(0) = \{e^{i\theta} f(e^{i\theta});\ f \in \mathscr{A}\}.$$

Then, by the same argument as used in the above proof, we see that $\mathscr{C}/\mathscr{A}(0)$ has the dual H_1. Thus:

$\mathscr{C}/\mathscr{A}(0)$ *has dual* H_1 *which has dual* $L_\infty/H_\infty(0)$.

In particular, H_1 *is a dual, whereas the larger space* L_1 *is not*. This has an important consequence — *the unit sphere in* H_1 is (sequentially) w* *compact over* $\mathscr{C}$ for instance, whilst the *unit sphere* in L_1 has its w* (sequential) *closure equal to* $\mathscr{M}$ (measures!). H_1 also has *some other* properties resembling those of the (reflexive) spaces L_p, $1 < p < \infty$, rather than those of L_1. There is a paper of D. Newman in the *Proc. A.M.S.* from 1963 on that.

Here is a table summarising the above duality results, together with some others, which are established in exactly the same way:

For Unit Circle

Space	Dual
$H_p, \ 1 \le p < \infty$	$L_q/H_q(0), \ \dfrac{1}{q} = 1 - \dfrac{1}{p}$
$H_p(0), \ 1 \le p < \infty$	$L_q/H_q, \ \dfrac{1}{q} = 1 - \dfrac{1}{p}$
$L_p/H_p, \ 1 \le p < \infty$	$H_q(0), \ \dfrac{1}{q} = 1 - \dfrac{1}{p}$
$L_p/H_p(0), \ 1 \le p < \infty$	$H_q, \ \dfrac{1}{q} = 1 - \dfrac{1}{p}$
$\mathscr{C}/\mathscr{A}$	$H_1(0)$
$\mathscr{C}/\mathscr{A}(0)$	H_1

Very similar (and more symmetric) results hold in connection with the H_p spaces for the *upper half plane*, introduced in Chapter VI. Instead of $\mathscr{C}$ and $\mathscr{A}$, the relevant spaces are here

$$\mathscr{C}_0 = \{f \text{ continuous on } \mathbb{R}; \ F(x) \to 0 \text{ as } x \to \pm\infty\}$$

$$\mathscr{A}_0 = \mathscr{C}_0 \cap H_\infty(\Im z > 0);$$

$\mathscr{A}_0$ is the space of functions *analytic* in $\Im z > 0$, having a *continuous* extension up to $\mathbb{R}$, and *vanishing at* ∞ *in the closed upper half plane.*

Both $\mathscr{C}_0$ and $\mathscr{A}_0$ are equipped with the sup-norm.

Proofs of the following results, presented in tabular form, are very much like the corresponding ones for the unit circle.

For Upper Half Plane

Space	Dual
$H_p, \ 1 \le p < \infty$	$L_q/H_q, \ \dfrac{1}{q} = 1 - \dfrac{1}{p}$
$L_p/H_p, \ 1 \le p < \infty$	$H_q, \ \dfrac{1}{q} = 1 - \dfrac{1}{p}$
$\mathscr{C}_0/\mathscr{A}_0$	H_1

2. Duality method of Havinson and of Rogosinski–Shapiro

The above duality results give us some theorems about approximation by H_p functions. We mostly just give results for the *unit circle*; analogous ones hold for the upper half plane.

Theorem *Let $F \in L_p(-\pi, \pi)$, $1 < p < \infty$, and call*

$$\|F - H_p\|_p = \inf\{\|F - h\|_p; \ h \in H_p\}.$$

Then, with $1/q = 1 - (1/p)$,

(i) $$\|F - H_p\|_p = \sup\left\{ \left|\int_{-\pi}^{\pi} F(e^{i\theta})g(e^{i\theta})\,d\theta\right| ;\quad g \in H_q(0)\ \&\ \|g\|_q = 1\right\}.$$

(ii) *There is an* $h_0 \in H_p$ *with* $\|F - H_p\|_p = \|F - h_0\|_p$ *(i.e., the minimum is* attained*).*
(iii) *There is a* $g_0 \in H_q(0)$, $\|g_0\|_q = 1$, *with*

$$\|F - H_p\|_p = \int_{-\pi}^{\pi} F(e^{i\theta})g_0(e^{i\theta})\,d\theta,$$

(i.e., the sup is attained*).*

Proof (i) is a metric restatement of the duality between L_p/H_p and $H_q(0)$.

To prove (ii), let $h_n \in H_p$ with $\|F - h_n\|_p \underset{n}{\longrightarrow} \|F - H_p\|_p$. Then $\|h_n\|_p$ is *bounded*, so there is a *subsequence* $\{h_{n_j}\}$ *converging weakly* in L_p to a *limit* h_0; one easily checks that $h_0 \in H_p$. Then $F - h_{n_j} \underset{j}{\longrightarrow} F - h_0$ (weakly). From *this*, it easily follows that

$$\|F - h_0\|_p \le \liminf_{j\to\infty} \left\|F - h_{n_j}\right\|_p = \|F - H_p\|_p,$$

so $\|F - h_0\|_p = \|F - H_p\|_p$. For the beginner, *here is the proof*: take any $\epsilon > 0$, and take g, $g \in L_q$, $\|g\|_q = 1$ with

$$\|F - h_0\|_p \le \left|\int_{-\pi}^{\pi} (F(e^{i\theta}) - h_0(e^{i\theta}))g(e^{i\theta})\,d\theta\right| + \epsilon.$$

The expression on the right, *by weak convergence* of $F - h_{n_j}$ to $F - h_0$, is

$$\lim_{j\to\infty} \left|\int_{-\pi}^{\pi} (F(e^{i\theta}) - h_{n_j}(e^{i\theta}))g(e^{i\theta})\,d\theta\right| + \epsilon.$$

But

$$\left|\int_{-\pi}^{\pi} (F(e^{i\theta}) - h_{n_j}(e^{i\theta}))g(e^{i\theta})\,d\theta\right| \le \left\|F - h_{n_j}\right\|_p,$$

since $\|g\|_q = 1$.

By going to another subsequence, we get $\|F - h_0\|_p \le \liminf_{j\to\infty} \left\|F - h_{n_j}\right\|_p + \varepsilon$. Squeeze ε.

Now we must prove (iii). By the Hahn–Banach theorem, *there is a linear functional* Λ *of norm* 1 on L_p/H_p such that $\Lambda(F + H_p) = \|F - H_p\|_p$. By a previous theorem and its proof,

$$\Lambda(f + H_p) = \int_{-\pi}^{\pi} f(e^{i\theta})g_0(e^{i\theta})\,d\theta$$

for all $f \in L_p$ and some $g_0 \in H_q(0)$ of q-norm 1.

We are done.

Theorem *Let* $F \in L_1(-\pi, \pi)$. *Then there is an* $h_0 \in H_1$ *with*

$$\|F - H_1\|_1 = \|F - h_0\|_1,$$

and there is a $g_0 \in H_\infty(0)$ *with* $\|g_0\|_\infty = 1$ *and*

$$(F(e^{i\theta}) - h_0(e^{i\theta}))\, g_0(e^{i\theta}) = |F(e^{i\theta}) - h_0(e^{i\theta})|$$

almost everywhere.

Proof According to a previous remark, the unit sphere in H_1 *is* w* *compact over* $\mathscr{C}$, so we can show the *existence* of an $h_0 \in H_1$ which minimises $\|F - h_0\|_1$ as in the proof of the last theorem.

As in the proof of (iii) in that theorem, we get a $g_0 \in H_\infty(0)$ with $\|g_0\|_\infty = 1$ and

$$\|F - H_1\|_1 = \|F - h_0\|_1 = \int_{-\pi}^{\pi} F(e^{i\theta})g_0(e^{i\theta})\, d\theta = \int_{-\pi}^{\pi} (F(e^{i\theta}) - h_0(e^{i\theta}))g_0(e^{i\theta})\, d\theta.$$

Since $|g_0(e^{i\theta})| \leq 1$, *equality* of

$$\int_{-\pi}^{\pi} |F(e^{i\theta}) - h_0(e^{i\theta})|\, d\theta \quad \text{and} \quad \int_{-\pi}^{\pi} (F(e^{i\theta}) - h_0(e^{i\theta}))g_0(e^{i\theta})\, d\theta$$

shows that $g_0(e^{i\theta})$ does the job required of it.

Theorem *Let $F \in \mathscr{C}$. Then:*

(i) $\|F - \mathscr{A}\|_\infty = \|F - H_\infty\|_\infty = \sup\left\{\left|\int_{-\infty}^{\infty} F(e^{i\theta})g(e^{i\theta})\, d\theta\right|;\quad g \in H_1(0)\ \&\ \|g\|_1 = 1\right\}.$

(ii) *There is a $g_0 \in H_1(0)$, $\|g_0\|_1 = 1$, with*

$$\|F - \mathscr{A}\|_\infty = \int_{-\pi}^{\pi} F(e^{i\theta})g_{(}e^{i\theta})\, d\theta.$$

(iii) *There is an $h_0 \in H_\infty$ with*

$$|F(e^{i\theta}) - h_0(e^{i\theta})| = \|F - \mathscr{A}\|_\infty \quad \text{a.e.}$$

Proof $H_\infty \supset \mathscr{A}$ so surely $\|F - H_\infty\|_\infty \leq \|F - \mathscr{A}\|_\infty$. Since $H_1(0)$ is the dual of $\mathscr{C}/\mathscr{A}$, we get a $g_0 \in H_1(0)$, $\|g_0\|_1 = 1$, so that (ii) holds. But

$$\int_{-\pi}^{\pi} h(e^{i\theta})g_0(e^{i\theta})\, d\theta = 0$$

for $h \in H_\infty$, so clearly

$$\left|\int_{-\pi}^{\pi} F(e^{i\theta})g_0(e^{i\theta})\, d\theta\right| \leq \|g_0\|_1\, \|F - H_\infty\|_\infty,$$

proving, by choice of g_0, that $\|F - \mathscr{A}\|_\infty \leq \|F - H_\infty\|_\infty$, yielding (i).

Now H_∞ is a *dual* (of $L_1/H_1(0)$), so *by the argument used above* (w* compactness here), there *is* an $h_0 \in H_\infty$ with $\|F - H_\infty\|_\infty = \|F - h_0\|_\infty$. Then

$$\|F - h_0\|_\infty = \|F - H_\infty\|_\infty = \|F - \mathscr{A}\|_\infty = \int_{-\pi}^{\pi} F(e^{i\theta})g_0(e^{i\theta})\, d\theta$$

$$= \int_{-\pi}^{\pi} [F(e^{i\theta}) - h_0(e^{i\theta})]g_0(e^{i\theta})\, d\theta.$$

That is, since

$$\int_{-\pi}^{\pi} |g_0(e^{i\theta})|\, d\theta = 1,$$

$$\int_{-\pi}^{\pi} [F(e^{i\theta}) - h_0(e^{i\theta})]g_0(e^{i\theta})\, d\theta = \|F - h_0\|_\infty \int_{-\pi}^{\pi} |g_0(e^{i\theta})|\, d\theta \quad !$$

So $|F(e^{i\theta}) - h_0(e^{i\theta})| = \|F - h_0\|_\infty$ a.e. *on the support* of g_0. *Since* $g_0 \in H_1(0)$, $|g_1(e^{i\theta})| > 0$ a.e. by an old theorem (in Chapter III, Section B). So $|F(e^{i\theta}) - h_0(e^{i\theta})| \equiv \|F - h_0\|_\infty$ a.e.

Q.E.D.

Scholium If $F \in \mathscr{C}$, there is *only one* $h \in H_\infty$ with $\|F - h\|_\infty = \|F - H_\infty\|_\infty$. For, *suppose there were two*, say, h_1 and h_2. *Take* the $g_0 \in H_1(0)$ of which the last theorem affirms the existence. Then we get as in the above proof

$$\int_{-\pi}^{\pi} (F(e^{i\theta}) - h_k(e^{i\theta}))g_0(e^{i\theta})\, d\theta = \|F - h_k\|_\infty \int_{-\pi}^{\pi} |g_0(e^{i\theta})|\, d\theta$$

with $k = 1$ and 2 so, *since* $|g_0(e^{i\theta})| > 0$ a.e., we *must have*

$$F(e^{i\theta}) - h_1(e^{i\theta}) = \|F - h_1\|_\infty \frac{|g_0(e^{i\theta})|}{g_0(e^{i\theta})} \quad \text{a.e.,}$$

$$F(e^{i\theta}) - h_2(e^{i\theta}) = \|F - h_2\|_\infty \frac{|g_0(e^{i\theta})|}{g_0(e^{i\theta})} \quad \text{a.e.,}$$

or, since

$$\|F - h_1\|_\infty = \|F - h_2\|_\infty = \|F - H_\infty\|_\infty,$$

$$h_1(e^{i\theta}) = F(e^{i\theta}) - \|F - H_\infty\|_\infty \frac{|g_0(e^{i\theta})|}{g_0(e^{i\theta})} = h_2(e^{i\theta}) \quad \text{a.e.,}$$

as required.

If $F \notin \mathscr{C}$, all we have is the

Theorem *Let $F \in L_\infty$. Then there is an $h_0 \in H_\infty$ with*

$$\|F - h_0\|_\infty = \|F - H_\infty\|_\infty = \sup\left\{\left|\int_{-\pi}^{\pi} F(e^{i\theta})g(e^{i\theta})\, d\theta\right| ; \quad g \in H_1(0) \;\&\; \|g\|_1 = 1\right\}.$$

Proof Existence of h_0 follows by the w* compactness of H_∞, and the rest by previous duality results.

Remark In general, the sup in the above theorem is *not attained* if $F \in H_\infty$ is *not continuous*. The work in the above scholium shows that *it cannot be attained* (by a $g_0 \in H_1(0)$) with $\|g_0\|_1 = 1$) if $F - H_\infty$ contains *more than one element* of norm *equal* to $\|F - H_\infty\|_\infty$, i.e., if there is *more than one* $h \in H_\infty$ which *minimizes* $\|F - h\|_\infty$. Cases where there are *several* $h \in H_\infty$ *minimizing* $\|F - h\|_\infty$ are *of practical importance* in various questions.

Historical remark The above (duality) approach is due to Rogosinski and H. S. Shapiro and, independently, to Havinson. It has been often rediscovered by various people who *didn't know* about prior work (e.g., by me!). There is a historical note by A. Shields in the *A.M.S. Translations*, Ser. 2, Vol. 32, 1963.

These results are obviously of great importance in studying approximation problems of various kinds, and continue to have many applications.

If $F \in H_\infty$ is *not continuous*, the sup in the above theorem *may not be attained even though there is only one $h \in H_\infty$ minimizing $\|F - h\|_\infty$.*

Example *Let E_1 and E_2 be two disjoint sets of positive measure on $\{|z| = 1\}$ adding up to $\{|z| = 1\}$, and let*

$$F(e^{i\theta}) = \begin{cases} 1 & \text{on } E_1 \\ -1 & \text{on } E_2. \end{cases}$$

Then there is no nonzero $h \in H_\infty$ with $\|F - h\|_\infty \le 1$. Neither is there any $g_0 \in H_1(0)$ with

$$\int_{-\pi}^{\pi} F(e^{i\theta})g_0(e^{i\theta})\, d\theta = \|g_0\|_1 = 1.$$

(E_1 and E_2 can even be two complementary arcs, so that F has only two points of discontinuity of very simple form.)

Proof There is no such g_0. *Suppose there were.* Then we'd have to have

$$g_0(e^{i\theta}) > 0 \quad \text{a.e. on } E_1$$

$$g_0(e^{i\theta}) < 0 \quad \text{a.e. on } E_2,$$

so $g_0(e^{i\theta})$ *is real* a.e. So since $g_0 \in H_1$, a generalization of the Schwarz reflection principle given in Chapter III shows that, if we put

$$g_0(z) = \overline{g_0\left(\frac{z}{|z|^2}\right)}$$

for $|z| > 1$, $g_0(z)$ *becomes regular everywhere in* $\mathbb{C}$ (even on $|z| = 1$). Also, $g_0(0) = 0$ makes $g_0(\infty) = 0$, so $g_0(z)$ is *bounded everywhere, hence constant, and the constant must be zero*, so $g_0 \equiv 0$. (*Another* way of seeing this is to observe that $\Im g_0(z)$ must *vanish identically*, so $g_0(z)$ must be *constant* and hence equal to 0.)

Suppose now that $h \not\equiv 0$, $h \in H_\infty$ and $|F(e^{i\theta}) - h(e^{i\theta})| \le 1$ a.e.

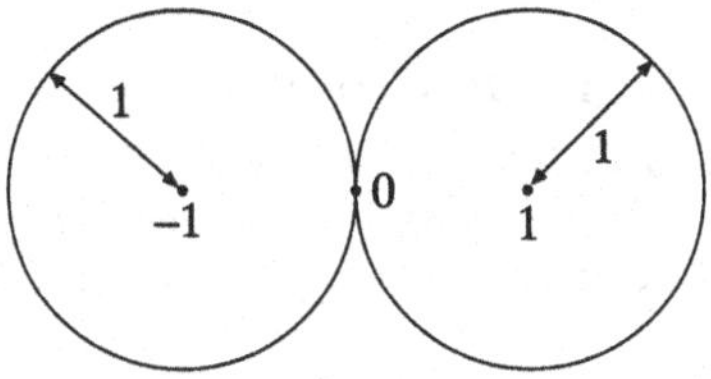

Then, *for* $e^{i\theta} \in E_1$, $h(e^{i\theta})$ *is within the (closed) right hand circle*, and *for* $e^{i\theta} \in E_2$, *within the (closed) left hand one*. By a well-known theorem (Chapter III, Section B), $h(e^{i\theta}) = 0$ only on a set of measure zero, so that $h(e^{i\theta})$ takes some values *really in* the right hand circle (away from 0) and *really in the left hand circle.*

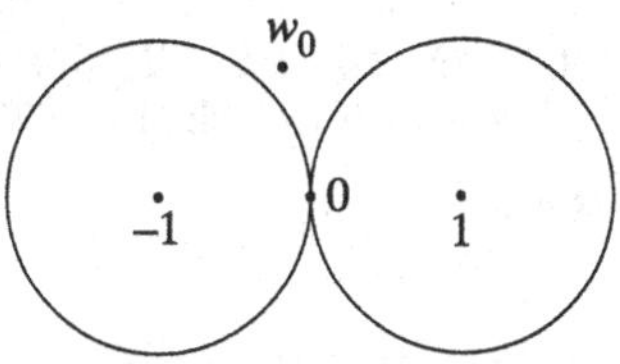

All the values of $h(z)$, for $|z| < 1$, *are in these two circles. Indeed*, let w_0 be *outside* them, and *suppose* $h(z_0) = w_0$. By Runge's theorem, we can get a polynomial $P(w)$ with $|P(w)| \le 1$ if $|w - 1| \le 1$ or $|w + 1| \le 1$ but $P(w_0) = 2$.

$P(h(z)) \in H_\infty$, and by construction $|P(h(e^{i\theta}))| \le 1$. Therefore *by the Poisson representation for H_∞ functions*, $|P(h(z_0))| \le 1$. But $h(z_0) = w_0$ and $P(w_0) = 2$, yielding a contradiction. So $h(z)$ takes $\{|z| < 1\}$ into the shaded region,

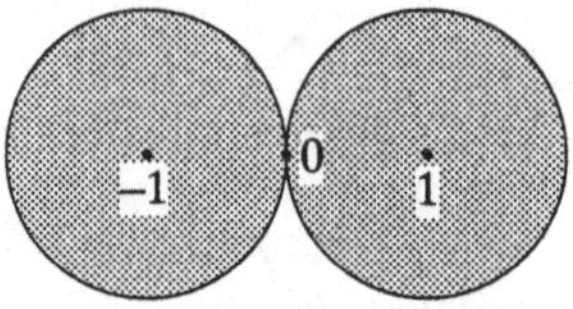

and since it has *some* boundary values *to the right* of 0 and *some* boundary values *to the left* of 0, it must, in $\{|z| < 1\}$, take values in *each* of the two circles. Therefore, *by connectivity*, there must be a z_0, $|z_0| < 1$, with $h(z_0) = 0$. By the *principle of conservation of domain*, $h(z)$ takes values filling out a neighbourhood of 0 for $\{|z| < 1\}$. *But it doesn't!*

So $h \not\equiv 0$ cannot exist, and *we are done.*

3. Sarason's theorem

Let us go back to the chain

$$\mathscr{C}/\mathscr{A}(0) \text{ has dual } H_1 \text{ which has dual } L_\infty/H_\infty(0).$$

Thus, if B is the Banach space $\mathscr{C}/\mathscr{A}(0)$, B^{**} is $L_\infty/H_\infty(0)$. Now B has a canonical isometric image in B^{**} obtained by *identifying linear functionals over* B^*. In the present case, the element of $L_\infty/H_\infty(0)$ corresponding to $F + \mathscr{A}(0)$, $F \in \mathscr{C}$, is the coset $\Phi + H_\infty(0)$, $\Phi \in L_\infty$, determined by

$$\int_{-\pi}^{\pi} F(e^{i\theta})g(e^{i\theta})\,d\theta = \int_{-\pi}^{\pi} g(e^{i\theta})\Phi(e^{i\theta})\,d\theta$$

for *all* $g \in H_1$. We see that this holds iff $\Phi \in F + H_\infty(0)$, which is to say that in the canonical embedding of $\mathscr{C}/\mathscr{A}(0)$ in $L_\infty/H_\infty(0)$, $F + \mathscr{A}(0)$, $F \in \mathscr{C}$, *corresponds* to $F + H_\infty(0)$. The *image* of $\mathscr{C}/\mathscr{A}(0)$ in $L_\infty/H_\infty(0)$ under this embedding is thus

$$\mathscr{E} = \{F + H_\infty(0);\ F \in \mathscr{C}\}.$$

In particular, $\mathscr{E}$ *is* $\|\ \|_\infty$*-closed in the quotient space* $L_\infty/H_\infty(0)$.

Now the canonical homomorphism $\Theta : L_\infty \mapsto L_\infty/H_\infty(0)$ *is continuous. Therefore* $\Theta^{-1}(\mathscr{E})$ is $\|\ \|_\infty$-closed in L_∞. But $\Theta^{-1}(\mathscr{E})$ is $\mathscr{C} + H_\infty(0)$ which also $= \mathscr{C} + H_\infty$ since $1 \in \mathscr{C}$. Therefore:

Theorem (Sarason) $\mathscr{C} + H_\infty$ *is* $\|\ \|_\infty$*-closed.*

From this we can prove:

Theorem (Sarason) *If* F *and* $G \in \mathscr{C} + H_\infty$, $FG \in \mathscr{C} + H_\infty$, *i.e.,* $\mathscr{C} + H_\infty$ *is an* algebra.

Proof It is enough to show that if $F \in \mathscr{C}$ and $G \in H_\infty$, $FG \in \mathscr{C} + H_\infty$. Let the $F_N(e^{i\theta})$ be of the form

$$\sum_{-N}^{N} A_k(N)e^{ik\theta}$$

such that $F_N \to F$ *uniformly*, then $F_N G \to FG$ *uniformly*, so if each $F_N G \in \mathscr{C} + H_\infty$, $FG \in \mathscr{C} + H_\infty$ by the above theorem.

But if

$$G(z) = \sum_{0}^{\infty} a_n z^n \in H_\infty,$$

clearly $G_N(z) = \sum_{N}^{\infty} a_n z^n \in H_\infty$, and

$$F_N(e^{i\theta})G_N(e^{i\theta}) = \sum_{-N}^{N} A_n(N)e^{in\theta}G_N(e^{i\theta}) \in H_\infty$$

because $e^{in\theta}G_N(e^{i\theta}) \in H_\infty$ for $n = -N, -N+1, -N+2, \ldots$. *Finally*, $F_N(G - G_N)$, *a trigonometric polynomial*, is in $\mathscr{C}$. So $F_N G = F_N(G - G_N) + F_N G_N \in \mathscr{C} + H_\infty$, and we're done.

This proof depends on the fact that H_1 is the dual of $\mathscr{C}/\mathscr{A}(0)$ and thus, *ultimately*, on the F. and M. Riesz theorem. As Sarason first showed, *norm closure of* $\mathscr{C} + H_\infty$ *is independent of the F. and M. Riesz theorem*. (In fact, Sarason presented this in his 1973 *Bull. A.M.S.* survey, but the *idea itself* comes from Zalcman.) In a 1974 lecture at McGill University, *Walter Rudin* abstracted the Zalcman argument into a general theorem, worthy in its own right:

Theorem *Let* B, $\|\ \|$ *be a Banach space, with* E *and* F *norm-closed subspaces of* B. *Suppose there is a family* $\mathfrak{T}$ *of linear operators on* B *with the following properties:*

(i) $|||T||| \le M < \infty$ *for all* $T \in \mathfrak{T}$.

(ii) $TB \subseteq E$ *for each* $T \in \mathfrak{T}$.

(iii) $TF \subseteq F$ *for each* $T \in \mathfrak{T}$.

(iv) *If* $u \in E$ *and* $\epsilon > 0$, *there is a* $T \in \mathfrak{T}$ *with* $\|Tu - u\| < \epsilon$.

Then $E + F$ *is norm-closed in* B.

Proof Let $x \in$ norm closure of $E+F$. Then we can find $u_n \in E$, $v_n \in F$, with $\|u_n + v_n\| \le 2^{-n}$ for $n \ge 2$ and

$$x = \sum_{n=1}^{\infty} (u_n + v_n).$$

Write $x_n = u_n + v_n$. Then $x = \sum_{n=1}^{\infty} x_n$, and we have

$$x_n = (u_n - T_n u_n + T_n x_n) + (v_n - T_n v_n)$$

where, for each n, $T_n \in \mathfrak{T}$ is chosen so that $\|u_n - T_n u_n\| < 2^{-n}$. Now $\tilde{u}_n = u_n - T_n u_n + T_n x_n \in E$ and $\|\tilde{u}_n\| \le 2^{-n} + \|T_n x_n\| \le 2^{-n}(1+M)$ for $n \ge 2$, since $\|x_n\| \le 2^{-n}$ for $n \ge 2$. Also, $v_n - T_n v_n = \tilde{v}_n \in F$, and $\|\tilde{v}_n\| \le \|\tilde{u}_n\| + \|x_n\| \le (2+M)2^{-n}$ for $n \ge 2$. *Therefore* $\sum_1^\infty \tilde{u}_n$ converges, to, say, $u \in E$ because E is closed, and $\sum_1^\infty \tilde{v}_n$ converges to $v \in F$ because F is closed. So

$$x = \sum_{1}^{\infty} x_n = \sum_{1}^{\infty} (\tilde{u}_n + \tilde{v}_n) = u + v \in E + F. \qquad \text{Q.E.D.}$$

Corollary $\mathscr{C} + H_\infty$ *is* $\|\ \|_\infty$*-closed.*

Proof Take $B = L_\infty$, $E = \mathscr{C}$, $F = H_\infty$, and let $\mathfrak{T} = \{T_N\ ;\ N = 1,2,\ldots\}$ where, if $F \in L_\infty$ and $F(e^{i\theta}) \sim \sum_{-\infty}^{\infty} A_n e^{in\theta}$,

$$(T_N F)(e^{i\theta}) = \sum_{-N}^{N} \left(1 - \frac{|n|}{N}\right) A_n e^{in\theta}$$

(the Nth Fejer partial sum of the Fourier series of F). Then $|||T_N||| \le 1$ and for each $F \in \mathscr{C}$, $||T_N F - F||_\infty \underset{N}{\longrightarrow} 0$. Clearly $T_N L_\infty \subseteq \mathscr{C}$ and $T_N H_\infty \subseteq H_\infty$. We have the desired result.

Many other similar applications of the general theorem can be given.

B. Elements of constant modulus in cosets of L_∞/H_∞. Marshall's theorem

1. A result of Adamian, Arov and Krein

Given $F \in L_\infty$, we have seen in the previous section that under certain circumstances (e.g., *F continuous*), the coset $F + H_\infty$ contains an element of constant modulus equal to $||F - H_\infty||_\infty$.

We are interested in seeing the extent to which elements of constant modulus occur in $F+H_\infty$. Around 1920, Nevanlinna proved some deep results about this, using very "hard" methods. Complete and definitive results were arrived at some 40 years later by Adamian, Arov and Krein with the help of operator theory, and Garnett found functional-analytic proofs of them towards the end of the 1970s. Here is one:

Theorem *If $F \in L_\infty$ and $||F - H_\infty||_\infty < 1$, $F + H_\infty$ contains an element ω with $|\omega(e^{i\theta})| \equiv 1$ a.e.*

Proof (Garnett) The idea is to look for the $\omega \in F + H_\infty$ with $||\omega||_\infty \le 1$ which *maximises*

$$\left| \int_{-\pi}^{\pi} \omega(e^{i\theta})\, d\theta \right|$$

and show that such an ω does the job.

Let

$$a = \sup\left\{ \left| \int_{-\pi}^{\pi} \omega(e^{i\theta})\, d\theta \right| ;\ \omega \in F + H_\infty,\ ||\omega||_\infty \le 1 \right\}.$$

There is indeed an $\omega \in F + H_\infty$, $||\omega||_\infty \le 1$, with

$$\left| \int_{-\pi}^{\pi} \omega(e^{i\theta})\, d\theta \right| = a.$$

For if we take $\omega_n \in F + H_\infty$, $||\omega_n||_\infty \le 1$, with

$$\left| \int_{-\pi}^{\pi} \omega_n(e^{i\theta})\, d\theta \right| \underset{n}{\longrightarrow} a,$$

we can let ω be a w* limit (in L_∞) of some w* *convergent subsequence* of the ω_n, and then $||\omega||_\infty \le 1$ while

$$\left| \int_{-\pi}^{\pi} \omega(e^{i\theta})\, d\theta \right| = a$$

since $1 \in L_1$!

We shall henceforth suppose that

$$\int_{-\pi}^{\pi} \omega(e^{i\theta})\,d\theta = a,$$

which is no restriction, since we can attain this by working with $e^{i\gamma}F$ instead of F, where γ is a real constant.

Now *firstly*, $\|\omega\|_\infty = 1$. For if $\|\omega\|_\infty = 1-\epsilon$ with $\epsilon > 0$, $\omega + \epsilon \in \omega + H_\infty = F + H_\infty$ (N.B. $\omega \in F + H_\infty$ because the above ω_n are, and H_∞ is w* closed!), $\|\omega + \epsilon\|_\infty \le 1$, and

$$\int_{-\pi}^{\pi} [\omega(e^{i\theta}) + \epsilon]\,d\theta = a + 2\pi\epsilon > a,$$

a contradiction to the choice of ω.

Secondly, $\|\omega - H_\infty(0)\|_\infty = 1$. For otherwise $\|\omega - H_\infty(0)\|_\infty < 1$, and then there is an $h \in H_\infty(0)$ with $\|\omega - h\|_\infty = 1-\epsilon$, $\epsilon > 0$. Then $-h+\epsilon \in H_\infty$, so

$$\omega - h + \epsilon \in F + H_\infty, \quad \|\omega - h + \epsilon\|_\infty \le 1,$$

and because $h \in H_\infty(0)$,

$$\int_{-\pi}^{\pi} (\omega(e^{i\theta}) - h(e^{i\theta}) + \epsilon)\,d\theta = \int_{-\pi}^{\pi} \omega(e^{i\theta})\,d\theta + 2\pi\epsilon = a + 2\pi\epsilon > a,$$

again contradicting the choice of $\omega \in F + H_\infty$.

Now, since $\|\omega - H_\infty(0)\|_\infty = 1$, by the above duality theorems, there is a sequence of $f_n \in H_1$, $\|f_n\|_1 = 1$, with

$$\int_{-\pi}^{\pi} \omega(e^{i\theta}) f_n(e^{i\theta})\,d\theta \underset{n}{\longrightarrow} 1.$$

We must eventually have $|f_n(0)| \ge c$ *for some* $c > 0$. *Indeed, if not*, then, without loss of generality, $f_n(0) = c_n$ with $c_n \underset{n}{\longrightarrow} 0$. Then $f_n - c_n \in H_1(0)$, $\|f_n - c_n\|_1 \underset{n}{\longrightarrow} 1$, whilst

$$\int_{-\pi}^{\pi} \omega(e^{i\theta})(f_n(e^{i\theta}) - c_n)\,d\theta \quad \textit{also} \underset{n}{\longrightarrow} 1.$$

Therefore, by the above duality theorems, $\|\omega - H_\infty\|_\infty \ge 1$, i.e., $\|F - H_\infty\|_\infty \ge 1$ since $\omega \in F + H_\infty$, *contradicting our hypothesis* that $\|F - H_\infty\|_\infty < 1$. So $|f_n(0)| \ge c > 0$.

We now show that this last inequality implies $|\omega(e^{i\theta})| \equiv 1$ a.e. *Assume not*. Then, for some $\lambda < 1$ there is a measurable E, $|E| > 0$, with $|\omega(e^{i\theta})| \le \lambda$ on E. Without loss of generality, $|E| < 2\pi$ so that also $|\sim E| > 0$. Since in any case $|\omega(e^{i\theta})| \le 1$, we have, because $\|f_n\|_1 = 1$,

$$\left|\int_{-\pi}^{\pi} \omega(e^{i\theta}) f_n(e^{i\theta})\,d\theta\right| \le \lambda \cdot \int_E |f_n(e^{i\theta})|\,d\theta + \left(1 - \int_E |f_n(e^{i\theta})|\,d\theta\right),$$

and in order for the *left hand side* to $\underset{n}{\longrightarrow} 1$ we *must have*

$$\int_E |f_n(e^{i\theta})|\,d\theta \underset{n}{\longrightarrow} 0.$$

Therefore

$$\frac{1}{|E|}\int_E \log|f_n(e^{i\theta})|\,d\theta \le \log\left(\frac{1}{|E|}\int_E |f_n(e^{i\theta})|\,d\theta\right) \underset{n}{\longrightarrow} -\infty.$$

At the same time,

$$\frac{1}{|\sim E|}\int_{\sim E} \log|f_n(e^{i\theta})|\,d\theta \le \log\left(\frac{\|f_n\|_1}{|\sim E|}\right) = \log\left(\frac{1}{|\sim E|}\right) < \infty,$$

so finally

$$\int_{-\pi}^{\pi} \log|f_n(e^{i\theta})|\,d\theta \underset{n}{\longrightarrow} -\infty.$$

Therefore the *outer factors* of the f_n *already* tend to 0 at the origin, so surely $|f_n(0)| \underset{n}{\longrightarrow} 0$. This contradicts $|f_n(0)| \ge c > 0$ just proven, so we must have $|\omega(e^{i\theta})| \equiv 1$ a.e., Q.E.D.

Corollary *Let $f \in H_\infty$, let $\|f\|_\infty < 1$, and let Ω be any* inner function. *Then there is another* inner function, $\omega \in f + \Omega H_\infty$.

Proof Apply the above theorem with $F(e^{i\theta}) = f(e^{i\theta})/\Omega(e^{i\theta})$. Then $\|F\|_\infty < 1$, so $\|F - H_\infty\|_\infty < 1$, and there is an $h \in H_\infty$ with $|F(e^{i\theta}) + h(e^{i\theta})| \equiv 1$ a.e. Then $f + \Omega h \in H_\infty$ and $|f(e^{i\theta}) + \Omega(e^{i\theta})h(e^{i\theta})| \equiv 1$ a.e., so $f + \Omega h$ is also an inner function.

Remark Adamian, Arov and Krein also proved that if $F \in L_\infty$ and $\|F - H_\infty\|_\infty = 1$ but $F - H_\infty$ contains *more than one element of norm* 1, then *there is* an $\omega \in F - H_\infty$ with $|\omega(e^{i\theta})| \equiv 1$ a.e. Garnett has a functional-analytic proof of *this* fact also, but we do not give it in this book.

2. Marshall's theorem

Around 1975, D. Marshall verified a long-standing conjecture about the uniform approximation of H_∞ functions by linear combinations of *Blaschke products*. In order to situate his result, let us first establish a much easier and very well-known analogous proposition about L_∞.

Theorem *Let $f \in L_\infty(-\pi,\pi)$ and $\|f\|_\infty \le 1$. Then, given any $\epsilon > 0$ we can find $u_1, \ldots u_n \in L_\infty$ with $|u_k(\theta)| \equiv 1$ a.e. (so-called* unimodular functions*) and numbers $\lambda_1, \ldots \lambda_n \ge 0$, $\sum_k \lambda_k = 1$, with*

$$\left\|f - \sum_k \lambda_k u_k\right\|_\infty < \epsilon.$$

In other words:

The norm-closed convex hull of the set of unimodular functions in L_∞ is precisely the unit sphere (ball) of L_∞.

Proof f can first be uniformly approximated by a measurable function of norm ≤ 1 *taking only a finite number of values*. This *new* function can then be uniformly approximated by convex linear combinations of unimodular functions, as *elementary duality theory* (separation theorems for convex bodies) for *finite dimensional spaces* shows.

There is *another* proof, which is more instructive here. Without loss of generality, $\|f\|_\infty \le 1 - (\epsilon/2)$, say; otherwise work with $(1 - (\epsilon/2))f$ instead of f and observe that $\|f - (1 - (\epsilon/2))f\|_\infty \le \epsilon/2$. Then we may assume $|f(\theta)| \le 1 - (\epsilon/2)$ *everywhere*, so, by *Cauchy's theorem* (!),

$$f(\theta) = \frac{1}{2\pi}\int_{-\pi}^{\pi} \frac{e^{it} + f(\theta)}{1 + \overline{f(\theta)}e^{it}}\,dt.$$

Now each of the functions

$$u_t = \frac{e^{it} + f(\theta)}{1 + \overline{f(\theta)}e^{it}}$$

is in L_∞, and $|u_t(\theta)| \equiv 1$! Also, since $\|f\| \le 1 - (\epsilon/2)$, we have

$$\|u_t - u_{t'}\|_\infty < \frac{\epsilon}{2}$$

if $|t - t'|$ is less than some positive number δ depending on ϵ. So for large N,

$$\left\| \frac{1}{2\pi}\int_{-\pi}^{\pi} u_t\, dt \; - \; \frac{1}{N}\sum_{k=1}^{N} u_{2\pi k/N} \right\|_\infty < \frac{\epsilon}{2}.$$

Therefore

$$\left\| f \; - \; \frac{1}{N}\sum_{k=1}^{N} u_{2\pi k/N} \right\|_\infty < \frac{\epsilon}{2},$$

and we're done.

The unimodular functions in H_∞ are just the *inner functions*, i.e., those having modulus 1 a.e. on the unit circumference. (See Chapter IV, Sections D, E, F and G.) It is very natural to conjecture that, in H_∞, the *norm-closed convex hull of the set of inner functions is the unit sphere of* H_∞. *Marshall* proved this.

Lemma (Douglas and Rudin) *Let* $u \in L_\infty$, $|u(\zeta)| \equiv 1$ *a.e. for* $|\zeta| = 1$. *Then there are* inner functions ω *and* Ω *in* H_∞ *with*

$$\left| u(\zeta) \; - \; \frac{\omega(\zeta)}{\Omega(\zeta)} \right| < \epsilon \quad \text{a.e.,} \quad |\zeta| = 1.$$

Proof (Marshall) If, for $k = 1, \ldots, N$, E_k denotes the subset of $\{|\zeta| = 1\}$ where

$$\frac{2\pi}{N}(k-1) \le \arg u(\zeta) < \frac{2\pi}{N}k,$$

the E_k are disjoint and add up to the unit circumference. Put

$$u_k(\zeta) = \begin{cases} e^{2\pi k i/N}, & \zeta \in E_k \\ 1, & \zeta \notin E_k, \; |\zeta| = 1. \end{cases}$$

Then each u_k is unimodular and *takes only two values*, and

$$\|u \; - \; u_1 \cdot u_2 \cdots u_N\|_\infty \; \le \; \frac{2\pi}{N}.$$

So it *clearly suffices* to establish the result for *each* factor u_k, i.e., for *unimodular functions taking only two values*, because the *product* of inner functions is surely an inner function.

Thus, let the unimodular function u take only two values, say 1 and γ, $|\gamma| = 1$, $\gamma \ne 1$, with

$$u(\zeta) = 1, \quad \zeta \in E$$

$$u(\zeta) = \gamma, \quad \zeta \in \sim E \quad \text{(complement in unit circumference).}$$

Using Poisson's formula, construct $V(z)$ bounded and harmonic in $\{|z| < 1\}$, having $\angle$ boundary values

$$V(\zeta) = 0 \; \text{ a.e. on } E$$

$$V(\zeta) = -K \; \text{ a.e. on} \sim E.$$

Then $h = \exp(V + i\tilde{V})$ is analytic and bounded in $|z| < 1$, and there takes values in the open ring $e^{-K} < |h| < 1$. On E, $|h(\zeta)| = 1$ a.e., and on $\sim E$, $|h(\zeta)| = e^{-K}$ a.e.

Let Φ_K be a conformal mapping of the ring $e^{-K} < |w| < 1$ onto the infinite domain (including ∞) obtained from $\mathbb{C} \cup \{\infty\}$ by removing therefrom the two segments $[-\epsilon, 0]$ and $[l, l']$, with

$$l = i\frac{1-\gamma}{1+\gamma}$$

and $l' > l$. We furthermore take Φ_K so as to map $|w| = 1$ onto $[-\epsilon, 0]$ and $|w| = e^{-K}$ onto $[l, l']$.

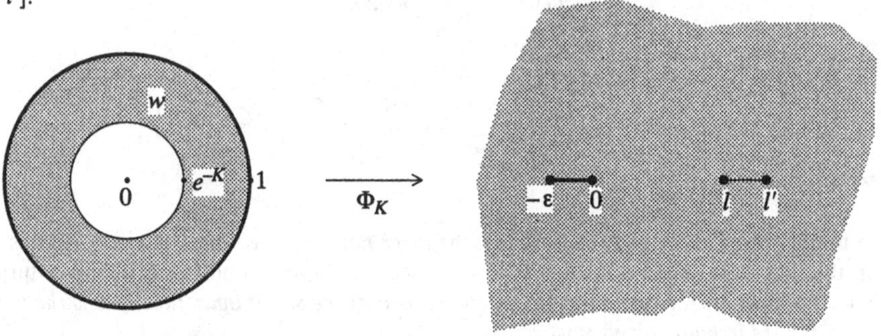

Here, $\epsilon > 0$ can be *chosen* as small as we like, but then $l' > l$ *is not free*, but *depends* on the *radius* e^{-K} *of the smaller circle*. It is, however, *true* that $l' \to l$ as $K \to \infty$, for in that limit Φ_K clearly tends to a conformal mapping of $\{|w| < 1\}$ onto $(\mathbb{C} \cup \{\infty\}) \sim [-\epsilon, 0]$ which takes 0 to l. Given $\epsilon > 0$, let us therefore *fix* K so large that $l' < l + \epsilon$.

Having determined K and Φ_K in the manner described, put

$$\Psi(w) = \frac{i - \Phi_K(w)}{i + \Phi_K(w)}$$

for $e^{-K} \le |w| \le 1$. Ψ takes the ring $e^{-K} < |w| < 1$ conformally onto the *complement*, in $\mathbb{C} \cup \{\infty\}$, of the two arcs

$$\widehat{\frac{i+\epsilon}{i-\epsilon},\, 1} \quad \text{and} \quad \widehat{\gamma, \gamma'}$$

lying on the unit circle.

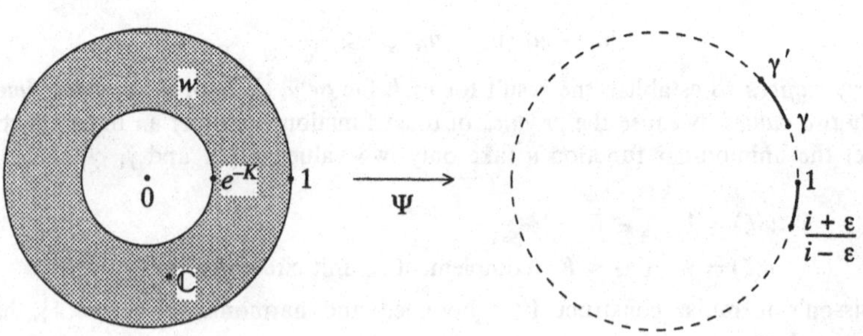

Here,

$$\gamma' = \frac{i - l'}{i + l'}$$

is close to

$$\gamma = \frac{i - l}{i + l}$$

by choice of K; in fact, *each of the two arcs*

$$\widehat{\frac{i+\epsilon}{i-\epsilon},\ 1} \quad \text{and} \quad \widehat{\gamma,\ \gamma'}$$

has diameter $< 2\epsilon$.

By construction of the function h we now see that $\Psi(h(\zeta))$ *lies on the arc*

$$\widehat{\frac{i+\epsilon}{i-\epsilon},\ 1}$$

for almost all ζ *in* E, and *on the arc*

$$\widehat{\gamma,\ \gamma'}$$

for almost all ζ *in* $\sim E$. *Therefore* $|\Psi(h(\zeta)) - u(\zeta)| < 2\epsilon$ a.e. on $|\zeta| = 1$, and $\Psi(h(\zeta))$ is *unimodular*, being a.e. in modulus equal to 1 there.

$\Psi(h(z))$ is *meromorphic* for $|z| < 1$. Indeed, Ψ being *conformal*, there is precisely *one point*, say c, $e^{-K} < |c| < 1$, with $\Psi(c) = \infty$, and $\Psi(w)$ has *just a simple pole* at c. Elsewhere in $e^{-K} < |w| < 1$, $\Psi(w)$ *is regular*. So $\Psi(h(z))$ is *regular* at the points z, $|z| < 1$, where $h(z) \neq c$, and has a *pole* at any z where $h(z) = c$, the *order* of this *pole* being the *order* of the *zero* that $h(z) - c$ has there.

The function $h(z) - c$ belongs to H_∞; by Chapter IV, Section D we can write

$$h(z) - c = \Omega(z)O(z)$$

with $\Omega(z)$ the *inner factor* of $h(z) - c$ and $O(z)$ its *outer factor*. We have:

$$|O(\zeta)| = |h(\zeta) - c| \geq 1 - |c| > 0 \quad \text{a.e. for } \zeta \text{ in } E,$$

$$|O(\zeta)| = |h(\zeta) - c| \geq |c| - e^{-K} > 0 \quad \text{a.e. for } \zeta \text{ in } \sim E.$$

Thus, $|O(\zeta)|$ is *bounded below* a.e. on $|\zeta| = 1$, so, by Chapter IV, Subsection E.2, $1/O(z) \in H_\infty$. In other words, $|\Omega(z)/(h(z) - c)|$ is *bounded above* in $|z| < 1$.

From what we have just seen, it follows easily that $\omega(z) = \Omega(z)\Psi(h(z))$ belongs to H_∞. This function is surely *analytic* in $|z| < 1$ because any *poles* of $\Psi(h(z))$ will be *cancelled* by corresponding *zeros* of $\Omega(z)$, the inner factor of $h(z) - c$. It is *bounded* in $|z| < 1$. Indeed, take any small $\delta > 0$; if $e^{-K} < |w| < 1$ and $|w - c| \geq \delta$, $\Psi(w)$ is *bounded*, hence $\Psi(h(z))$, and therefore $\Omega(z)\Psi(h(z))$, is *bounded* on the set of z where $|h(z) - c| \geq \delta$. For $|w - c| < \delta$, we have $|\Psi(w)| \leq A/|w - c|$, say, so if $|h(z) - c| < \delta$,

$$|\Omega(z)\Psi(h(z))| \ \leq\ A|\Omega(z)/(h(z) - c)|.$$

We have, however, just seen that the *expression* on the *right* is *bounded* in $|z| < 1$.

For almost all ζ, $|\zeta| = 1$, we now have

$$|\omega(\zeta)| = |\Omega(\zeta)|\,|\Psi(h(\zeta))| = 1.$$

Therefore $\omega \in H_\infty$ is *also an inner function*, like Ω. Since

$$\Psi(h(\zeta)) \ =\ \frac{\omega(\zeta)}{\Omega(\zeta)},$$

we have

$$\left|u(\zeta) \ -\ \frac{\omega(\zeta)}{\Omega(\zeta)}\right| \ =\ |u(\zeta) \ -\ \Psi(h(\zeta))| \ <\ 2\epsilon \quad \text{a.e.,} \quad |\zeta| = 1.$$

The lemma is proved with 2ϵ instead of ϵ.

Corollary *Let $f \in L_\infty$, $\|f\|_\infty \le 1$, and let $\epsilon > 0$. Then we can find* inner functions $\omega_1, \ldots, \omega_n$, $\Omega_1, \ldots, \Omega_n$, *and numbers* $\lambda_k > 0$, $\sum_1^n \lambda_k = 1$, *with*

$$\left| f(\zeta) - \sum_1^n \lambda_k \omega_k(\zeta)/\Omega_k(\zeta) \right| < \epsilon \quad \text{a.e.,} \quad |\zeta| = 1.$$

Proof Use the lemma and the first theorem of this subsection.

Remark *All the Ω_k can be taken equal* — just use a common denominator! For the product of inner functions is an inner function.

Lemma *Let $f \in H_\infty$. Then we can find* inner functions $\Omega, \omega, \omega_1, \ldots, \omega_n$ *and real constants* $a, a_1, \ldots, a_n$ *such that*

(i) $g = (a\omega + a_1\omega_1 + \ldots + a_n\omega_n)/\Omega$ *is in* H_∞

(ii) $\|f - g\|_\infty < 2\epsilon$.

Proof Without loss of generality, $\|f\|_\infty \le 1$. Then, by the preceding corollary and remark, we get inner functions $\omega_1, \ldots, \omega_n, \Omega$, and numbers $\lambda_k > 0$ with

$$\left| f(\zeta) - \sum_1^n \lambda_k \omega_k(\zeta)/\Omega(\zeta) \right| < \epsilon/2 \text{ a.e. for } |\zeta| = 1.$$

TRICK. Call

$$F(\zeta) = \sum_1^n \lambda_k \omega_k(\zeta)/\Omega(\zeta).$$

Since $f \in H_\infty$ (!), *the previous inequality shows that* $\|F - H_\infty\|_\infty < \epsilon$. *Therefore by the theorem in Subsection* 1, there is a $g \in H_\infty$ with $|g(\zeta) - F(\zeta)| \equiv \epsilon$ a.e.! Clearly

$$\|f - g\|_\infty \le \|f - F\|_\infty + \|F - g\|_\infty < 2\epsilon.$$

Also,

$$\Omega(\zeta)g(\zeta) - \Omega(\zeta)F(\zeta) \in H_\infty,$$

because

$$\Omega F = \sum_k \lambda_k \omega_k \in H_\infty.$$

So, since $|\Omega(\zeta)| \equiv 1$ a.e. for $|\zeta| = 1$, $\Omega(\zeta)g(\zeta) - \Omega(\zeta)F(\zeta)$ *must equal* $\epsilon\omega(\zeta)$ *with an inner function* ω, being in H_∞ *and of constant modulus* ϵ a.e. *on* $|\zeta| = 1$. So

$$g(\zeta) = \epsilon\frac{\omega(\zeta)}{\Omega(\zeta)} + \sum_1^n \lambda_k \frac{\omega_k(\zeta)}{\Omega(\zeta)}.$$

We are done.

Theorem (Marshall) *Let $f \in H_\infty$ and $\|f\|_\infty \le 1$. Given $\epsilon > 0$, we can find* inner functions, $u_1, \ldots, u_n$ *and* positive *numbers*, $\lambda_1, \ldots, \lambda_n$, $\sum_1^n \lambda_k = 1$, *with*

$$\left\| f - \sum_k \lambda_k u_k \right\|_\infty < 4\epsilon.$$

Proof Without loss of generality, we may suppose $\|f\|_\infty \le 1 - 2\epsilon$, otherwise we may work with $(1 - 2\epsilon)f$ instead of f. By the preceding lemma, we can find a $g \in H_\infty$ *of the very special form*

$$\sum_k a_k \omega_k / \Omega$$

with real constants a_k and *inner functions* ω_k, Ω, such that $\|f - g\|_\infty < \epsilon$. In particular, $\|g\|_\infty < 1 - \epsilon$. Now since, for $|\zeta| = 1$, we have, almost everywhere, $|\omega_k(\zeta)| = 1$, $|\Omega(\zeta)| = 1$, it follows that

$$\overline{g(\zeta)} = \sum_k a_k \frac{\Omega(\zeta)}{\omega_k(\zeta)} \quad \text{a.e.}$$

Let Ω_1 be the *product* of the ω_k — *it is an inner function.* Then, although $\overline{g(\zeta)}$ is *usually not* in H_∞, $\Omega_1(\zeta)\overline{g(\zeta)}$ *is!*

To functions $g \in H_\infty$ *having this property,* we can apply a *modification* of the Cauchy integral argument used to prove the *first* theorem of the present subsection.

$|g(\zeta)| \le 1 - \epsilon$ a.e., so, by Cauchy's theorem,

$$g(\zeta) = \frac{1}{2\pi} \int_{-\pi}^{\pi} \frac{\gamma e^{it} + g(\zeta)}{1 + \overline{g(\zeta)}\gamma e^{it}} \, dt,$$

γ *being any number of modulus* 1.

Now apply Bernard's trick, and take $\gamma = \Omega_1(\zeta)$!

Then we get

$$g(\zeta) = \frac{1}{2\pi} \int_{-\pi}^{\pi} \frac{\Omega_1(\zeta) e^{it} + g(\zeta)}{1 + \Omega_1(\zeta)\overline{g(\zeta)} e^{it}} \, dt.$$

Each of the functions

$$u_t(\zeta) = \frac{\Omega_1(\zeta) e^{it} + g(\zeta)}{1 + \Omega_1(\zeta)\overline{g(\zeta)} e^{it}}$$

belongs to H_∞, *because* Ω_1, g *and* $\Omega_1 \bar{g}$ *do, and because* $\|\Omega_1 \bar{g}\|_\infty \le 1 - \epsilon < 1$! Since $|\Omega_1(\zeta)| = 1$ a.e. for $|\zeta| = 1$, we have $|u_t(\zeta)| = 1$, a.e., $|\zeta| = 1$, i.e., *the functions* u_t *are inner.*

Because $\|g\|_\infty \le 1 - \epsilon$, we can apply the argument used to prove the first theorem of this subsection (approximation of the preceding integral by a Riemann sum) to conclude that

$$\left\| g - \frac{1}{N} \sum_{k=1}^{N} u_{2\pi k/N} \right\|_\infty < \epsilon$$

if N is large.

Therefore

$$\left\| f - \frac{1}{N} \sum_{k=1}^{N} u_{2\pi k/N} \right\|_\infty < 2\epsilon$$

for our f with $\|f\|_\infty < 1 - 2\epsilon$; if we only know $\|f\|_\infty \le 1$ we get a similar approximation to f to within 4ϵ instead of 2ϵ. Q.E.D.

Recall now *Frostman's theorem* from Chapter IV, Section G. *This says that any inner function can be uniformly approximated by Blaschke products**. Combining this fact with the previous result, we immediately have

Marshall's Theorem *Let* $f \in H_\infty$, $\|f\|_\infty \le 1$. *Then there are Blaschke products* $B_1, \ldots, B_n$

* N.B. The Blaschke products spoken of here and in the next result are understood to contain *arbitrary constant factors of modulus* 1.

and positive numbers λ_k, $\sum_1^n \lambda_k = 1$, *with*

$$\left\| f - \sum_k \lambda_k B_k \right\|_\infty < \epsilon,$$

$\epsilon > 0$ *being arbitrary.*

Thus, *the unit sphere in* H_∞ *is the norm-closed convex hull of the set of Blaschke products.* Truly a most beautiful result.

C. Szegö's theorem

Let μ be a finite *positive* measure on $[-\pi, \pi]$, and let $\mathscr{P}(0)$ denote the *class of polynomials* $P(z)$ with $P(0) = 0$, i.e., *polynomials without constant term*. If $1 \le p < \infty$, we are interested in *how small* we can make

$$\int_{-\pi}^{\pi} |1 - P(e^{i\theta})|^p \, d\mu(\theta) \quad \text{for } P \in \mathscr{P}(0).$$

Theorem *If* σ *is a positive finite* singular *measure,*

$$\inf \left\{ \int_{-\pi}^{\pi} |1 - P(e^{i\theta})|^p \, d\sigma(\theta); \; P \in \mathscr{P}(0) \right\}$$

is zero.

Proof Assume the inf is strictly positive. Then, for

$$\frac{1}{q} + \frac{1}{p} = 1,$$

there is a $G \in L_q(d\sigma)$ with

$$\int_{-\pi}^{\pi} G(e^{i\theta}) e^{in\theta} \, d\sigma(\theta) = 0, \quad n = 1, 2, \ldots,$$

but

$$\int_{-\pi}^{\pi} G(e^{i\theta}) \cdot 1 \, d\sigma(\theta) > 0.$$

By Hölder, $G \in L_1(d\sigma)$, so $ds(\theta) = G(e^{i\theta}) \, d\sigma(\theta)$ is a *finite Radon measure* on $[-\pi, \pi]$, whose *support is contained in that of* $d\sigma(\theta)$, i.e., *in a set of Lebesgue measure zero*. So $ds(\theta)$ is *singular*. But

$$\int_{-\pi}^{\pi} e^{in\theta} \, ds(\theta) = 0, \quad n = 1, 2, \ldots,$$

so by the theorem of F. and M. Riesz, $ds(\theta)$ *is absolutely continuous!*

Therefore $\mathrm{d}s(\theta)$ *must* $= 0$. But on the other hand,

$$\int_{-\pi}^{\pi} 1 \cdot \mathrm{d}s(\theta) > 0.$$

We have a contradiction. The theorem is proven.

Theorem (Kolmogorov) *Let* $\mathrm{d}\mu(\theta) = w(\theta)\,\mathrm{d}\theta + \mathrm{d}\sigma(\theta)$, *with* $w \in L_1$, $w \geq 0$, $\mathrm{d}\sigma \geq 0$ *and* σ singular. *Let* $1 \leq p < \infty$.

Then

$$\inf_{P \in \mathscr{P}(0)} \int_{-\pi}^{\pi} |1 - P(e^{i\theta})|^p w(\theta)\,\mathrm{d}\theta = \inf_{P \in \mathscr{P}(0)} \int_{-\pi}^{\pi} |1 - P(e^{i\theta})|^p\,\mathrm{d}\mu(\theta).$$

Remark Thus, *only the absolutely continuous part of* μ *matters.*

Proof of theorem Clearly

$$\inf_{P \in \mathscr{P}(0)} \int_{-\pi}^{\pi} |1 - P(e^{i\theta})|^p w(\theta)\,\mathrm{d}\theta \leq \inf_{P \in \mathscr{P}(0)} \int_{-\pi}^{\pi} |1 - P(e^{i\theta})|^p\,\mathrm{d}\mu(\theta)$$

so it suffices to prove the reverse inequality.

Let

$$K = \inf_{P \in \mathscr{P}(0)} \int_{-\pi}^{\pi} |1 - P(e^{i\theta})|^p w(\theta)\,\mathrm{d}\theta.$$

Then there is a $P \in \mathscr{P}(0)$ with

$$\int_{-\pi}^{\pi} |1 - P(e^{i\theta})|^p w(\theta)\,\mathrm{d}\theta < K + \epsilon.$$

It suffices to find a $Q \in \mathscr{P}(0)$ with

$$\int_{-\pi}^{\pi} |1 - Q(e^{i\theta})|^p (w(\theta)\,\mathrm{d}\theta + \mathrm{d}\sigma(\theta)) < K + 4\epsilon,$$

say.

First of all, by the initial theorem of this section, we can find a $P_1 \in \mathscr{P}(0)$ with

$$\int_{-\pi}^{\pi} |1 - P_1(e^{i\theta})|^p\,\mathrm{d}\sigma(\theta) < \epsilon.$$

Let $P_2(e^{i\theta}) = P_1(e^{i\theta}) - P(e^{i\theta})$; then we have $P_2 \in \mathscr{P}(0)$ and

$$\int_{-\pi}^{\pi} |1 - P(e^{i\theta}) - P_2(e^{i\theta})|^p\,\mathrm{d}\sigma(\theta) < \epsilon.$$

Take a closed $E \subseteq$ support of σ with

$$\int_{\sim E} \left(1 + |P(e^{i\theta})| + |P_2(e^{i\theta})|\right)^p \mathrm{d}\sigma(\theta) < \epsilon;$$

this is of course possible. Because $\mathrm{d}\sigma$ is *singular*, $|E| = 0$. Therefore by a construction used in the first proof of the F. and M. Riesz theorem given in Chapter II, we can find an $h \in \mathscr{A}$ (see beginning of Subsection A.1) with $h(e^{i\theta}) \equiv 1$ for $e^{i\theta} \in E$ and $|h(e^{i\theta})| < 1$ if $e^{i\theta} \notin E$. So for $n = 1, 2, \ldots,$

$$\int_E |1 - P(e^{i\theta}) - [h(e^{i\theta})]^n P_2(e^{i\theta})|^p \, d\sigma(\theta) = \int_E |1 - P(e^{i\theta}) - P_2(e^{i\theta})|^p \, d\sigma(\theta) \; < \; \epsilon.$$

Since in any case $|h(e^{i\theta})| \le 1$, we see, by choice of E, that

$$\int_{-\pi}^{\pi} |1 - P(e^{i\theta}) - [h(e^{i\theta})]^n P_2(e^{i\theta})|^p \, d\sigma(\theta) \; \le \; \epsilon + \int_{\sim E} \left(1 + |P(e^{i\theta})| + |P_2(e^{i\theta})|\right)^p d\sigma(\theta) \; < \; 2\epsilon.$$

Since $|h(e^{i\theta})| < 1$ outside E, hence a.e.,

$$1 - P(e^{i\theta}) - (h(e^{i\theta}))^n P_2(e^{i\theta}) \underset{n}{\longrightarrow} 1 - P(e^{i\theta}) \quad \text{a.e.,}$$

whence, since $w(\theta) \in L_1$, by Lebesgue's dominated convergence theorem,

$$\int_{-\pi}^{\pi} |1 - P(e^{i\theta}) - (h(e^{i\theta}))^n P_2(e^{i\theta})|^p w(\theta) \, d\theta \underset{n}{\longrightarrow} \int_{-\pi}^{\pi} |1 - P(e^{i\theta})|^p w(\theta) \, d\theta \; < \; K + \epsilon.$$

There is thus an n sufficiently large for

$$\int_{-\pi}^{\pi} |1 - P(e^{i\theta}) - (h(e^{i\theta}))^n P_2(e^{i\theta})|^p w(\theta) \, d\theta \quad \text{to be} \quad < \; K + 2\epsilon.$$

Thence, *finally*,

$$\int_{-\pi}^{\pi} |1 - P(e^{i\theta}) - (h(e^{i\theta}))^n P_2(e^{i\theta})|^p \, d\mu(\theta) \; < \; K + 2\epsilon + 2\epsilon = K + 4\epsilon.$$

Now $h(e^{i\theta}) \in \mathscr{A}$, so $h(e^{i\theta})$ can be *uniformly* approximated as *closely* as we want by a *polynomial*

$$\sum_0^N a_k e^{ik\theta}.$$

We have $\left(\sum_0^N a_k e^{ik\theta}\right)^n P_2(e^{i\theta}) \in \mathscr{P}(0)$ since $P_2 \in \mathscr{P}(0)$, and if $\sum_0^N a_k e^{ik\theta}$ is *close enough* to $h(e^{i\theta})$, we will *still* have

$$\int_{-\pi}^{\pi} |1 - Q(e^{i\theta})|^p \, d\mu(\theta) < K + 4\epsilon$$

with $Q \in \mathscr{P}(0)$ given by

$$Q(e^{i\theta}) = P(e^{i\theta}) + \left(\sum_0^N a_k e^{ik\theta}\right)^n P_2(e^{i\theta}).$$

This does it.

Our study thus reduces to the determination of

$$\inf_{P \in \mathscr{P}(0)} \int_{-\pi}^{\pi} |1 - P(e^{i\theta})|^p w(\theta) \, d\theta$$

with $w \in L_1(-\pi,\pi)$, $w \geq 0$. This problem is completely solved by the most beautiful and elegant

Theorem (of Szegö) *If* $1 \leq p < \infty$,

$$\inf_{P\in\mathscr{P}(0)} \frac{1}{2\pi}\int_{-\pi}^{\pi} |1-P(e^{i\theta})|^p w(\theta)\,d\theta \;=\; \exp\left(\frac{1}{2\pi}\int_{-\pi}^{\pi}\log w(\theta)\,d\theta\right).$$

Proof Suppose first of all that

$$\int_{-\pi}^{\pi}\log^- w(\theta)\,d\theta \;<\; \infty,$$

so that

$$\log w(\theta) \in L_1(-\pi,\pi).$$

(*Surely* $\int_{-\pi}^{\pi}\log^+ w(\theta)\,d\theta < \infty$ because $w \in L_1$.) It is convenient to work with

$$w_1(\theta) = w(\theta)\cdot\exp\left(-\frac{1}{2\pi}\int_{-\pi}^{\pi}\log w(t)\,dt\right)$$

and

$$K = \exp\left(\frac{1}{2\pi}\int_{-\pi}^{\pi}\log w(t)\,dt\right),$$

so as to have $w(\theta) = Kw_1(\theta)$ with

$$\int_{-\pi}^{\pi}\log w_1(\theta)\,d\theta = 0.$$

In the present case, *the desired result will follow* if we show that

$$\inf_{P\in\mathscr{P}(0)}\int_{-\pi}^{\pi}|1-P(e^{i\theta})|^p w_1(\theta)\,d\theta = 2\pi.$$

Because $\log w_1(\theta) \in L_1(-\pi,\pi)$, we can form the analytic function

$$f(z) = \frac{1}{2\pi}\int_{-\pi}^{\pi}\frac{e^{it}+z}{e^{it}-z}\log w_1(t)\,dt, \quad |z|<1;$$

we have

$$f(0) = \frac{1}{2\pi}\int_{-\pi}^{\pi}\log w_1(t)\,dt = 0.$$

Since $\log w_1 \in L_1$, by the computation in Chapter IV, Subsection E.2, $\exp(f(z)/p)$ belongs to H_p. It is *outer*, and by the material in Chapter I, Section D,

$$|\exp(f(e^{i\theta})/p)| = (w_1(\theta))^{1/p} \quad \text{a.e.}$$

If $P \in \mathscr{P}(0)$, $G(z) = (1-P(z))\exp(f(z)/p)$ is *also* in H_p, and $G(0) = 1$ since $f(0) = 0$. So for $r < 1$,

$$\frac{1}{2\pi}\int_{-\pi}^{\pi} |G(re^{i\theta})|^p \, d\theta \geq 1.$$

Since $G(re^{i\theta}) \to G(e^{i\theta})$ in L_p norm as $r \to 1$ (Chapter IV, Section C), we get

$$\frac{1}{2\pi}\int_{-\pi}^{\pi} |G(e^{i\theta})|^p \, d\theta \geq 1,$$

i.e.,

$$\int_{-\pi}^{\pi} |1 - P(e^{i\theta})|^p w_1(\theta) \, d\theta \geq 2\pi$$

for any $P \in \mathscr{P}(0)$, in view of the relation of $f(e^{i\theta})$ to $w_1(\theta)$. *We have thus shown that the* inf *in question is* $\geq 2\pi$.

To prove the reverse inequality, observe that, since $\exp(f(z)/p)$ is *outer*, there is, *by Beurling's theorem* (Chapter IV, Section E) a sequence of polynomials $Q_n(z)$ with

$$Q_n(z)\exp(f(z)/p) \xrightarrow[n]{} 1$$

in H_p norm. Because $\exp(f(0)/p) = 1$, we must have $Q_n(0) \xrightarrow[n]{} 1$, so also

$$\frac{Q_n(z)}{Q_n(0)} \exp \frac{f(z)}{p} \xrightarrow[n]{} 1$$

in H_p norm. We can write $Q_n(z)/Q_n(0) = 1 - P_n(z)$ with $P_n \in \mathscr{P}(0)$, so we surely have

$$\int_{-\pi}^{\pi} |1 - P_n(e^{i\theta})|^p w_1(\theta) \, d\theta = \int_{-\pi}^{\pi} |1 - P_n(e^{i\theta})|^p |e^{f(e^{i\theta})}| \, d\theta \xrightarrow[n]{} \int_{-\pi}^{\pi} 1^p \, d\theta = 2\pi,$$

showing that *the desired* inf *is* $\leq 2\pi$.

The formula in question is thus proved in the present case.

If now

$$\int_{-\pi}^{\pi} \log w(\theta) \, d\theta = -\infty,$$

let us take

$$w_n(\theta) = \max\left(w(\theta),\ 1/n\right);$$

then each $\log w_n \in L_1$. We have

$$\int_{-\pi}^{\pi} \log w_n(\theta) \, d\theta \xrightarrow[n]{} -\infty.$$

For each n, $w_n(\theta) \geq w(\theta)$, so that

$$\inf_{P \in \mathscr{P}(0)} \int_{-\pi}^{\pi} |1 - P(e^{i\theta})|^p w(\theta) \, d\theta \leq \inf_{P \in \mathscr{P}(0)} \int_{-\pi}^{\pi} |1 - P(e^{i\theta})|^p w_n(\theta) \, d\theta$$

$$= 2\pi \exp\left(\frac{1}{2\pi}\int_{-\pi}^{\pi} \log w_n(\theta) \, d\theta\right)$$

by what has already been done. Since this holds for every n, we see, making $n \to \infty$, that

$$\inf_{P \in \mathscr{P}(0)} \int_{-\pi}^{\pi} |1 - P(e^{i\theta})|^p w(\theta)\,\mathrm{d}\theta = 0.$$

But this equals

$$2\pi \exp\left(\frac{1}{2\pi} \int_{-\pi}^{\pi} \log w(\theta)\,\mathrm{d}\theta \right)$$

in this case.

Szegö's theorem is completely proved. Q.E.D.

Remark The connection with Beurling's theorem (Chapter IV, Section E) is seen to be *very close*, and the preceding discussion could just as well have been placed at the end of Chapter IV.

D. The Helson–Szegö theorem

In a 1960 Bologna *Annali* paper, Helson and Szegö give a characterization of the finite positive measures μ on $[-\pi, \pi]$ having the property that

$$\int_{-\pi}^{\pi} |\tilde{T}(\theta)|^2\,\mathrm{d}\mu(\theta) \le \text{const.} \int_{-\pi}^{\pi} |T(\theta)|^2\,\mathrm{d}\mu(\theta)$$

for all *trigonometric polynomials* $T(\theta)$.

A trigonometric polynomial $T(\theta)$ is simply a *finite sum* of the form $\sum_n a_n e^{in\theta}$, and for such a function $T(\theta)$, the *harmonic conjugate* $\tilde{T}(\theta)$ (Chapter I, Section E) is

$$-i \sum_n (\operatorname{sgn} n) a_n e^{in\theta},$$

where we put $\operatorname{sgn} 0 = 0$.

Definition A positive measure μ is called a *Helson–Szegö measure* if

$$\int_{-\pi}^{\pi} |\tilde{T}(\theta)|^2\,\mathrm{d}\mu(\theta) \le \text{const.} \int_{-\pi}^{\pi} |T(\theta)|^2\,\mathrm{d}\mu(\theta)$$

for all trigonometric polynomials T.

Simple direct computation shows that $\mathrm{d}\mu(\theta) = \mathrm{d}\theta$ is a Helson–Szegö measure.

Theorem *A Helson–Szegö measure is absolutely continuous.*

Proof Suppose E is closed, $|E| = 0$, but $\mu(E) > 0$. *Fatou's construction*, used in one of the proofs of the theorem of the brothers Riesz (Chapter II, Section A) gives us an $h \in \mathscr{A}$ (for definition see Subsection A.1) with $h(e^{i\theta}) \equiv 1$ on E and $|h(e^{i\theta})| < 1$ elsewhere. We have, surely, $h(0) = a$ with $|a| < 1$. Put $F_n(z) = (h(z))^n - a^n$. Then $F_n \in \mathscr{A}$ and $F_n(0) = 0$ so $\Re F_n = -\widetilde{\Im F_n}$. If μ is Helson–Szegö, we have therefore

$$\int_{-\pi}^{\pi} |(\Re F_n)(e^{i\theta})|^2\,\mathrm{d}\mu(\theta) \le C \int_{-\pi}^{\pi} |(\Im F_n)(e^{i\theta})|^2\,\mathrm{d}\mu(\theta)$$

since, because $F_n \in \mathscr{A}$, there is, for each n, a sequence of trigonometric polynomials T_m with $T_m(\theta) \underset{m}{\longrightarrow} (\Im F_n)(e^{i\theta})$ *uniformly* and $\tilde{T}_m(\theta) \underset{m}{\longrightarrow} -(\Re F_n)(e^{i\theta})$ *uniformly*. Now, *on* E,

$$(\Re F_n)(e^{i\theta}) = 1 - \Re a^n \underset{n}{\longrightarrow} 1,$$

so

$$\liminf_{n\to\infty} \int_{-\pi}^{\pi} |(\Re F_n)(e^{i\theta})|^2 \, d\mu(\theta) \geq \mu(E) > 0.$$

However,

$$|(\Im F_n)(e^{i\theta})| \leq 1 + |a|^n < 2,$$

whilst, *for* $e^{i\theta} \in E$,

$$(\Im F_n)(e^{i\theta}) = -\Im a^n \underset{n}{\longrightarrow} 0,$$

and for $e^{i\theta} \notin E$,

$$|(\Im F_n)(e^{i\theta})| \leq |h(e^{i\theta})|^n + |\Im a^n| \underset{n}{\longrightarrow} 0.$$

So $(\Im F_n)(e^{i\theta}) \underset{n}{\longrightarrow} 0$ *boundedly* and *everywhere*, and hence

$$\int_{-\pi}^{\pi} |(\Im F_n)(e^{i\theta})|^2 \, d\mu(\theta) \underset{n}{\longrightarrow} 0.$$

We have reached a contradiction, and are done.

Theorem *A nonzero Helson–Szegö measure is necessarily of the form* $d\mu(\theta) = w(\theta)\, d\theta$ *with* $w \geq 0$ *in* $L_1(-\pi, \pi)$ *and* $\int_{-\pi}^{\pi} \log w(\theta)\, d\theta > -\infty$.

Proof By the previous theorem $d\mu(\theta) = w(\theta)\, d\theta$ with $w \in L_1$ and $w \geq 0$. Suppose

$$\int_{-\pi}^{\pi} \log w(\theta)\, d\theta = -\infty;$$

then we use Szegö's theorem (Section C) to get a contradiction. For *then* we have a *sequence* of trigonometric polynomials P_n belonging to our old friend $\mathscr{P}(0)$ of Section C (the set of finite sums of the form $\sum_{n>0} a_n e^{in\theta}$) such that

$$\int_{-\pi}^{\pi} |1 - P_n(\theta)|^2 w(\theta)\, d\theta \underset{n}{\longrightarrow} 0.$$

Put $T_n(\theta) = 1 - P_n(\theta)$. Then $P_n = \tilde{\tilde{T}}_n$, so if $w(\theta)\, d\theta$ *is* a Helson–Szegö measure,

$$\int_{-\pi}^{\pi} |P_n(\theta)|^2 w(\theta)\, d\theta = \int_{-\pi}^{\pi} |\tilde{\tilde{T}}_n(\theta)|^2 w(\theta)\, d\theta \leq C \int_{-\pi}^{\pi} |\tilde{T}_n(\theta)|^2 w(\theta)\, d\theta \leq C^2 \int_{-\pi}^{\pi} |T_n(\theta)|^2 w(\theta)\, d\theta \underset{n}{\longrightarrow} 0.$$

So

$$\int_{-\pi}^{\pi} |P_n(\theta)|^2 w(\theta)\, d\theta \underset{n}{\longrightarrow} 0,$$

and, finally,

$$\int_{-\pi}^{\pi} 1^2 w(\theta)\, d\theta = 0,$$

or $w(\theta) = 0$ a.e. We are done.

> Our determination of all Helson–Szegö measures *is thus reduced* to the examination of those of the form $w(\theta)\,d\theta$ with positive $w \in L_1$ and $\int_{-\pi}^{\pi} \log w(\theta)\,d\theta > -\infty$.

Helson and Szegö introduced an auxiliary operation related to $\sim$.

Definition If $T(\theta) = \sum_n a_n e^{in\theta}$ is a trigonometric polynomial, call

$$(\Pi T)(\theta) = \sum_{n>0} a_n e^{in\theta}.$$

We use also the following notations:

$$\langle f, g\rangle_w = \int_{-\pi}^{\pi} f(\theta)\overline{g(\theta)}w(\theta)\,d\theta$$

$$\|f\|_w = \sqrt{\int_{-\pi}^{\pi} |f(\theta)|^2 w(\theta)\,d\theta}.$$

Lemma

$$\|\tilde{T}\|_w \le K\,\|T\|_w$$

for all trigonometric polynomials T *and some* K *iff*

$$\|\Pi T\|_w \le C\,\|T\|_w$$

for all such T *and some* C.

Proof

Step 1 If $\|\Pi T\|_w \le C\,\|T\|_w$ for all T, observe that for $T(\theta) = \sum_n a_n e^{in\theta}$,

$$\overline{T(\theta)} = \sum_n \overline{a_{-n}} e^{in\theta},$$

so $\overline{(\Pi \overline{T})} = \sum_{n<0} a_n e^{in\theta}$, and finally

$$\tilde{T} = -i\Pi T + i\overline{(\Pi \overline{T})}.$$

Clearly $\|\overline{T}\|_w = \|T\|_w$, so

$$\|\tilde{T}\|_w \le \|\Pi T\|_w + \|\Pi \overline{T}\|_w \le C\,\|T\|_w + C\,\|T\|_w = 2C\,\|T\|_w.$$

Step 2 If $\|\tilde{T}\|_w \le K\,\|T\|_w$ for all T, observe that $2\Pi T = -\tilde{\tilde{T}} + i\tilde{T}$, so

$$\|\Pi T\|_w \le \frac{K^2 + K}{2}\,\|T\|_w. \qquad \text{Q.E.D.}$$

Lemma $w(\theta)\,d\theta$ *is a Helson–Szegö measure iff there is a* $\rho < 1$ *such that, whenever* $P, Q \in \mathscr{P}(0)$,

$$\left| \Re \int_{-\pi}^{\pi} P(e^{i\theta}) \cdot e^{-i\theta} Q(e^{i\theta}) w(\theta)\,d\theta \right| \le \rho\,\|P\|_w\,\|Q\|_w.$$

Proof By the above lemma, $w(\theta)\,d\theta$ is Helson–Szegö iff $\|\Pi T\|_w \le C\|T\|_w$ for each trigonometric polynomial T. Any such T can be written as $P(\theta) + e^{i\theta}\overline{Q(e^{i\theta})}$ where $P, Q \in \mathscr{P}(0)$ and $P = \Pi T$. Because $|e^{i\theta}| \equiv 1$, we have for such T,

$$\|T\|_w^2 = \|P\|_w^2 + \|Q\|_w^2 + 2\Re \int_{-\pi}^{\pi} P(e^{i\theta})e^{-i\theta}Q(e^{i\theta})w(\theta)\,d\theta.$$

If, first of all, the inequality in the statement of the lemma holds, this last expression is

$$\ge \|P\|_w^2 + \|Q\|_w^2 - 2\rho\|P\|_w\|Q\|_w = (1-\rho^2)\|P\|_w^2 + (\rho\|P\|_w - \|Q\|_w)^2$$
$$\ge (1-\rho^2)\|P\|_w^2 = (1-\rho^2)\|\Pi T\|_w^2,$$

proving

$$\|\Pi T\|_w^2 \le \frac{1}{1-\rho^2}\|T\|_w^2.$$

Conversely, if $\|P\|_w^2 \le C^2\|T\|_w^2$ (where, without loss of generality, $C > 1$!), we have, when $\|P\|_w = \|Q\|_w = 1$, $\|T\|_w^2 \ge 1/C^2$, so that, using the formula for $\|T\|_w^2$ given above,

$$1 + 1 + 2\Re \int_{-\pi}^{\pi} P(e^{i\theta})e^{-i\theta}Q(e^{i\theta})w(\theta)\,d\theta \ge \frac{1}{C^2},$$

$$\Re \int_{-\pi}^{\pi} P(e^{i\theta})e^{-i\theta}Q(e^{i\theta})w(\theta)\,d\theta \ge -\left(1 - \frac{1}{2C^2}\right).$$

Repeating the argument with $-P$ instead of P we get, after changing signs,

$$\Re \int_{-\pi}^{\pi} P(e^{i\theta})e^{-i\theta}Q(e^{i\theta})w(\theta)\,d\theta \le 1 - \frac{1}{2C^2}.$$

These two inequalities are the same as the one asserted with

$$\rho = 1 - \frac{1}{2C^2}.$$

The lemma is proved.

Theorem (Helson and Szegö) *A measure* $d\mu$ *such that*

$$\int_{-\pi}^{\pi} |\tilde{T}(\theta)|^2\,d\mu(\theta) \le C \int_{-\pi}^{\pi} |T(\theta)|^2\,d\mu(\theta)$$

for all trigonometric polynomials T *is necessarily of the form*

$$d\mu(\theta) = e^{u(\theta)+\tilde{v}(\theta)}\,d\theta$$

where u *and* v *are* real valued, $|u(\theta)|$ *is* bounded, *and*

$$|v(\theta)| \le \frac{\pi}{2} - \epsilon, \quad \epsilon > 0.$$

Conversely, if $d\mu(\theta) = e^{u(\theta)+\tilde{v}(\theta)}\,d\theta$

with u *and* v as stated, *the above inequality holds with some* C.

Proof By the first two theorems of this section, we may restrict ourselves to examination of $d\mu(\theta) = w(\theta)\,d\theta$ with $w \in L_1$, and

$$\int_{-\pi}^{\pi} \log w(\theta)\,d\theta > -\infty.$$

Take any such w, and form

$$\phi(z) = \exp\left(\frac{1}{2\pi}\int_{-\pi}^{\pi} \frac{e^{it}+z}{e^{it}-z} \log\sqrt{w(t)}\,dt\right);$$

because $w \in L_1$, $\phi \in H_2$ (*and is outer*!), and by Chapter I, Section D,

$$|\phi(e^{i\theta})|^2 = w(\theta) \quad \text{a.e.}$$

Call

$$\frac{w(\theta)}{(\phi(e^{i\theta}))^2} = e^{iv(\theta)}.$$

By the previous lemma, $w(\theta)\,d\theta$ is Helson–Szegö iff, for some $\rho < 1$, $P, Q \in \mathscr{P}(0)$ and $\|P\|_w \le 1$, $\|Q\|_w \le 1$ imply

$$\left|\Re \int_{-\pi}^{\pi} \phi(e^{i\theta})P(e^{i\theta}) \cdot \phi(e^{i\theta})e^{-i\theta}Q(e^{i\theta})e^{iv(\theta)}\,d\theta\right| \le \rho.$$

By replacing P with $e^{i\gamma}P$ and varying γ through all real values, the latter condition is seen to be equivalent to

$$\left|\int_{-\pi}^{\pi} \phi(e^{i\theta})P(e^{i\theta}) \cdot \phi(e^{i\theta})e^{-i\theta}Q(e^{i\theta}) \cdot e^{iv(\theta)}\,d\theta\right| \le \rho\,\|\phi P\|_2\,\|\phi Q\|_2$$

for all $P, Q \in \mathscr{P}(0)$.

Now by Beurling's theorem (Chapter IV, Section E), $\phi\mathscr{P}(0)$ *is dense in* $H_2(0)$ *and* $e^{-i\theta}\phi(e^{i\theta}) \cdot \mathscr{P}(0)$ *is dense in* H_2.

So the previous condition is *equivalent* to

$$\left|\int_{-\pi}^{\pi} e^{iv(\theta)}F(e^{i\theta})G(e^{i\theta})\,d\theta\right| \le \rho\,\|F\|_2\,\|G\|_2$$

for *all* $F \in H_2(0)$ and $G \in H_2$.

By Chapter IV, *any* $f \in H_1(0)$ can be written as FG with $F \in H_2(0)$, $G \in H_2$, and $\|F\|_2 = \|G\|_2 = \sqrt{\|f\|_1}$.

So *finally*, $w(\theta)\,d\theta$ is a Helson–Szegö measure iff, for some $\rho < 1$

$$\left|\int_{-\pi}^{\pi} e^{iv(\theta)}f(\theta)\,d\theta\right| \le \rho\,\|f\|_1, \quad f \in H_1(0).$$

By Subsection A.2, this is equivalent to $\left\|e^{iv(\theta)} - H_\infty\right\|_\infty \le \rho < 1.$

Suppose now that $w_0(\theta) = e^{\tilde{v}(\theta)}$ with

$$|v(\theta)| \le \frac{\pi}{2} - \epsilon, \quad \epsilon > 0.$$

Then, as we see at once, the corresponding outer function $\phi_0(e^{i\theta})$ satisfies $(\phi_0(e^{i\theta}))^2 = e^{\tilde{v}(\theta)-iv(\theta)}$, so $w_0(\theta)/(\phi_0(e^{i\theta}))^2 = e^{iv(\theta)}$.

Since $|v(\theta)| \leq (\pi/2) - \epsilon$, we have

$$|e^{iv(\theta)} - \sin\epsilon| \leq \cos\epsilon < 1,$$

and the *constant* $\sin\epsilon$ is *in* H_∞, so the above boxed relation *is* fulfilled with $\rho = \cos\epsilon < 1$. *Then*, $||\tilde{T}||_{w_0} \leq K_0 ||T||_{w_0}$ as we saw above. If now $w(\theta) = e^{u(\theta)} w_0(\theta)$, and $-c \leq u(\theta) \leq c$, we *still have*

$$||\tilde{T}||_w \leq e^{2c} ||T||_w .$$

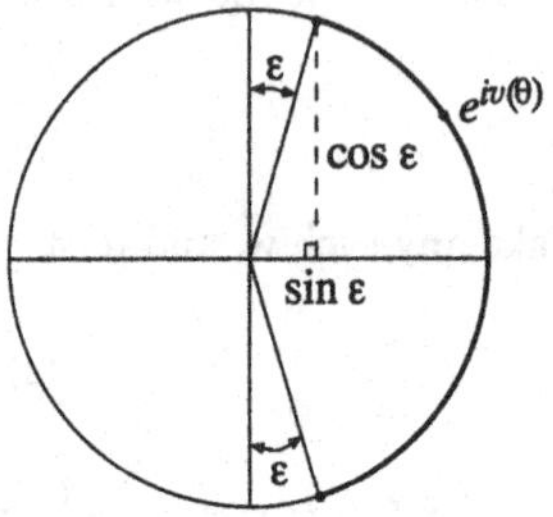

So our condition on w is sufficient.

It is necessary. For if the above boxed relation *holds*, we get an $h \in H_\infty$ with

$$\left|\frac{w(\theta)}{(\phi(e^{i\theta}))^2} - h(e^{i\theta})\right| = |e^{iv(\theta)} - h(e^{i\theta})| \leq \rho < 1.$$

Multiplying by $|\phi(e^{i\theta})|^2 = w(e^{i\theta})$, we get, a.e.,

$$|(\phi(e^{i\theta}))^2 h(e^{i\theta}) - w(\theta)| \leq \rho w(\theta),$$

where $w(\theta) > 0$ a.e. (because $\int_{-\pi}^{\pi} \log w(\theta)\, d\theta > -\infty$.) *So the boundary values* $(\phi(e^{i\theta}))^2 h(e^{i\theta})$ *of the* H_1 *function* $\phi^2 h$ *belong* a.e. *to the sector* $|\arg W| \leq \arcsin\rho < \pi/2$, and, *by Poisson's formula for* H_1 *functions* (Chapter II), *the values* $(\phi(z))^2 h(z)$ *also belong to that sector* for $|z| < 1$. *In particular*, $\Re\phi^2 h \geq 0$ there, so, by a previous exercise, $\phi^2 h$ is *outer*. So $\phi^2 h$ is of the form $e^{i\gamma} e^{u+i\tilde{u}}$ where γ is *real*. We *now* have $|\tilde{u}(e^{i\theta}) + \gamma| \leq \arcsin\rho$, so by Chapter V we certainly can take the harmonic conjugate

$$\widetilde{\tilde{u} + \gamma} = -u + u(0).$$

Call $-\tilde{u}(e^{i\theta}) - \gamma = v_1(\theta)$. *Then* $|v_1(\theta)| \leq \arcsin\rho$ and $u(e^{i\theta}) = \tilde{v}_1(\theta) + c$, c a real constant, so

$$|h(e^{i\theta})||\phi(e^{i\theta})|^2 = e^{\tilde{v}_1(\theta)+c}.$$

Finally,

$$|e^{iv(\theta)} - h(e^{i\theta})| \leq \rho < 1$$

makes

$$\frac{1}{1+\rho} \leq \frac{1}{|h(e^{i\theta})|} \leq \frac{1}{1-\rho},$$

so we can call

$$\frac{e^c}{|h(e^{i\theta})|} = e^{u_1(\theta)}$$

with a bounded (*above* **and** *below*!) real u_1. *Then*

$$w(\theta) = |\phi(e^{i\theta})|^2 = e^{u_1(\theta)+\tilde{v}_1(\theta)}$$

with $|v_1(\theta)| \leq \arcsin\rho < \pi/2$, Q.E.D.

Remark This is a most satisfying result. The above line of investigation has been continued by Helson and Sarason, and also by me.

Remark In 1970 or 71, Hunt, Muckenhoupt and Wheeden *determined completely all weights* $w(\theta)$ *for which*

$$\int_{-\pi}^{\pi} |\tilde{T}(\theta)|^p w(\theta)\, d\theta \leq c \int_{-\pi}^{\pi} |T(\theta)|^p w(\theta)\, d\theta,$$

where $1 < p < \infty$. *Their solution looks entirely different from that of Helson and Szegö, and, for* $p = 2$, *is as follows:*

> *$w(\theta)\,d\theta$ is a Helson–Szegö measure iff, for all intervals I,*
>
> $$\left\{\frac{1}{|I|}\int_I w(\theta)\,d\theta\right\}\left\{\frac{1}{|I|}\int_I \frac{d\theta}{w(\theta)}\right\} \leq c,$$
>
> *a finite constant independent of I.*

Coifman and Fefferman recently published a simplification of Hunt, Muckenhoupt and Wheeden's work in *Studia Mathematica.*

The above boxed condition must be *equivalent* to the fact that

$$\log w(\theta) - \tilde{v}(\theta) \in L_\infty(-\pi, \pi)$$

for some v with $\|v\|_\infty < \pi/2$. *Why*, is not so evident. The matter is involved with the theory of *BMO* (the class of *functions of bounded mean oscillation*), to be treated in Chapter X. At about the time I was giving the course on which the first edition of this book was based, Garnett and Jones found a *direct proof* of the equivalence, which, however, does not quite give the precise upper bound $\pi/2$ for $\|v\|_\infty$. Their work is published in the *Annals of Math.* for 1978.

Problem 7 Let $\omega(\theta) \in L_\infty(-\pi, \pi)$ and $|\omega(\theta)| \equiv 1$ a.e. The problem is to show that there is a *nonzero* $h \in H_\infty$ such that

(1) $$|\omega(\theta) - h(\theta)| \leq 1 \quad \text{a.e.}$$

if and only if there is a nonzero $f \in H_1$ *with*

(2) $$\omega(\theta) = \frac{f(\theta)}{|f(\theta)|} \quad \text{a.e.}$$

(a) If there *is* an $f \in H_1$ satisfying (2), show that $h = f/(P + i\tilde{P})$ is in H_∞ and satisfies (1), where $P(\theta) = |f(\theta)|$.

(*b*)* If there is a nonzero $h \in H_\infty$ satisfying (1), show that $he^{\tilde{\psi} - i\psi} \in H_1$, where ψ,

$$-\frac{\pi}{2} \leq \psi(\theta) \leq \frac{\pi}{2},$$

is such that

$$e^{-i\psi(\theta)}\overline{\omega(\theta)}h(\theta) > 0 \quad \text{a.e.}$$

Hence get an $f \in H_1$ satisfying (2).

HINT:

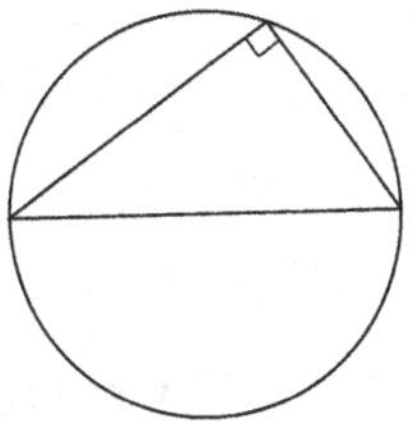

VIII

Application of the Hardy–Littlewood Maximal Function

The present chapter is based on some fairly old work in analysis, but nevertheless marks the beginning of the more recent developments in the theory of H_p spaces.

A. Use of the distribution function

In this chapter, we mostly consider only functions which are finite almost everywhere.

Definition If $f(x)$ is complex-valued on $\mathbb{R}$, and finite a.e., we write, for $\lambda > 0$,

$$m_f(\lambda) = |\{x;\ |f(x)| > \lambda\}|.$$

$m_f(\lambda)$ is called the *distribution function** of f.

The function $m_f(\lambda)$ is clearly *decreasing*; there is, of course, nothing to prevent its being *infinite* for some or all values of $\lambda > 0$.

By definition (!) *of the Lebesgue integral*, we clearly have

$$\boxed{\int_{-\infty}^{\infty} |f(x)|^p \, dx = \int_{0+}^{\infty} \lambda^p(-dm_f(\mu))}$$

for $p > 0$, provided that $m_f(\lambda) \longrightarrow 0$ for $\lambda \longrightarrow \infty$. We have, moreover, the

Lemma *For* $0 < p < \infty$,

$$\int_{-\infty}^{\infty} |f(x)|^p \, dx = p \int_{0}^{\infty} \lambda^{p-1} m_f(\lambda) \, d\lambda.$$

Proof By partial integration. There is a slight complication due to the fact that $\mathbb{R}$ has infinite measure.

---

* In Chapter V, Subsection C.1 we used $|h(\theta)| \geq \lambda$ in the definition of $m_h(\lambda)$. *Here* we prefer to use $>$ in order to involve *open sets* in our discussion; see Subsection B.1 below.

If, for some $c > 0$, $m_f(\lambda) \geq c$ for *all* $\lambda > 0$, then

$$\int_0^\infty \lambda^{p-1} m_f(\lambda)\, d\lambda$$

is clearly *infinite*, but $\int_{-\infty}^{\infty} |f(x)|^p\, dx$ is clearly $\geq \lambda^p \cdot c$ for all $\lambda > 0$, hence *also infinite*.

So we need only consider the case where $m_f(\lambda) \to 0$ as $\lambda \to \infty$.

Let $0 < \epsilon < R < \infty$. Then

$$(*) \qquad -\int_\epsilon^R \lambda^p dm_f(\lambda) = \epsilon^p m_f(\epsilon) - R^p m_f(R) + p\int_\epsilon^R \lambda^{p-1} m_f(\lambda)\, d\lambda.$$

(i) If

$$\int_0^\infty \lambda^{p-1} m_f(\lambda)\, d\lambda < \infty,$$

then

$$p\int_0^\epsilon \lambda^{p-1} m_f(\lambda)\, d\lambda \to 0$$

as $\epsilon \to 0$, i.e., since $m_f(\lambda)$ *decreases*,

$$m_f(\epsilon)\int_0^\epsilon p\lambda^{p-1}\, d\lambda \to 0$$

as $\epsilon \to 0$; that is, $\epsilon^p m_f(\epsilon) \to 0$ for $\epsilon \to 0$.

Again, given $\delta > 0$, there is an A so large that for all $R > A$,

$$p\int_A^R \lambda^{p-1} m_f(\lambda)\, d\lambda < \delta.$$

Therefore

$$m_f(R)\int_A^R p\lambda^{p-1}\, d\lambda = (R^p - A^p) m_f(R) < \delta,$$

so, for $R > A$ *large enough*, $R^p m_f(R) < 2\delta$. This proves $R^p m_f(R) \to 0$ as $R \to \infty$.

Making $\epsilon \to 0$ and $R \to \infty$ in (*), we thus find

$$p\int_0^\infty \lambda^{p-1} m_f(\lambda)\, d\lambda = -\int_{0+}^\infty \lambda^p dm_f(\lambda) = \int_{-\infty}^\infty |f(x)|^p\, dx,$$

in case the extreme left-hand member is finite.

(ii) Suppose *now* that

$$-\int_{0+}^\infty \lambda^p dm_f(\lambda) < \infty.$$

As remarked above, we may assume $m_f(\lambda) \to 0$ for $\lambda \to \infty$, and then

$$R^p m_f(R) = -R^p\int_R^\infty dm_f(\lambda) \leq -\int_R^\infty \lambda^p dm_f(\lambda),$$

so $R^p m_f(R) \to 0$ as $R \to \infty$. Making $R \to \infty$ in (*) gives, for $\epsilon > 0$,

$$\epsilon^p m_f(\epsilon) + p\int_\epsilon^\infty \lambda^{p-1} m_f(\lambda)\, d\lambda = -\int_\epsilon^\infty \lambda^p dm_f(\lambda),$$

i.e.,

$$p\int_\epsilon^\infty \lambda^{p-1} m_f(\lambda)\, d\lambda \leq -\int_\epsilon^\infty \lambda^p dm_f(\lambda).$$

Making $\epsilon \to 0$, we see that

$$\int_0^\infty \lambda^{p-1} m_f(\lambda)\,d\lambda < \infty,$$

and we are back in case (i). So the lemma holds.

B. The Hardy–Littlewood maximal function

1. Theorem of Hardy and Littlewood

Definition Let $f(x)$ be measurable on $\mathbb{R}$. The *Hardy–Littlewood maximal function*, $f^M(x)$, is

$$f^M(x) = \sup_{\xi<x<\xi'} \left\{ \frac{1}{\xi'-\xi} \int_\xi^{\xi'} |f(t)|\,dt \right\}.$$

Evidently, if f is *bounded*, $|f^M(x)| \le \|f\|_\infty$.

Theorem of Hardy and Littlewood

$$\boxed{m_{f^M}(\lambda) \le \frac{2}{\lambda} \int_{\{f^M(x)>\lambda\}} |f(x)|\,dx, \quad \lambda > 0.}$$

Proof Without loss of generality, $f(x) \ge 0$. Write

$$f_1(x) = \sup_{h>0} \frac{1}{h} \int_x^{x+h} f(t)\,dt, \qquad f_2(x) = \sup_{h>0} \frac{1}{h} \int_{x-h}^{x} f(t)\,dt.$$

For given $\lambda > 0$, call $E_1 = \{x;\ f_1(x) > \lambda\}$, $E_2 = \{x;\ f_2(x) > \lambda\}$, $E = \{x;\ f^M(x) > \lambda\}$; *since*

$$\frac{1}{h_1+h_2}\int_{x-h_1}^{x+h_2} f(t)\,dt = \frac{h_1}{h_1+h_2}\left(\frac{1}{h_1}\int_{x-h_1}^{x} f(t)\,dt\right) + \frac{h_2}{h_1+h_2}\left(\frac{1}{h_2}\int_{x}^{x+h_1} f(t)\,dt\right),$$

we have $E \subseteq E_1 \cup E_2$, so that $m_{f^M}(\lambda) \le m_{f_1}(\lambda) + m_{f_2}(\lambda)$, and the theorem will *clearly follow* as soon as we show that

$$m_{f_1}(\lambda) = \frac{1}{\lambda}\int_{E_1} f(x)\,dx, \qquad m_{f_2}(\lambda) = \frac{1}{\lambda}\int_{E_2} f(x)\,dx.$$

We content ourselves with the proof of the *first* relation, that of the *second* being very similar. Write

$$F(x) = \int_0^x f(t)\,dt,$$

using the usual convention if $x < 0$; it suffices to consider the case where $F(x)$ is everywhere finite and hence *continuous*, since otherwise both sides of the above boxed relation will be infinite for $\lambda > 0$. In that event, $F(x)$ is *increasing* and E_1 is a countable union of disjoint intervals J_k (one *may* be of infinite length), obtained from the graph of $F(x)$ vs. x by the following construction, due to F. Riesz:

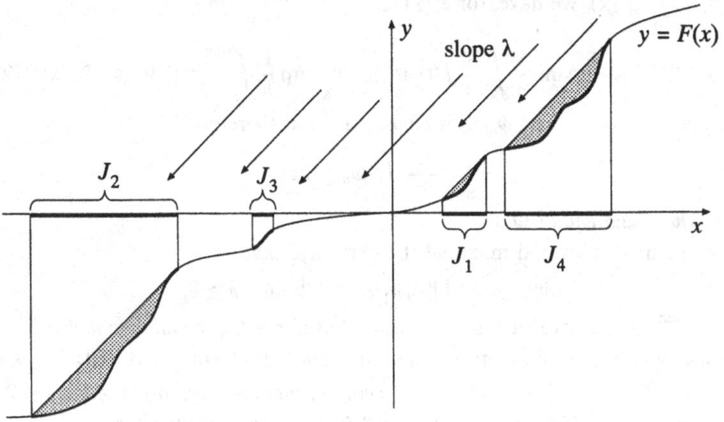

One *shines light* down along the *top* of the graph of $F(x)$ vs. x, in the *direction opposite* to that of the lines of slope λ. The J_k are the *intervals* of the x-axis *directly below* (or above) the *portions of the curve left in shadow*.

From the figure, it is manifest that for each J_k,

$$\lambda|J_k| \;=\; \text{increase of } F \text{ on } J_k \;=\; \int_{J_k} f(t)\,dt.$$

Hence, adding over the J_k,

$$\lambda|E_1| = \int_{E_1} f(t)\,dt,$$

i.e.,

$$m_{f_1}(\lambda) \;=\; |E_1| \;=\; \lambda^{-1}\int_{E_1} f(t)\,dt,$$

as required.

We are done.

Scholium The theorem of Hardy and Littlewood can be applied to yield a quick proof of the fact that if $f \in L_1$ and $F(x) = \int_0^x f(t)\,dt$, then $F'(x)$ exists and equals $f(x)$ almost everywhere (Lebesgue's theorem).

Indeed, if $f \in L_1$ we can construct a sequence of *continuous* functions $\phi_n(x)$ with $\|\phi_n\|_1 \le 4^{-n}$ for $n \ge 2$, such that the series $\sum_0^\infty \phi_n(x)$ converges to $f(x)$ *both* in L_1-norm *and* a.e. For each $n \geq 2$, let $\mathcal{O}_n = \{x;\ \phi_n^M(x) > 1/2^n\}$, and put $G_m = \bigcup_{n>m}\mathcal{O}_n$. If $x \notin G_m$,

$$\sup_{h,h'>0} \frac{1}{h+h'} \int_{x-h'}^{x+h} \left|\sum_{n>m} \phi_n(t)\right| dt \;\leq\; \sum_{n>m} \phi_n^M(x) \;\leq\; 2^{-m},$$

so, if $\Phi_m(x) = \sum_0^m \phi_n(x)$, we have, for $x \notin G_m$,

$$\Phi_m(x) - 2^{-m} \;\leq\; \liminf_{h\to 0} \frac{1}{h}\int_x^{x+h} f(t)\,dt \;\leq\; \limsup_{h\to 0} \frac{1}{h}\int_x^{x+h} f(t)\,dt \;\leq\; \Phi_m(x) + 2^{-m},$$

because $\Phi_m(x) = f(x) - \sum_{n>m}\phi_n(x)$ is *continuous* and therefore

$$\frac{\mathrm{d}}{\mathrm{d}x}\int_0^x \Phi_m(t)\,\mathrm{d}t$$

exists everywhere and equals $\Phi_m(x)$.

Now the Hardy–Littlewood maximal theorem says that

$$|\mathcal{O}_n| \leq 2^{n+1}\,\|\phi_n\|_1 \leq 2^{-n+1} \quad \text{for } n \geq 2,$$

so $|G_m| \leq 2^{-m+1}$. The above calculation shows that if $x \notin G_m$ *for all m sufficiently large* and $\lim_{m\to\infty}\Phi_m(x)$ *exists, then* $F'(x)$ *also* exists and *equals* that limit. But $\Phi_m(x) \underset{m}{\longrightarrow} f(x)$ a.e., and $G_1 \supset G_2 \supset G_3 \supset \ldots$, with $G = \bigcap_{m>1} G_m$ *being of measure zero because* $|G_m| \leq 2^{-m+1}$. So $F'(x)$ exists and equals $f(x)$ for almost all $x \notin G$, i.e., almost everywhere.

2. Norm inequalities for f^M

Theorem *If* $p > 1$ *(sic!),* $\|f^M\|_p \leq C_p\,\|f\|_p$ *with a constant* C_p *depending only on p.*

Proof One can use the obvious relation $\|f^M\|_\infty \leq \|f\|_\infty$ and the Hardy–Littlewood maximal theorem to set up a Marcinkiewicz argument like the one used in Chapter V, Subsection C.2 in order to prove corresponding inequalities for the harmonic conjugate $\tilde{f}$.

Here, however, a simple trick works. Without loss of generality, $f(x) \geq 0$. We can *approximate* $f(x)$ from *below* by a sequence of *bounded functions of compact support*. For each ϕ of the *latter*, $\phi^M(x)$ is bounded and $O(1/|x|)$ at ∞. By making the sequence of ϕ's monotone increasing, we get a sequence of $\phi^M(x)$ tending to $f^M(x)$ *from below*, in monotone increasing fashion, so that $\|f^M\|_p$ is the limit of the $\|\phi^M\|_p$. It thus clearly suffices to show that $\|\phi^M\|_p \leq C_p\,\|\phi\|_p$ for each ϕ.

Now, however, we are *assured* that $\|\phi^M\|_p < \infty$ because $p > 1$, $\phi^M(x)$ being bounded and $O(1/|x|)$ at ∞, and we proceed as follows. From Section A,

$$\|\phi^M\|_p^p \;=\; p\int_0^\infty \lambda^{p-1} m_{\phi^M}(\lambda)\,\mathrm{d}\lambda$$

which, in turn, is

$$\leq 2p\int_0^\infty \left\{\int_{\{\phi^M(x)>\lambda\}} \phi(x)\,\mathrm{d}x\right\} \lambda^{p-2}\,\mathrm{d}\lambda$$

by the Hardy–Littlewood maximal theorem. On *changing the order of integration* (!), this last is seen to equal

$$2p\int_{-\infty}^{\infty}\int_0^{\phi^M(x)} \lambda^{p-2}\phi(x)\,\mathrm{d}\lambda\,\mathrm{d}x \;=\; \frac{2p}{p-1}\int_{-\infty}^{\infty} \phi(x)(\phi^M(x))^{p-1}\,\mathrm{d}x.$$

Hölder's inequality now shows that the *last* expression is

$$\leq \frac{2p}{p-1}\,||\phi||_p\,||\phi^M||_p^{p-1},$$

so we have

$$||\phi^M||_p^p \;\leq\; \frac{2p}{p-1}\,||\phi||_p\,||\phi^M||_p^{p-1}.$$

Since $||\phi^M||_p^{p-1} < \infty$, *it can be cancelled from both sides* (!), yielding

$$||\phi^M||_p \leq \frac{2p}{p-1}\,||\phi||_p,$$

and proving the theorem, with $C_p = 2p/(p-1)$. Q.E.D.

3. Criterion for integrability of f^M on sets of finite measure

On $\mathbb{R}$, there is *no substitute* for the theorem of the preceding subsection in the case $p = 1$. That's because $\mathbb{R}$ has *infinite measure.*

Take, for instance

$$f(x) = \begin{cases} 1, & |x| \leq 1 \\ 0, & |x| > 1; \end{cases}$$

then, as is easily seen,

$$f^M(x) \geq \frac{\text{const.}}{|x|}$$

for large $|x|$, so $f^M \notin L_1(\mathbb{R})$, despite the fact that f belongs to *all* L_p spaces and is of compact support.

The *correct* L_1 substitute is for $L_1(E)$ with $|E| < \infty$, and it is established by a Marcinkiewicz argument like the one used in proving Zygmund's theorem (Chapter V, Subsection C.3).

Theorem *If* $|E| < \infty$,

$$\int_E f^M(x)\,\mathrm{d}x \leq 2|E| + 4\int_{-\infty}^{\infty} |f(x)| \log^+ |2f(x)|\,\mathrm{d}x.$$

Proof Let $\mu(\lambda) = \big|\{x \in E;\ f^M(x) > \lambda\}\big|$, then

$$\int_E f^M(x)\,\mathrm{d}x \leq |E| + \int_1^{\infty} \lambda(-\,\mathrm{d}\mu(\lambda))$$

which, by integration by parts, is seen to be less than or equal (in fact *equal*) to

$$|E| + \mu(1) + \int_1^{\infty} \mu(\lambda)\,\mathrm{d}\lambda \;\leq\; 2|E| + \int_1^{\infty} \mu(\lambda)\,\mathrm{d}\lambda \;\leq\; 2|E| + \int_1^{\infty} m_{f^M}(\lambda)\,\mathrm{d}\lambda.$$

To estimate $m_{f^M}(\lambda)$, we use Marcinkiewicz' trick and write, for given $\lambda > 0$,

$$f(x) = g_\lambda(x) + h_\lambda(x),$$

where

$$g_\lambda(x) = \begin{cases} f(x), & |f(x)| \leq \lambda/2 \\ 0, & \text{otherwise,} \end{cases} \qquad h_\lambda(x) = \begin{cases} 0, & |f(x)| \leq \lambda/2 \\ f(x), & |f(x)| > \lambda/2. \end{cases}$$

Clearly $f^M(x) \leq g_\lambda^M(x) + h_\lambda^M(x)$, and $||g_\lambda||_\infty \leq \lambda/2$, so that $||g_\lambda^M||_\infty \leq \lambda/2$. Therefore $f^M(x) > \lambda$ implies $h_\lambda^M(x) > \lambda/2$, so

$$m_{f^M}(x) \leq |\{x;\ h_\lambda^M(x) > \lambda/2\}|$$

which, by the Hardy–Littlewood maximal theorem, is

$$\leq \frac{2}{\lambda/2}\int_{-\infty}^{\infty} |h_\lambda(x)|\,dx = \frac{4}{\lambda}\int_{\{|f(x)|>\lambda/2\}} |f(x)|\,dx = \frac{4}{\lambda}\int_{\lambda/2}^{\infty} s(-dm_f(s)).$$

Therefore,

$$\int_E f^M(x)\,dx \leq 2|E| + \int_1^\infty m_{f^M}(\lambda)\,d\lambda \leq 2|E| + \int_1^\infty \left\{\frac{4}{\lambda}\int_{\lambda/2}^{\infty} s(-dm_f(s))\right\} d\lambda$$

$$= 2|E| + \int_{1/2}^{\infty}\int_1^{2s} \frac{4s}{\lambda}\,d\lambda(-dm_f(s)) = 2|E| + \int_{1/2}^{\infty} (4s\log 2s)(-dm_f(s))$$

$$= 2|E| + 4\int_{\{|f(x)|>1/2\}} |f(x)|\log(2|f(x)|)\,dx = 2|E| + 4\int_{-\infty}^{\infty} |f(x)|\log^+(2|f(x)|)\,dx,$$

Q.E.D.

Since f^M only depends on the *modulus*, $|f(x)|$, of f, the *form* of the known partial converse to Zygmund's theorem (Chapter V, Subsection C.4) should lead us to suspect that the result just proven *also* has a *converse*. Despite, however, the length of time that the above theorem has been known (it is in the *first edition* of Zygmund's book!) its converse was not noticed until 1969, when it was published by E. M. Stein. *In fact*, the *steps* in the *above proof* can *practically* be *reversed*!

Theorem *Let $f \in L_1(\mathbb{R})$ be of* compact support. *Then*

$$\int_E f^M(x)\,dx$$

is finite for every E of finite measure only if

$$\int_{-\infty}^{\infty} |f(x)|\log^+|f(x)|\,dx < \infty.$$

Proof If f is of compact support, clearly $f^M(x) \leq$ const./$|x|$ for large $|x|$, so $E = \{x;\ f^M(x) > 1\}$ is of finite measure. Finiteness of

$$\int_E f^M(x)\,dx$$

with this E will lead to the desired conclusion.

Let us take the partial maximal function

$$f_1(x) = \sup_{h>0} \frac{1}{h}\int_x^{x+h} |f(t)|\,dt$$

already used in the proof of the Hardy–Littlewood theorem (Subsection 1). Then $f_1(x) \leq f^M(x)$, so

$$\int_{\{f_1(x)>1\}} f_1(x)\,dx \leq \int_E f^M(x)\,dx < \infty.$$

Now, however,

$$\int_{\{f_1(x)>1\}} f_1(x)\,dx = -\int_1^\infty \lambda\, dm_{f_1}(\lambda) = m_{f_1}(\lambda) + \int_1^\infty m_{f_1}(\lambda)\,d\lambda \geq \int_1^\infty m_{f_1}(\lambda)\,d\lambda,$$

so the last quantity must be finite.

In proving the Hardy–Littlewood theorem, we *actually showed* that

$$m_{f_1}(\lambda) = \frac{1}{\lambda}\int_{\{f_1(x)>\lambda\}} |f(x)|\,dx,$$

so

$$\int_1^\infty m_{f_1}(\lambda)\,d\lambda = \int_1^\infty \int_{\{f_1(x)>\lambda\}} |f(x)|\,dx\frac{d\lambda}{\lambda} = \int_{\{f_1(x)>1\}} \int_1^{f_1(x)} |f(x)|\frac{d\lambda}{\lambda}\,dx$$

$$= \int_{\{f_1(x)>1\}} |f(x)| \log f_1(x)\,dx.$$

But $f_1(x) \geq |f(x)|$ a.e., so that the last integral is

$$\geq \int_{\{f(x)>1\}} |f(x)| \log f_1(x)\,dx \geq \int_{\{f(x)>1\}} |f(x)| \log |f(x)|\,dx = \int_{-\infty}^\infty |f(x)| \log^+ |f(x)|\,dx.$$

In short,

$$\int_{\{f_1(x)>1\}} f_1(x)\,dx \geq \int_{-\infty}^\infty |f(x)| \log^+ |f(x)|\,dx,$$

and if the quantity on the *left* is finite, so is the one on the *right*. We are done.

C. Application to functions analytic or harmonic in the upper half plane, or in unit circle

1. Estimate of Poisson's integral

Lemma *Let*

$$\int_{-\infty}^\infty \frac{|f(t)|\,dt}{1+t^2} < \infty,$$

and put for $\Im z > 0$,

$$V(z) = \frac{1}{\pi}\int_{-\infty}^\infty \frac{y}{(x-t)^2+y^2} f(t)\,dt.$$

Then

$$|V(x+iy)| \le \left(\frac{|x|}{y}+2\right) f^M(0).$$

Proof By a calculation which is essentially the same as one already carried out in Chapter I, Subsection D.3. We may assume $f^M(0) < \infty$, since otherwise there is nothing to prove.

Integration by parts yields:

$$|V(x+iy)| \le \frac{1}{\pi}\int_{-\infty}^{\infty} \frac{y}{(x-t)^2+y^2}|f(t)|\,dt$$

$$= \left(\frac{1}{\pi}\frac{y}{(x-t)^2+y^2}\int_0^t |f(s)|\,ds\right]_{-\infty}^{\infty} + \frac{1}{\pi}\int_{-\infty}^{\infty}\left\{\frac{2y(t-x)}{\left((x-t)^2+y^2\right)^2}\int_0^t |f(s)|\,ds\right\} dt.$$

Because $f^M(0) < \infty$, the *integrated* term is *zero*. The *second* term can be rewritten as

$$\frac{1}{\pi}\int_{-\infty}^{\infty} \frac{2y(t-x)t}{\left((x-t)^2+y^2\right)^2}\cdot\left(\frac{1}{t}\int_0^t |f(s)|\,ds\right) dt,$$

with

$$t^{-1}\int_0^t |f(s)|\,ds \le f^M(0)$$

by definition. We have

$$\frac{2y(t-x)t}{\left((x-t)^2+y^2\right)^2} = \frac{2y(t-x)^2}{\left((t-x)^2+y^2\right)^2} + \frac{2y(t-x)\cdot x}{\left((t-x)^2+y^2\right)^2},$$

and this is in absolute value

$$\le \frac{2y+|x|}{(t-x)^2+y^2}.$$

Since

$$\frac{1}{\pi}\int_{-\infty}^{\infty} \frac{y}{(t-x)^2+y^2}\,dt = 1,$$

substitution into the previous expression yields

$$|V(x+iy)| \le \left(2+\frac{|x|}{y}\right) f^M(0).$$

Q.E.D.

2. The non-tangential maximal function $F^*(x)$, $-\infty < x < \infty$

Definition If $F(z)$ is defined for $\Im z > 0$, put, for $x \in \mathbb{R}$,

$$F^*(x) = \sup\{|F(\xi+i\eta)|;\ 0 \le |\xi - x| < \eta\}.$$

Thus, $F^*(x)$ is the sup of $|F(\zeta)|$ in the $90°$ sector S_x, symmetric about the vertical, with vertex at x:

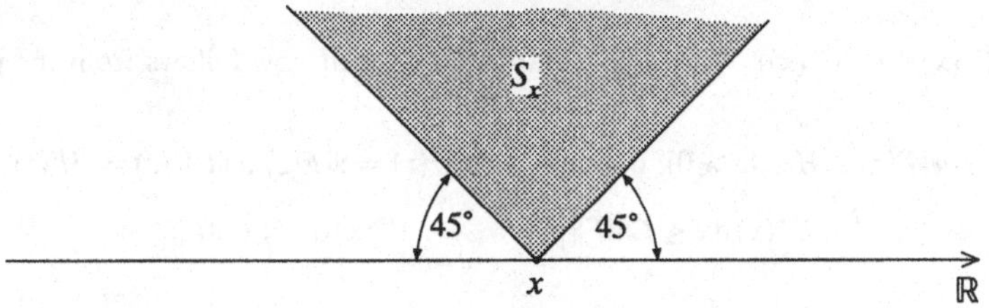

Any other angle less that 180° could have been chosen instead of the 90°, which we take here merely for convenience in writing.

$F^*(x)$ is frequently referred to as the *non-tangential maximal function* of $F(z)$.

Theorem *Let $p > 1$ (sic!), let $V(z)$ be harmonic in $\Im z > 0$, and suppose that*

$$\|V\|_p^p = \sup_{h>0} \int_{-\infty}^{\infty} |V(x+ih)|^p \, dx$$

is finite. *Then*

$$\int_{-\infty}^{\infty} (V^*(x))^p \, dx \leq K_p \|V\|_p^p,$$

with a constant K_p depending only on p.

Proof By Chapter VI, Section C there is an $f \in L_p(\mathbb{R})$, with $\|f\|_p$ equal to the norm $\|V\|_p$ defined above (see very end of Section B in Chapter VI), such that

$$V(z) = \frac{1}{\pi} \int_{-\infty}^{\infty} \frac{y}{|z-t|^2} f(t) \, dt.$$

By the lemma of Subsection 1, we have, making a translation along the x-axis:

$$V^*(x) \leq 3f^M(x).$$

But $\|f^M\|_p \leq C_p \|f\|_p$ by Subsection B.2. So

$$\|V^*\|_p \leq 3C_p \|f\|_p = 3C_p \|V\|_p.$$

Q.E.D.

Theorem *Let $p > 0$ and $F(z) \in H_p(\Im z > 0)$. Then*

$$\int_{-\infty}^{\infty} (F^*(x))^p \, dx \leq K_p \|F\|_p^{(p)}$$

with a K_p depending only on p, where (in order to follow the usage of Chapter IV), we put

$$(p) = \begin{cases} 1, & 0 < p < 1 \\ p, & p \geq 1. \end{cases}$$

Proof By Chapter VI, Section C we can write $F(z) = B(z)G(z)$, where $B(z)$ is a *Blaschke product* for the upper half plane (so that, in particular, $F^*(x) \leq G^*(x)$!), $G(z)$ *has no zeros* in $\Im z > 0$, and $G \in H_p(\Im z > 0)$ with $\|G\|_p = \|F\|_p$.

Therefore $V(z) = (G(z))^{p/2}$ can be defined so as to be analytic, *hence harmonic* (complex valued) in $\Im z > 0$, and

$$\|V\|_2^2 = \|G\|_p^{(p)} = \|F\|_p^{(p)}.$$

We have $(F^*(x))^p \le (G^*(x))^p = (V^*(x))^2$. Our desired result now follows from the previous theorem.

Corollary *For $F(z) \in H_1(\Im z > 0)$, we have, with $U(z) = \Re F(z)$ and $V(z) = \Im F(z)$,*

$$\int_{-\infty}^{\infty} U^*(x)\,dx \le C\,\|F\|_1, \qquad \int_{-\infty}^{\infty} V^*(x)\,dx \le C\,\|F\|_1.$$

3. The non-tangential maximal function $f^*(\theta)$, $-\pi \le \theta \le \pi$

Definition If $f(z)$ is defined in $\{|z| < 1\}$ and θ is real we put

$$f^*(\theta) = \sup\{|f(z)|;\ z \in S_\theta\},$$

where S_θ is the shaded region shown here:

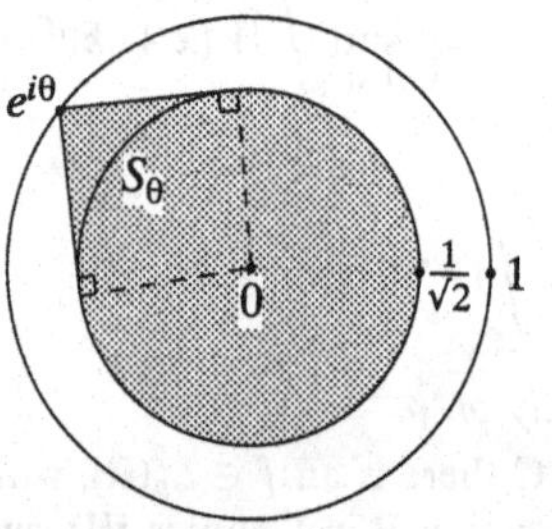

By arguments similar to those used in Subsection 2, we can establish* the following results:

Theorem *Let $p > 1$ (sic!), let $v(z)$ be harmonic in $\{|z| < 1\}$ and suppose that*

$$\|v\|_p^p = \sup_{r<1} \int_{-\pi}^{\pi} |v(re^{i\theta})|^p\,d\theta$$

is finite. Then, with a k_p depending only on p,

$$\int_{-\pi}^{\pi} (v^*(\theta))^p\,d\theta \le k_p\,\|v\|_p^p.$$

Theorem *Let $f(z) \in H_p(|z| < 1)$ for some $p > 0$. Then, with a constant K_p depending only on p,*

$$\int_{-\pi}^{\pi} (f^*(\theta))^p\,d\theta \le K_p\,\|f\|_p^{(p)},$$

where $(p) = \sup(1, p)$.

Corollary *For $f \in H_1$ and $u = \Re f$,*

$$\int_{-\pi}^{\pi} u^*(\theta)\,d\theta \le C\,\|f\|_1.$$

---

* In doing this, one works with the Hardy–Littlewood maximal function corresponding to $\chi(\theta)v(e^{i\theta})$, where χ is the characteristic function of $[-2\pi, 2\pi]$ (sic!) and $v(e^{i\theta})$ the boundary value of the harmonic function involved.

4. The maximal Hilbert transform

Lemma

$$\left|\int_{|t|>y} \frac{f(t)}{t}\,dt - \int_{-\infty}^{\infty} \frac{tf(t)}{t^2+y^2}\,dt\right| \leq (1+2\pi)f^M(0).$$

Proof The expression on the left equals

$$\left|\int_{|t|>y} \frac{y^2}{t(y^2+t^2)} f(t)\,dt - \int_{-y}^{y} \frac{tf(t)\,dt}{y^2+t^2}\right|$$
$$\leq \frac{1}{2y}\int_{-y}^{y} |f(t)|\,dt + \int_{-\infty}^{\infty} \frac{y}{t^2+y^2}|f(t)|\,dt$$
$$\leq f^M(0) + \sup_{y>0} \int_{-\infty}^{\infty} \frac{y}{t^2+y^2}|f(t)|\,dt \leq (1+2\pi)f^M(0)$$

by the lemma of Subsection 1. Q.E.D.

Theorem *Let $1 < p < \infty$ (sic!), and $f \in L_p(\mathbb{R})$. The so-called* maximal Hilbert transform

$$\check{f}(x) = \sup_{\epsilon>0} \left|\frac{1}{\pi}\int_{|t-x|>\epsilon} \frac{f(t)}{x-t}\,dt\right|$$

satisfies $\|\check{f}\|_p \leq K_p \|f\|_p$.

Proof For $\Im z > 0$, put

$$V(z) = \frac{1}{\pi}\int_{-\infty}^{\infty} \frac{x-t}{|z-t|^2} f(t)\,dt,$$

then $V(z)$ is harmonic for $\Im z > 0$, and by Chapter VI, Section D, for each $h > 0$,

$$\int_{-\infty}^{\infty} |V(x+ih)|^p\,dx \leq C_p \|f\|_p^p,$$

with a C_p depending only on p (M. Riesz' theorem). Hence, by Subsection 2 above, $\|V^*\|_p \leq K_p \|f\|_p$ with some constant K_p. Making a translation, we see by the lemma just proven that

$$\left|\frac{1}{\pi}\int_{|t-x|>y} \frac{f(t)}{x-t}\,dt - V(x+iy)\right| \leq \left(\frac{1}{\pi}+2\right) f^M(x)$$

for each $y > 0$.

Therefore,

$$\check{f}(x) \leq \left(\frac{1}{\pi}+2\right) f^M(x) + V^*(x).$$

But we know that $\|f^M\|_p \leq \tilde{C}_p \|f\|_p$ here, according to Subsection B.2. Since $\|V^*\|_p \leq K_p \|f\|_p$, we're done.

Remark An analogous result holds for the unit circle.

D. Maximal function characterization of $\Re H_1$

If $F(z) \in H_1(\Im z > 0)$ and $U(z) = \Re F(z)$, a corollary of Subsection C.2 says that

$U^*(x) \in L_1(\mathbb{R})$. In 1971, Burkholder, Gundy and Silverstein discovered (with the help of probability theory) that the *converse* result holds!

1. Theorem of Burkholder, Gundy and Silverstein

Theorem *If $U(z)$ is real valued and harmonic in $\Im z > 0$ and*

$$\int_{-\infty}^{\infty} U^*(x)\,dx < \infty,$$

then $U = \Re F$ for an $F \in H_1(\Im z > 0)$.

Proof Based on two ideas in Fefferman and Stein's 1972 *Acta* paper, and worked up especially for the course on which this book was based.

For $h > 0$ and $\Im z \geq 0$ write $U_h(x) = U(z + ih)$. Note that since $U^* \in L_1$, $\|U_h\|_1$ (notation of Subsection C.2) is surely bounded for $h > 0$. This being the case, U has a *harmonic conjugate* V, given, for *any* half plane $\Im z > h > 0$, by

$$V(z) = \frac{1}{\pi}\int_{-\infty}^{\infty} \frac{x-t}{(x-t)^2 + (y-h)^2} U_h(t)\,dt.$$

The *claim* is that $U + iV \in H_1$.

In order to prove this, we will show that

$$\int_{-\infty}^{\infty} |V_h(x)|\,dx \leq 4\int_{-\infty}^{\infty} U^*(x)\,dx$$

for each $h > 0$. *In fact*, we will show that

$$\int_{-\infty}^{\infty} |V_h(x)|\,dx \leq 4\int_{-\infty}^{\infty} U_h^*(x)\,dx$$

which is *stronger*, since $U_h^*(x) \leq U^*(x)$ as the following diagram makes clear:

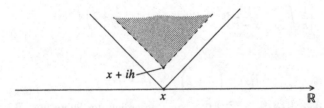

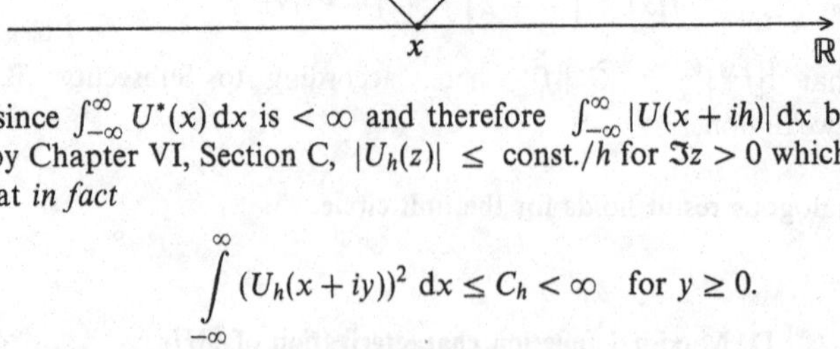

Again, since $\int_{-\infty}^{\infty} U^*(x)\,dx$ is $< \infty$ and therefore $\int_{-\infty}^{\infty} |U(x + ih)|\,dx$ bounded for $h > 0$, we have, by Chapter VI, Section C, $|U_h(z)| \leq$ const./h for $\Im z > 0$ which, since $\|U\|_1 < \infty$, implies that *in fact*

$$\int_{-\infty}^{\infty} (U_h(x+iy))^2\,dx \leq C_h < \infty \quad \text{for } y \geq 0.$$

(C_h may be *enormous* if $h > 0$ is *small*, but that is of *no concern* to us here!) By Chapter

VI, Section D we now see that

$$\int_{-\infty}^{\infty} (V_h(x+iy))^2 \, dx \le C_h$$

for $y \ge 0$, so that in fact

$$U_h + iV_h \in H_2(\Im z > 0)$$

for each $h > 0$. The function $U_h(z) + iV_h(z)$ is continuous in $\Im z \ge 0$ and tends to *zero* when $z \to \infty$ there, since (Chapter VI, Section D)

$$U_h(z) + iV_h(z) = \frac{i}{\pi} \int_{-\infty}^{\infty} \frac{1}{z - t + (ih/2)} U_{h/2}(t) \, dt$$

for $\Im z \ge 0$.

Let, for $\lambda > 0$,

$$m(\lambda) = |\{x;\ U_h^*(x) > \lambda\}|,$$

$$\mu(\lambda) = |\{x;\ |V_h(x)| > \lambda\}|.$$

We proceed to estimate $\mu(\lambda)$ *in terms of* $m(\lambda)$. Call $\mathcal{O}_\lambda = \{x;\ U_h^*(x) > \lambda\}$, and $E_\lambda = \mathbb{R} \sim \mathcal{O}_\lambda$. We have $|\mathcal{O}_\lambda| = m(\lambda)$, and clearly

$$\mu(\lambda) \ \le\ |\{x \in E_\lambda;\ V_h(x) > \lambda\}| + |\mathcal{O}_\lambda| \ = \ m(\lambda) + |\{x \in E_\lambda;\ |V_h(x)| > \lambda\}|.$$

We proceed to estimate the *second term* on the *right*.

$\mathcal{O}_\lambda$ is a bounded open set, hence a disjoint union of finite open intervals J_k:

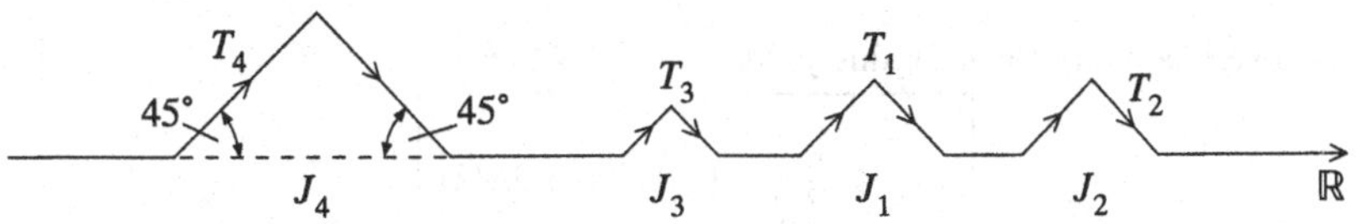

Above J_k, let T_k be the 45° roof constructed as shown, let $T = \bigcup_k T_k$, and let Γ be the *curve* consisting of the T_k and $E_\lambda = \mathbb{R} \sim \mathcal{O}_\lambda$, *so oriented* that $x = \Re z$ *increases* as z *moves along* Γ in the *positive sense*:

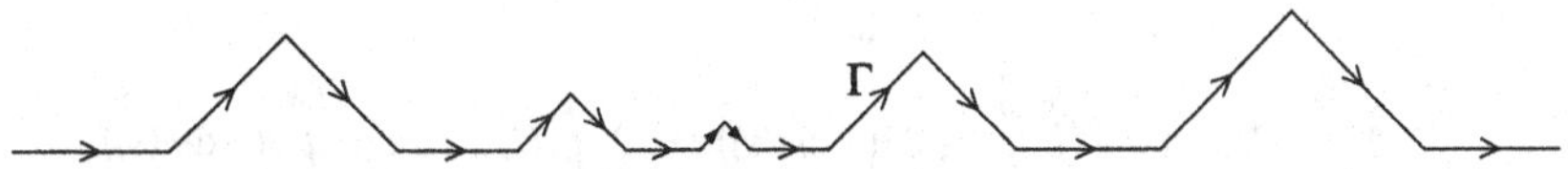

Since $U_h + iV_h \in H_2$ and $\mathcal{O}_\lambda$, hence T, is bounded, by Cauchy's theorem and the lemma at the beginning of Chapter VI, Section D:

$$\int_\Gamma (U_h(z) + iV_h(z))^2 \, dz \ = \ 0.$$

On taking real parts, this becomes

$$\int_{E_\lambda} (U_h^2 - V_h^2) \, dx + \int_T (U_h^2 - V_h^2) \, dx - 2 \int_T U_h V_h \, dy \ = \ 0.$$

Now on each piece T_k of T, $\mathrm{d}y = \pm \mathrm{d}x$, so

$$\left|2\int_T U_h V_h \,\mathrm{d}y\right| \leq \int_T (U_h^2 + V_h^2)\,\mathrm{d}x$$

which, substituted in the previous, yields, on transposition:

$$\int_{E_\lambda} V_h^2 \,\mathrm{d}x \leq \int_{E_\lambda} U_h^2 \,\mathrm{d}x + 2\int_T U_h^2 \,\mathrm{d}x.$$

But on each segment of T, $|U_h(z)| \leq \lambda$, *since any such such segment has a foot in* $E_\lambda = \{x;\ U_h^*(x) \leq \lambda\}$.
(This is the *first* idea of Fefferman and Stein.)

Therefore

$$\int_{E_\lambda} V_h^2 \,\mathrm{d}x \leq \int_{E_\lambda} (U_h^*)^2 \,\mathrm{d}x + 2\lambda^2 \int_T \mathrm{d}x = \int_{E_\lambda} (U_h^*)^2 \,\mathrm{d}x + 2\lambda^2 m(\lambda),$$

since

$$\int_T \mathrm{d}x = |\mathcal{O}_\lambda| = m(\lambda).$$

We have now

$$\int_{E_\lambda} (U_h^*)^2 \,\mathrm{d}x = \int_{\{U_h^*(x)\leq\lambda\}} (U_h^*(x))^2 \,\mathrm{d}x = \int_0^\lambda s^2(-\mathrm{d}m(s)),$$

hence (and this is the *second* idea of Fefferman and Stein),

$$|\{x \in E_\lambda;\ |V_h(x)| > \lambda\}| \leq \frac{1}{\lambda^2}\int_{E_\lambda} (V_h)^2 \,\mathrm{d}x \leq \frac{1}{\lambda^2}\int_0^\lambda s^2(-\mathrm{d}m(s)) + 2m(\lambda).$$

By the above inequality for $\mu(\lambda)$, this yields

$$\mu(\lambda) \leq \frac{1}{\lambda^2}\int_0^\lambda s^2(-\mathrm{d}m(s)) + 3m(\lambda).$$

The lemma of Section A now gives

$$\int_{-\infty}^{\infty} |V_h(x)|\,\mathrm{d}x = \int_0^\infty \mu(\lambda)\,\mathrm{d}\lambda \leq \int_0^\infty\int_0^\lambda \frac{s^2}{\lambda^2}(-\mathrm{d}m(s))\,\mathrm{d}\lambda + 3\int_0^\infty m(\lambda)\,\mathrm{d}\lambda$$

$$= 3\int_{-\infty}^{\infty} U_h^*(x)\,\mathrm{d}x + \int_0^\infty\int_s^\infty \frac{s^2}{\lambda^2}\,\mathrm{d}\lambda(-\mathrm{d}m(s)) = 3\int_{-\infty}^{\infty} U_h^*(x)\,\mathrm{d}x + \int_0^\infty s(-\mathrm{d}m(s))$$

$$= 4\int_{-\infty}^{\infty} U_h^*(x)\,\mathrm{d}x,$$

as required.

The theorem is completely proved.

In like manner we can establish:

Theorem *Let $u(z)$ be harmonic in $\{|z| < 1\}$. Then $u + i\tilde{u} \in H_1(|z| < 1)$ provided that*

$$\int_{-\pi}^{\pi} u^*(\theta)\, d\theta < \infty.$$

(For definition of $u^*(\theta)$, see Subsection C.3.)

2. Characterization in terms of the radial maximal function

The theorem of Burkhölder, Gundy and Silverstein can be *sharpened*. In order to do this, we use a remarkable result as a *substitute* for (lacking) subharmonicity of $|u(z)|^p$ (u harmonic) when $0 < p < 1$.

Lemma (Fefferman and Stein, 1972) *Let $u(z)$ be harmonic in the disk $\{|z| < R\}$, and let $0 < p < 1$. There is a constant C_p depending* only *on p such that*

$$|u(0)|^p \le \frac{C_p}{R^2} \int_0^R \int_0^{2\pi} |u(re^{i\theta})|^p r\, dr\, d\theta.$$

Proof By homogeneity considerations, we first reduce the general situation to the case where $R = 1$ and

$$\int_0^1 \int_0^{2\pi} |u(re^{i\theta})|^p r\, dr\, d\theta = 1.$$

Write, for $0 < r < 1$,

$$M(r) = \sup_\theta |u(re^{i\theta})|,$$

and

$$I(r) = \int_0^{2\pi} |u(re^{i\theta})|^p\, d\theta.$$

According to our preliminary reductions, we have

$$\int_0^1 I(r) r\, dr = 1,$$

so surely

$$\int_{1/2}^1 I(r) \frac{dr}{r} \le 4,$$

and, by the inequality between geometric and arithmetic means,

$$\frac{1}{\log 2} \int_{1/2}^1 \log I(r) \frac{dr}{r} \le \log \left\{ \frac{\int_{1/2}^1 (I(r)/r)\, dr}{\log 2} \right\} \le \log\left(\frac{4}{\log 2}\right),$$

whence

$$\boxed{\int_{1/2}^1 \log I(r) \frac{dr}{r} \le K,}$$

a pure number whose exact value need not concern us here.

Let $\alpha > 1$. By Poisson's formula (Chapter I!), we easily see that for $0 < r < 1$,

$$M(r^\alpha) \le \frac{1}{\pi\left(1 - (r^\alpha/r)\right)} \int_0^{2\pi} |u(re^{i\theta})|\, d\theta \le \frac{1}{\pi(1 - r^{\alpha-1})}(M(r))^{1-p} I(r).$$

Taking logarithms, we get

$$\log M(r^\alpha) \le (1-p)\log M(r) + \log I(r) + \log\left\{\frac{1}{\pi(1 - r^{\alpha-1})}\right\}.$$

Now (TRICK!) *multiply both sides by*

$$\frac{dr}{r} = \frac{1}{\alpha}\frac{d(r^\alpha)}{r^\alpha}$$

and integrate from 1/2 *to* 1. Putting $r^\alpha = \rho$ on the *right*, and using the boxed inequality proved above, we obtain

$$\frac{1}{\alpha}\int_{(1/2)^\alpha}^{1} \log M(\rho)\frac{d\rho}{\rho} \le (1-p)\int_{1/2}^{1} \log M(r)\frac{dr}{r} + K + \int_{1/2}^{1} \log\left\{\frac{1}{\pi(1 - r^{\alpha-1})}\right\}\frac{dr}{r}.$$

Here,

$$K + \int_{1/2}^{1} \log\left\{\frac{1}{\pi(1 - r^{\alpha-1})}\right\}\frac{dr}{r} = C_\alpha$$

is *finite* and depends *only* on α. *Now choose* α *so close to* 1 *that*

$$\frac{1}{\alpha} > 1 - p;$$

we can *do* this since $0 < p < 1$. Then we find

$$\frac{1}{\alpha}\int_{(1/2)^\alpha}^{1/2} \log M(r)\frac{dr}{r} + \left(\frac{1}{\alpha} - (1-p)\right)\int_{1/2}^{1} \log M(r)\frac{dr}{r} \le C_\alpha.$$

From this we see that there is a number l_α, depending *only* on α, with $\log M(r) \le l_\alpha$ for some r, $(1/2)^\alpha \le r \le 1$. *Then, by the principle of maximum*, $|u(0)| \le \exp l_\alpha$. Taking $C_p = \exp pl_\alpha$ after having chosen α as above, we see that the lemma holds as stated. Q.E.D.

Definition If $U(z)$ is defined for $\Im z > 0$ and $x \in \mathbb{R}$, put

$$U^+(x) = \sup_{y>0} |U(x+iy)|.$$

$U^+(x)$ is the so-called *'radial' maximal function* of U. Clearly $U^+ \le U^*$.

Theorem (Fefferman and Stein, 1972) *If* $U(z)$ *is harmonic and real in* $\Im z > 0$ *and* $U^+(x) \in L_1(\mathbb{R})$, *then* $U = \Re F$ *for an* $F \in H_1(\Im z > 0)$.

Proof We show that $U^+ \in L_1$ implies that $U^* \in L_1$, which is enough by the theorem of Subsection 1.

Let, for $\Im z > 0$, $W(z) = |U(z)|^{1/2}$. The picture shows that if $x \in \mathbb{R}$ and $0 \le |\xi - x| < \eta$, the disc $\{Z\,;\ |Z - (\xi + i\eta)| < \eta\}$ lies in $\Im z > 0$, so, by the lemma, with $p = 1/2$,

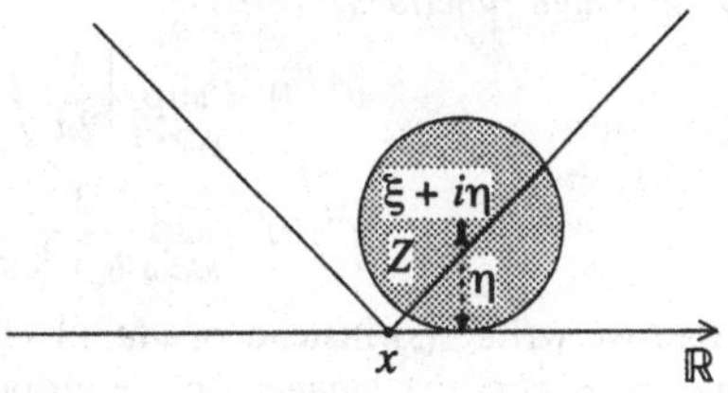

$$W(\xi + i\eta) \le \frac{C}{\eta^2} \iint\limits_{\{|Z-(\xi+i\eta)|<\eta\}} W(Z)\,dX\,dY,$$

where we write $Z = X + iY$.

Therefore, a fortiori,

$$W(\xi + i\eta) \ \le\ \frac{C}{\eta^2} \int_0^{2\eta} \int_{\xi-\eta}^{\xi+\eta} W(Z)\,dX\,dY \ \le\ \frac{2C}{\eta} \int_{\xi-\eta}^{\xi+\eta} W^+(X)\,dX,$$

since $W(Z) \le W^+(X)$.

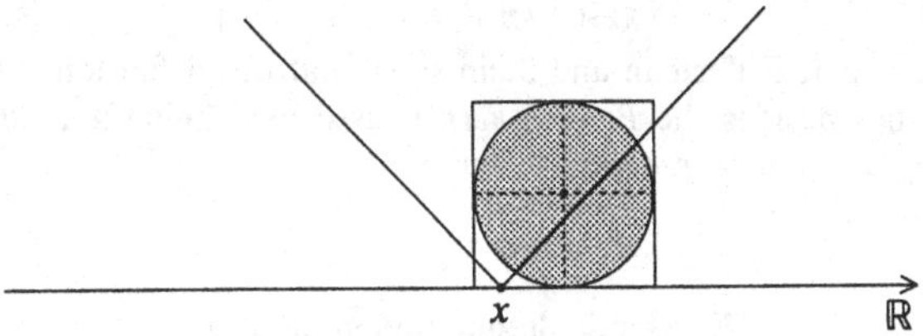

Thence, if $|\xi - x| < \eta$,

$$W(\xi + i\eta) \ \le\ \frac{2C}{\eta} \int_{x-2\eta}^{x+2\eta} W^+(X)\,dX \ \le\ 8C(W^+)^M(x),$$

by definition of the Hardy–Littlewood maximal function (for W^+*)!* Taking the sup over $\xi + i\eta$ with $0 \le |\xi - x| < \eta$, we see finally that $W^*(x) \le 8C(W^+)^M(x)$. Now, however, $W(z) = \sqrt{|U(z)|}$, so $W^+(x) = \sqrt{U^+(x)}$ is in L_2 by hypothesis. From Subsection B.2 we get $\|(W^+)^M\|_2 \le K\,\|W^+\|_2 = K\sqrt{\|U^+\|_1}$, so, by the previous relation,

$$\sqrt{\|U^*\|_1} \ = \ \|W^*\|_2 \ \le\ 8CK\sqrt{\|U^+\|_1},$$

proving $\|U^*\|_1 < \infty$ if $\|U^+\|_1 < \infty$.

We are done.

In the same way one can prove:

Theorem *Let* $u(z)$ *be harmonic in* $|z| < 1$ *and let*

$$u^+(\theta) \ = \ \sup_{0 \le r < 1} |u(re^{i\theta})|$$

belong to $L_1(-\pi, \pi)$*. Then* $u + i\tilde{u} \in H_1(|z| < 1)$*.*

3. Discussion

Remark Note the *different positions* of the *absolute value signs* in the *definitions* of the *two maximal functions*

$$u^+(\theta) = \sup_{0\le r<1} \left| \frac{1}{2\pi} \int_{-\pi}^{\pi} \frac{1-r^2}{1+r^2-2r\cos t} u(\theta - t)\,dt \right|,$$

$$u^M(\theta) = \sup_{h,k>0} \frac{1}{h+k} \int_{-h}^{k} |u(\theta - t)|\,dt.$$

(Here we write $u(s)$ instead of $u(e^{is})$.) Obviously, the *first* of these functions depends in a *much more sensitive manner* on the properties of u (possible oscillatory behaviour) than the *second*. It is *easy* to give *examples* of functions $u \in L_1(-\pi, \pi)$ for which $u^+(\theta) \in L_1(-\pi, \pi)$ but $u^M(\theta) \notin L_1(-\pi, \pi)$. *Indeed*, by the second theorem of Subsection B.3 together with a corollary in Subsection C.3, any $f \in H_1(|z| < 1)$ with $u(\theta) = \Re f(e^{i\theta})$ *of variable sign*, and such that $u \log^+ |u| \notin L_1(-\pi, \pi)$, provides us with such a u.

One is tempted to define a Hardy–Littlewood maximal function which is *more sensitively related* to u by putting, for instance.

$$\breve{u}(\theta) = \sup_{h,k>0} \left| \frac{1}{h+k} \int_{-h}^{k} u(\theta - t)\,dt \right|.$$

In their 1972 *Acta* paper, Fefferman and Stein show that this definition is *not useful*. *Other approximate* identities besides the *Poisson kernel*, used in defining u^+, *can* be used, but a certain *smoothness* seems to be *required* of them.

E. Atomic decomposition in $\Re H_1$

The sawtooth construction used in Subsection D.1 leads to a remarkable representation for the *real parts* of functions in H_1. That representation, due to Coifman but really based on an idea of Herz, will enable us to give a simple alternative proof of Fefferman's duality theorem in Section G of Chapter X.

What we have in mind is the expansion of any function $\Re F$, $F \in H_1$, in a series of constant multiples of so-called *atoms*. Such expansions are available for $\Re F$ with $F \in H_1(\Im z > 0)$ and for $F \in H_1(|z| < 1)$; we discuss the *former* space first, as the notation for it is a bit simpler. The treatment for the *second* space is analogous except for one point, and there we will omit a good part of the details.

I think the procedure followed here was first used by J.M. Wilson. He has adapted it to the treatment of H_1 functions of several variables.

1. Discussion of $\Re H_1(\Im z > 0)$

Various specifications of the notion of an atom are used in different circumstances. For the present treatment, the following one suffices.

Definition An *atom* of $\Re H_1(\Im z > 0)$ is a real valued measurable function $\phi(x)$, zero outside a certain finite interval $I \subseteq \mathbb{R}$, with $|\phi(x)| \le 1/|I|$ a.e. and $\int_I \phi(x)\,dx = 0$.

Atoms really are real parts of functions in H_1. Indeed, if $\phi(x)$ is an atom and

$$\tilde{\phi}(x) = \frac{1}{\pi} \int_{-\infty}^{\infty} \frac{\phi(t)\,dt}{x-t}$$

its Hilbert transform, $\phi(x) + i\tilde{\phi}(x) = F(x)$ a.e. for a certain $F \in H_1(\Im z > 0)$. Let us first see that that follows if $\phi + i\tilde{\phi} \in L_1(\mathbb{R})$.

In this circumstance, the function

$$U(z) = \frac{1}{\pi}\int_{-\infty}^{\infty} \frac{y\phi(t)}{(x-t)^2+y^2}\,dt,$$

harmonic in $\Im z > 0$, has there the harmonic conjugate

$$\tilde{U}(z) = \frac{1}{\pi}\int_{-\infty}^{\infty} \frac{(x-t)\phi(t)}{(x-t)^2+y^2}\,dt.$$

Here, where ϕ and hence $\tilde{\phi}$ both actually belong to $L_2(\mathbb{R})$, we have by a corollary near the end of Section D, Chapter VI,

$$\tilde{U}(z) = \frac{1}{\pi}\int_{-\infty}^{\infty} \frac{y\tilde{\phi}(t)}{(x-t)^2+y^2}\,dt$$

for $y > 0$, making

$$F(z) = \frac{1}{\pi}\int_{-\infty}^{\infty} \frac{y(\phi(t)+i\tilde{\phi}(t))}{(x-t)^2+y^2}\,dt$$

for the function $F(z) = U(z) + i\tilde{U}(z)$ analytic in $\Im z > 0$. Provided that $\phi + i\tilde{\phi} \in L_1(\mathbb{R})$, this makes

$$\int_{-\infty}^{\infty} |F(x+iy)|\,dx \;\leq\; \|\phi + i\tilde{\phi}\|_1$$

for each $y > 0$, and then $F \in H_1(\Im z > 0)$, and it has the non-tangential boundary value $\phi(x) + i\tilde{\phi}(x)$ at almost every $x \in \mathbb{R}$.

Now, however, we have the

Lemma *If ϕ is an atom, $\|\phi + i\tilde{\phi}\|_1 < \pi$.*

Proof Let I be the interval associated with ϕ in the above definition, and take I^* as the interval *with the same centre as* I, but having *thrice* its length. Since $|\phi(x)| \leq 1/|I|$ a.e. vanishes outside I, $\|\phi\|_1 \leq 1$.

Also,

$$\|\tilde{\phi}\|_1 = \int_{I^*} |\tilde{\phi}(x)|\,dx + \int_{\mathbb{R}\sim I^*} |\tilde{\phi}(x)|\,dx.$$

For the first integral on the right we use Schwarz and refer to a remark near the end of Section D in Chapter VI:

$$\int_{I^*} |\tilde{\phi}(x)|\,dx \;\leq\; |I^*|^{1/2}\,\|\tilde{\phi}\|_2 \;=\; (3|I|)^{1/2}\,\|\phi\|_2 \;\leq\; (3|I|)^{1/2}(|I|/|I|^2)^{1/2} \;=\; \sqrt{3}.$$

In order to estimate

$$\int_{\mathbb{R}\sim I^*} |\tilde{\phi}(x)|\,dx,$$

we resort to a *trick*. Because $\int_I \phi(t)\,dt = 0$ with ϕ vanishing outside I, we may write, for $x \notin I$,

$$\tilde{\phi}(x) = \frac{1}{\pi}\int_I \left(\frac{1}{x-t} - \frac{1}{x-c}\right)\phi(t)\,dt,$$

where c is the *midpoint* of I. Using $|\phi(t)| \leq 1/|I|$ a.e., we get from this

$$|\tilde{\phi}(x)| \leq \frac{|I|}{2\pi(\operatorname{dist}(x,I))^2}, \quad x \notin I,$$

and thence,

$$\int_{\mathbb{R}\sim I^*} |\tilde{\phi}(x)|\, dx \leq \frac{1}{\pi},$$

$\mathbb{R} \sim I^*$ consisting of the x with $\operatorname{dist}(x,I) \geq |I|$.

We thus have $||\tilde{\phi}||_1 \leq \sqrt{3} + (1/\pi)$, and finally

$$||\phi + i\tilde{\phi}||_1 \leq 1 + \sqrt{3} + (1/\pi) < \pi.$$

Done.

We see by this lemma and the observation preceding it that if the ϕ_n are atoms and $\sum_n |a_n| < \infty$ for numbers $a_n \in \mathbb{R}$, the series

$$\sum_n a_n \phi_n$$

converges in $L_1(\mathbb{R})$ to a function ϕ with $\phi + i\tilde{\phi} \in H_1(\Im z > 0)$ and

$$||\phi + i\tilde{\phi}||_1 \leq \pi \sum_n |a_n|.$$

What is remarkable is that the converse holds! Given $F \in H_1(\Im z > 0)$, we can *find* atoms ϕ_n and numbers $a_n \in \mathbb{R}$ with

$$\Re F(x) = \sum_n a_n \phi_n(x) \quad \text{a.e.},$$

and $\sum_n |a_n| \leq B\,||F||_1$. Here, B is a certain numerical constant independent of F. The proof of this result depends on the corollary at the very end of Subsection C.2, and one of its main ideas is the same as the one involving the 'tents' T_k in Subsection D.1.

Let, then $F \in H_1(\Im z > 0)$. For $x \in \mathbb{R}$, take, as in Subsection C.2, $F^*(x)$ to be the supremum of $|F(\zeta)|$ in the 90° sector

$$S_x = \{\zeta;\ 0 \leq |\Re\zeta - x| < \Im\zeta\}.$$

The set

$$\mathcal{O}_\lambda = \{x;\ F^*(x) > \lambda\}$$

is then open (in $\mathbb{R}$) for any $\lambda > 0$. Indeed, if $x \in \mathcal{O}_\lambda$ we have $F(\zeta) > \lambda$ for a certain $\zeta \in S_x$. Since F is *continuous* in the upper half-plane, we then have $|F(\zeta')| > \lambda$ for $\zeta' = \zeta + x' - x \in S_{x'}$ whenever $|x' - x|$ is sufficiently small, so such x' also belong to $\mathcal{O}_\lambda$.

It is important to realise that $\mathbb{R} \sim \mathcal{O}_\lambda$ is *always non-empty for* $\lambda > 0$. Each *interval component* of $\mathcal{O}_\lambda$ must in fact have *finite length*, for otherwise

$$\int_{\mathbb{R}} F^*(x)\, dx$$

would be *infinite*, contradicting the corollary at the end of Subsection C.2. There is, by the way, no valid analogue of the first statement for $H_1(|z| < 1)$, and that's the main reason why the treatment for that space differs from the present one.

Lemma *For each of the interval components I of* $\mathcal{O}_\lambda$, $\lambda > 0$, *we have*

$$\left|\int_I \Re F(x)\,\mathrm{d}x\right| \le \sqrt{2}\lambda|I|.$$

Proof Let $I = (a, b)$ with $-\infty < a < b < \infty$, and take the tent T lying above I,

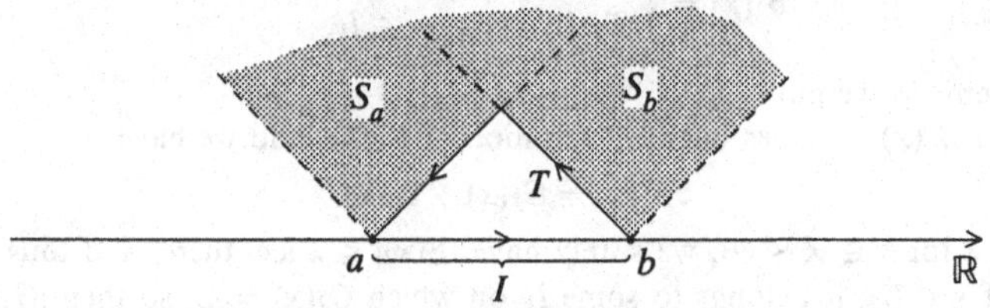

consisting of the segments from a and b making 45° angles with the real axis and meeting above I's midpoint. If $h > 0$, we have by Cauchy's theorem,

$$\int_I F(\zeta + ih)\,\mathrm{d}\zeta + \int_T F(\zeta + ih)\,\mathrm{d}\zeta = 0,$$

where, on I and T, we use the orientation indicated in the figure. Here, since $a \notin \mathcal{O}_\lambda$ and $b \notin \mathcal{O}_\lambda$ (!), $|F(\zeta + ih)| \le \lambda$ whenever $\zeta + ih$ belongs to the sectors S_a, S_b and, in particular, for ζ on T and *any* $h > 0$. Also, since $F \in H_1(\Im z > 0)$,

$$\int_{\mathbb{R}} |F(x + ih) - F(x)|\,\mathrm{d}x \to 0 \quad \text{as } h \to 0.$$

Using this property on I and bounded convergence on T we see, on making $h \to 0$ in the previous relation, that

$$\int_I F(x)\,\mathrm{d}x + \int_T F(\zeta)\,\mathrm{d}\zeta = 0.$$

Since $|F(\zeta)| \le \lambda$ on T and length T is $\le \sqrt{2}|I|$, the second integral on the left is in modulus $\le \sqrt{2}\lambda|I|$. Therefore

$$\left|\int_I F(x)\,\mathrm{d}x\right| \le \sqrt{2}\lambda|I|,$$

and on

$$\left|\int_I \Re F(x)\,\mathrm{d}x\right|$$

we have (at least) the same bound. Done.

This lemma gives us a *decomposition* of $\Re F(x)$ corresponding to any $\lambda > 0$. For such a λ, the set $\mathcal{O}_\lambda$ is a *countable disjoint union of* (finite) *open intervals* I_k, with

$$\left|\int_{I_k} \Re F(x)\,\mathrm{d}x\right| \le \sqrt{2}\lambda|I_k|$$

for each k. (It is of course possible that there are *no* I_k; then $\mathcal{O}_\lambda$ is empty and $F^*(x) \le \lambda$ on $\mathbb{R}$.) We can thus find numbers λ_k, $-\sqrt{2}\lambda \le \lambda_k \le \sqrt{2}\lambda$, with

$$\int_{I_k} (\Re F(x) - \lambda_k)\,\mathrm{d}x = 0.$$

Define then two functions, $G_\lambda(x)$ (thought of as the 'good part' of $\Re F(x)$ corresponding to the parameter λ) and $B_\lambda(x)$ (its 'bad part'), by putting

$$G_\lambda(x) = \begin{cases} \Re F(x), & x \in \mathbb{R} \sim \mathcal{O}_\lambda, \\ \lambda_k & x \in I_k, \end{cases}$$

and

$$B_\lambda(x) = \begin{cases} 0, & x \in \mathbb{R} \sim \mathcal{O}_\lambda, \\ \Re F(x) - \lambda_k & x \in I_k. \end{cases}$$

(In case $\mathcal{O}_\lambda$ is empty, we put $G_\lambda(x) = \Re F(x)$ and $B_\lambda(x) = 0$.)

The function $B_\lambda(x)$ vanishes outside the union of the I_k, and we have

$$\Re F(x) = G_\lambda(x) + B_\lambda(x).$$

Since $F^*(x) \leq \lambda$ for $x \in \mathbb{R} \sim \mathcal{O}_\lambda$, we surely have $|F(x)| \leq \lambda$ a.e. there, and thus $|G_\lambda(x)| \leq \lambda$ a.e., $x \notin \mathcal{O}_\lambda$. If $x \in \mathcal{O}_\lambda$, it belongs to some I_k on which $G_\lambda(x) = \lambda_k$, so then $|G_\lambda(x)| \leq \sqrt{2}\lambda$. Therefore

$$|G_\lambda(x)| \leq \sqrt{2}\lambda \quad \text{a.e.,} \quad x \in \mathbb{R}.$$

And our choice of the λ_k makes

$$\int_{I_k} B_\lambda(x)\, dx = 0$$

for each k.

From the G_λ and B_λ we now form what is called a *Calderón–Zygmund decomposition* of $\Re F(x)$. Taking $\lambda = 2^n$ with $n \in \mathbb{Z}$, $-\infty < n < \infty$ (sic!), we write

$$g_n(x) = G_{2^n}(x), \quad b_n(x) = B_{2^n}(x),$$

making

$$\Re F(x) = g_n(x) + b_n(x)$$

for each n. We have, by the above, $|g_n(x)| \leq \sqrt{2} \cdot 2^n$ a.e., so

$$g_n(x) \to 0 \quad \text{a.e. as } n \to -\infty.$$

By the corollary at the end of Subsection C.2, $F^*(x) < \infty$ a.e., so almost every real x belongs to

$$\mathbb{R} \sim \bigcap_n \mathcal{O}_{2^n}.$$

According to the above specification of $G_\lambda(x)$, this makes $g_n(x) = \Re F(x)$ at almost every $x \in \mathbb{R}$ for all sufficiently *large* n (depending on x). Thus,

$$\Re F(x) = \lim_{N \to \infty} \sum_{n=-N}^{N} (g_{n+1}(x) - g_n(x)) \quad \text{a.e., } x \in \mathbb{R}.$$

But

$$g_{n+1}(x) - g_n(x) = b_n(x) - b_{n+1}(x) \quad \text{a.e.}$$

(at least) by one of the above relations! Thence,

$$\Re F(x) = \lim_{N \to \infty} \sum_{n=-N}^{N} (b_n(x) - b_{n+1}(x)) \quad \text{a.e., } x \in \mathbb{R}.$$

The point here is that each *term* $b_n(x) - b_{n+1}(x)$ can be expanded in a series of real multiples of atoms. Denote, for each n, the *interval components* of $\mathcal{O}_{2^n}$ by $I_n(k)$. When λ gets *bigger*, $\mathcal{O}_\lambda$ gets *smaller*, so for any given n, *all* the $I_{n+1}(j)$ are included in the (disjoint!)

union of the $I_n(k)$ (for the different values of k). For each pair (n,k), let $\psi_{n,k}(x)$ be the function *equal* to $b_n(x) - b_{n+1}(x)$ on $I_n(k)$ and to *zero* outside that interval. Then,

$$\int_{I_n(k)} \psi_{n,k}(x)\,dx = 0.$$

To see this, denote the set of indices j for which $I_{n+1}(j) \subseteq I_n(k)$ by $S(n,k)$:

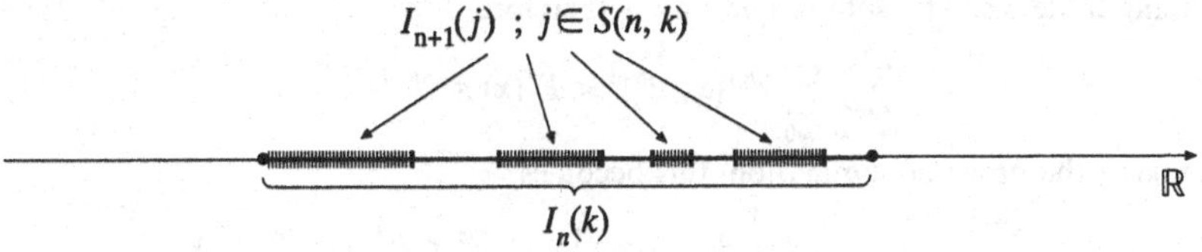

Then we can write

$$\int_{I_n(k)} \psi_{n,k}(x)\,dx \;=\; \int_{I_n(k)} b_n(x)\,dx \;-\; \sum_{j\in S(n,k)} \int_{I_{n+1}(j)} b_{n+1}(x)\,dx \;=\; 0.$$

(Should $S(n,k)$ be infinite, the expression of the second term on the right as a *sum of integrals* is justified by dominated convergence, $|b_{n+1}(x)|$ being $\le |\Re F(x)| + \sqrt{2}\cdot 2^{n+1}$.)

We also have

$$|b_n(x) - b_{n+1}(x)| = |g_{n+1}(x) - g_n(x)| \le \sqrt{2}\cdot 3\cdot 2^n \quad \text{a.e.}$$

Therefore, if we put

$$\phi_{n,k}(x) = \frac{\sqrt{2}}{3}\, 2^{-(n+1)}\, |I_n(k)|^{-1} \psi_{n,k}(x),$$

each function $\phi_{n,k}$ *is an atom* (associated with the interval $I_n(k)$).

Since $b_n(x) - b_{n+1}(x)$ vanishes outside $\mathcal{O}_{2^n}$, the *disjoint* union of the $I_n(k)$, we have

$$b_n(x) - b_{n+1}(x) \;=\; \sum_k \sqrt{2}\cdot 3\cdot 2^n\, |I_n(k)|\, \phi_{n,k}(x),$$

with the series on the right convergent a.e. (wherever the left-hand difference is defined and finite). Thence, by the previous,

$$\Re F(x) \;=\; \lim_{N\to\infty} \sum_{n=-N}^{N} \sum_k \sqrt{2}\cdot 3\cdot 2^n\, |I_n(k)|\, \phi_{n,k}(x) \quad \text{a.e., } x \in \mathbb{R},$$

and it is claimed that this is a decomposition of the kind sought. To verify that, it will be enough to show that

$$\sum_{n=-\infty}^{\infty} \sum_k \sqrt{2}\cdot 3\cdot 2^n\, |I_n(k)| \;<\; \infty.$$

Then, the $\phi_{n,k}(x)$ being *atoms* and hence satisfying $\|\phi_{n,k}\|_1 \le 1$ (!), it will follow that the series

$$\sum_{n=-\infty}^{\infty} \sum_k \sqrt{2}\cdot 3\cdot 2^n\, |I_n(k)|\, \phi_{n,k}(x)$$

converges unconditionally in $L_1(\mathbb{R})$ (and *necessarily* to $\Re F(x)$).

The sum

$$\sum_{n=-\infty}^{\infty}\sum_k 2^n\,|I_n(k)|$$

can be estimated in terms of $\|F\|_1$. For each n, we have

$$\sum_k 2^n\,|I_n(k)| \;=\; 2^n\,|\mathcal{O}_{2^n}| \;=\; 2^n\,|\{x;\; F^*(x) > 2^n\}| \;=\; 2^n \sum_{l=0}^{\infty} |\{x;\; 2^{n+l} < F^*(x) \le 2^{n+l+1}\}|,$$

$F^*(x)$ being finite a.e. The sum in question is therefore

$$\sum_{n=-\infty}^{\infty}\sum_{l=0}^{\infty} 2^n\,|\{x;\; 2^{n+l} < F^*(x) \le 2^{n+l+1}\}|.$$

On changing the order of summation, this becomes

$$\sum_{m=-\infty}^{\infty} 2^m\left(1+\frac{1}{2}+\frac{1}{4}+\cdots\right)|\{x;\; 2^m < F^*(x) \le 2^{m+1}\}|$$

which is obviously

$$\le 2\int_{\mathbb{R}} F^*(x)\,\mathrm{d}x.$$

By the corollary at the end of Subsection C.2, we see from this that

$$\sum_{n=-\infty}^{\infty}\sum_k \sqrt{2}\cdot 3\cdot 2^n\,|I_n(k)| \;\le\; B\,\|F\|_1$$

for a certain numerical constant B.

Putting $a_{n,k} = \sqrt{2}\cdot 3\cdot 2^n\,|I_n(k)|$, we thus have

$$\sum_{n,k} |a_{n,k}| \;\le\; B\,\|F\|_1$$

and

$$\Re F(x) = \sum_{n,k} a_{n,k}\phi_{n,k}(x) \quad \text{a.e., } x\in\mathbb{R},$$

with the atoms $\phi_{n,k}(x)$, the series converging unconditionally in $L_1(\mathbb{R})$. After converting the double indices to simple ones, we have the

Theorem (Coifman) *To any F in $H_1(\Im z > 0)$ there correspond atoms $\phi_n(x)$ and numbers $a_n \in \mathbb{R}$ with*

$$\sum_n |a_n| \le B\,\|F\|_1$$

and

$$\Re F(x) = \sum_n a_n\phi_n(x) \quad \text{a.e.,}$$

the series being unconditionally convergent in $L_1(\mathbb{R})$. Here, B is an absolute numerical constant.

2. Consideration of $\Re H_1(|z| < 1)$

When dealing with $H_1(|z| < 1)$, one uses the ice-cream cone shaped regions S_θ described at the beginning of Subsection C.3 to form the non-tangential maximal function (S_θ has its tip at $e^{i\theta}$). For $F \in H_1(|z| < 1)$, we put

$$F^*(\theta) = \sup\{|F(z)|;\; z\in S_\theta\}.$$

The principal difference between the present situation and the one of the preceding subsection is that here $F^*(\theta)$ has a *strictly positive infimum* (for $0 \le \theta \le 2\pi$), except for the case where $F(z) \equiv 0$. That is so because each S_θ includes the whole disc $\{|z| < 1/\sqrt{2}\}$, making $F^*(\theta)$ always $\ge$ the supremum of $|F(z)|$ on that disc.

It will be convenient in what follows to designate $\inf_\theta F^*(\theta)$ *by* λ_0. For $\lambda > 0$, we will take $\mathcal{O}_\lambda$ to be the set of $e^{i\theta}$ with $F^*(\theta) > \lambda$; *then* $\mathcal{O}_\lambda$ *can be a proper* (open) *subset of the unit circumference only for* $\lambda \geq \lambda_0$.

Let us now make the

Definition An *atom* for $\Re H_1(|z| < 1)$ is a real valued function $f(e^{i\theta})$ supported on an arc J of the unit circumference (N.B. J being perhaps *all* of the latter!), with $|f(e^{i\theta})| \leq 1/|J|$ a.e. and $\int_J f(e^{i\theta})\,d\theta = 0$.

An argument like the one at the beginning of the last subsection shows that every atom for $\Re H_1(|z| < 1)$ is equal a.e. to $\Re F(e^{i\theta})$ for some $F \in H_1(|z| < 1)$ with $\|F\|_1 \leq A$, an absolute constant. Here, however, we must have $\Re F(0) = 0$, *so atoms by themselves cannot suffice for the representation* (the 'building up') *of arbitrary functions in* $\Re H_1(|z| < 1)$. It will be seen that if we also throw in *arbitrary constants*, we will have all the building blocks we need.

Starting with a *non-zero* $F \in H_1(|z| < 1)$, we look at the set $\mathcal{O}_\lambda$ corresponding to any $\lambda > \lambda_0$ (here > 0 by the above). In these circumstances, $\mathcal{O}_\lambda$ is a *disjoint* union of *proper open arcs* on the unit circumference, and for *those* we have the

Lemma *If* J *is any component arc of* $\mathcal{O}_\lambda$ *for* $\lambda > \lambda_0$,

$$\left|\int_J \Re F(e^{i\theta})\,d\theta\right| \leq 2\lambda\,|J|.$$

Proof Is like that of the second lemma in the preceding subsection. One takes the curvilinear triangle or trapezoid T with 'base' J, constructed in remark (c) of Chapter III, Subsection D.1. By Cauchy's theorem,

$$\int_{\partial T} F(rz)\frac{dz}{z} = 0$$

for any $r < 1$, where ∂T denotes the oriented boundary of T. One observes that $|F(rz)| \leq \lambda$ for z on $\partial T \sim J$ and makes $r \to 1$; the desired inequality then follows from the facts that $|\partial T \sim J| \leq \sqrt{2}\,|J|$ (exercise: *check* this!), and that $|z| \geq 1/\sqrt{2}$ on ∂T.

We now follow the procedure of the last subsection as closely as we can. For $\lambda > \lambda_0$, $\mathcal{O}_\lambda$ is a disjoint union of certain arcs J_k on the unit circumference, and the lemma gives us numbers λ_k, $-2\lambda \leq \lambda_k \leq 2\lambda$, such that

$$\int_{J_k} \{\Re F(e^{i\theta}) - \lambda_k\}\,d\theta = 0$$

for each k. We can therefore put

$$G_\lambda(e^{i\theta}) = \begin{cases} \Re F(e^{i\theta}), & e^{i\theta} \notin \mathcal{O}_\lambda, \\ \lambda_k, & e^{i\theta} \in J_k, \end{cases}$$

and

$$B_\lambda(e^{i\theta}) = \begin{cases} 0, & e^{i\theta} \notin \mathcal{O}_\lambda, \\ \Re F(e^{i\theta}) - \lambda_k, & e^{i\theta} \in J_k \end{cases}$$

when $\lambda > \lambda_0$, and we will have

$$\Re F(e^{i\theta}) = G_\lambda(e^{i\theta}) + B_\lambda(e^{i\theta}),$$

with

$$|G_\lambda(e^{i\theta})| \leq 2\lambda \quad \text{and} \quad \int_{J_k} B_\lambda(e^{i\theta})\,d\theta = 0$$

for each k.

To start the *Caldéron–Zygmund decomposition* of $\Re F$, we take

$$g_n(e^{i\theta}) = G_{2^n}(e^{i\theta}), \qquad b_n(e^{i\theta}) = B_{2^n}(e^{i\theta})$$

for the $n \in \mathbb{Z}$ *with* $2^n > \lambda_0$. We need, however, to specify g_n and b_n for *one more* value of n.

Let n_0 be the largest $n \in \mathbb{Z}$ with $2^{n_0} \leq \lambda_0$ (which in the present circumstances is > 0). We certainly have

$$\int_0^{2\pi} \{\Re F(e^{i\theta}) - \Re F(0)\}\, d\theta = 0,$$

so, on putting

$$g_{n_0}(e^{i\theta}) = \Re F(0), \qquad b_{n_0}(e^{i\theta}) \;=\; \Re F(e^{i\theta}) - \Re F(0),$$

we see that

$$\int_0^{2\pi} b_{n_0}(e^{i\theta})\, d\theta = 0.$$

Again,

$$|F(0)| \leq \inf_\theta F^*(\theta) = \lambda_0$$

(see beginning of this subsection), so

$$|g_{n_0}(e^{i\theta})| \;=\; |\Re F(0)| \;\leq\; \lambda_0 \;<\; 2^{n_0+1}.$$

From here on, the work goes very much as it did for $\Re H_1(\Im z > 0)$. We have, for $n \geq n_0$,

$$g_{n+1}(e^{i\theta}) - g_n(e^{i\theta}) \;=\; b_n(e^{i\theta}) - b_{n+1}(e^{i\theta}) \text{ a.e.,}$$

so for almost all θ,

$$\Re F(e^{i\theta}) \;=\; \lim_{N\to\infty} g_N(e^{i\theta}) \;=\; g_{n_0}(e^{i\theta}) + \lim_{N\to\infty} \sum_{n=n_0}^{N} \{g_{n+1}(e^{i\theta}) - g_n(e^{i\theta})\}$$

$$= \;\Re F(0) + \lim_{N\to\infty} \sum_{n=n_0}^{N} \{b_n(e^{i\theta}) - b_{n+1}(e^{i\theta})\}.$$

Here, $b_{n_0}(e^{i\theta}) - b_{n_0+1}(e^{i\theta})$ is a constant multiple of *one* atom whose associated arc is the *whole* unit circumference. The atomic decomposition of the remaining terms $b_n(e^{i\theta}) - b_{n+1}(e^{i\theta})$ with $n > n_0$ is carried out as before; at the end one uses the relation

$$\int_0^{2\pi} F^*(\theta)\, d\theta \;\leq\; K\, \|F\|_1$$

proved in Subsection C.3, where K is an absolute constant. In this way we arrive at the

Theorem *Given* $F \in H_1(|z| < 1)$, *there are atoms* f_n *for* $\Re H_1(|z| < 1)$ *and real numbers* a_n, *such that*

$$\sum_n |a_n| \leq B\, \|F\|_1$$

and

$$\Re F(e^{i\theta}) \;=\; \Re F(0) + \sum_n a_n f_n(e^{i\theta}) \text{ a.e.,}$$

the series being unconditionally convergent in $L_1(0, 2\pi)$. Here, B is an absolute numerical constant.

F. Carleson measures

Definition A positive Borel measure μ (not necessarily finite) defined in $\Im z > 0$ is called a *Carleson measure* iff there is a constant K such that, for all $F \in H_1(\Im z > 0)$,

$$(*) \qquad \iint_{\Im z>0} |F(z)|\, d\mu(z) \;\le\; K\, \|F\|_1 .$$

Carleson measures are important in the next two chapters. Their geometric characterization is given by the

Theorem (Carleson) *μ is a Carleson measure iff, for all $x \in \mathbb{R}$ and all $h > 0$,*

$$(**) \qquad \boxed{\mu((x, x+h) \times (0, h)) \;\le\; Ch}$$

with a constant C independent of x and h.

Proof

Only if Let $x_0 \in \mathbb{R}$ and $h > 0$ be given. The test function

$$f(z) = \frac{h}{(z - x_0 + ih)^2}$$

belongs to $H_1(\Im z > 0)$ and $\|f\|_1 = \pi$.

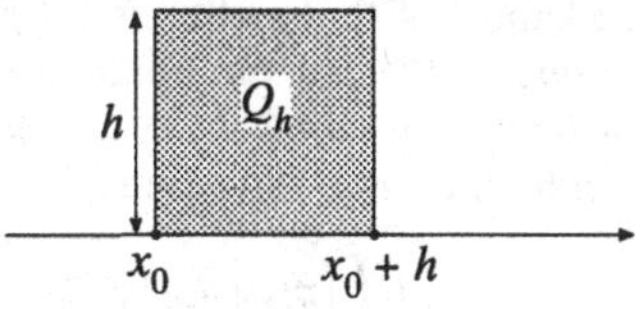

Also, $|f(z)| \ge 1/5h$ for $z \in Q_h = (x_0, x_0+h) \times (0, h)$, so if $(*)$ holds,

$$\frac{1}{5h}\mu(Q_h) \le \iint_{\Im z>0} |f(z)|\, d\mu(z) \le \pi K,$$

i.e., $\mu(Q_h) \le 5\pi K \cdot h$, proving $(**)$ with $C = 5\pi K$.

If It is enough to prove $(*)$ for all $F \in H_1$ of the special form $F(x) = f_h(z) = f(z + ih)$, where $f \in H_1(\Im z > 0)$ and $h > 0$. Take any such F; by the Poisson representation for f in the upper half-plane (Chapter VI) it is clear that $F(z)$ is *continuous* for $\Im z \ge 0$, and tends to zero as $z \to \infty$ in the upper half plane. We must establish $(*)$ with a K independent of F.

To this end, put, for $\lambda > 0$,

$$M(\lambda) = \mu(\{z;\ \Im z > 0 \ \&\ |F(z)| > \lambda\}).$$

Using an idea of Hörmander, we prove that $M(\lambda) \le C m_{F^*}(\lambda)$ with the C from $(**)$. Here, $F^*(x)$ is the non-tangential maximal function of F (Subsection C.2) and $m_{F^*}(\lambda)$ is the distribution function of F^* (Section A).

Let $E_\lambda = \{x \in \mathbb{R};\ F^*(x) \le \lambda\}$ and $\mathcal{O}_\lambda = \mathbb{R} \sim E_\lambda = \{x \in \mathbb{R};\ F^*(x) > \lambda\}$ so that $m_{F^*}(\lambda) = |\mathcal{O}_\lambda|$. If $x_0 \in E_\lambda$, *every* $z = x + iy$ *in the sector* $\overline{S}_{x_0} = \{z;\ |x - x_0| \le y \ \&\ y > 0\}$ must satisfy $|F(z)| \le \lambda$, therefore, $\Omega_\lambda = \{z;\ \Im z > 0, |F(z)| > \lambda\}$ *must be contained in* Ω'_λ, *the* COMPLEMENT, *in* $\Im z > 0$, *of* $\bigcup_{x_0 \in E_\lambda} \overline{S}_{x_0}$.

In the present case, since $F(z)$ is continuous in $\Im z \geq 0$ and 0 at ∞ there, $\mathcal{O}_\lambda = \mathbb{R} \sim E_\lambda$ is a *bounded open set on* $\mathbb{R}$, hence a *disjoint union* of *finite open intervals* J_k. Let, for each k, Δ_k be the open 45° isosceles triangle lying in $\Im z > 0$ with base J_k.

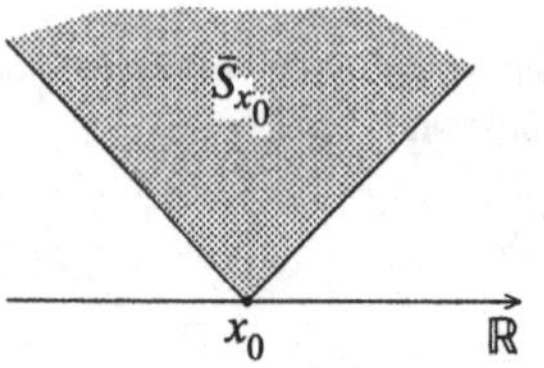

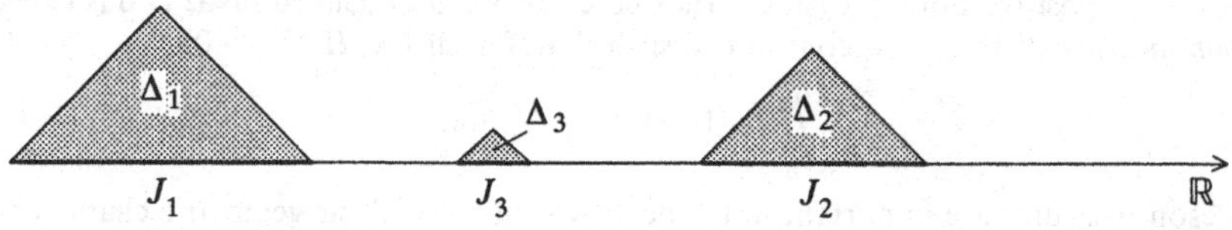

What we have here is the analogue for the upper half plane of *Privalov's construction*, described in Subsection D.1 of Chapter III. Arguing as in that place, we see that the set Ω'_λ introduced above is equal to the *union* of the Δ_k. Therefore $\Omega_\lambda \subseteq \bigcup_k \Delta_k$, and $M(\lambda) = \mu(\Omega_\lambda) \leq \sum_k \mu(\Delta_k)$. But for each k, by (**), $\mu(\Delta_k) \leq C|J_k|$, so

$$M(\lambda) \;\leq\; C\sum_k |J_k| \;=\; C|\mathcal{O}_\lambda| \;=\; Cm_{F^*}(\lambda).$$

Now we just bring in the material of Section A. The lemma of that section is *not particular* to *Lebesgue measure on the line*, but holds (with the same proof) for *distribution functions defined by means of quite general* σ-finite *measures*. Therefore

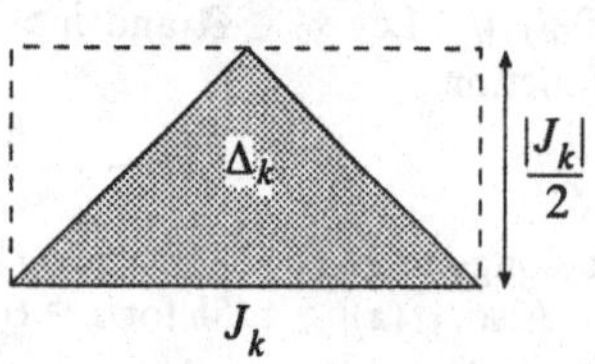

$$\iint_{\Im z>0} |F(z)|\,d\mu(z) \;=\; \int_0^\infty M(\lambda)\,d\lambda.$$

By the above work, the integral on the right is

$$\leq C\int_0^\infty m_{F^*}(\lambda)\,d\lambda \;=\; C\int_{-\infty}^\infty F^*(x)\,dx,$$

again by the lemma of Section A.

But here ($F \in H_1$!),

$$\int_{-\infty}^\infty F^*(x)\,dx \;\leq\; C_1\,\|F_1\|$$

by the second theorem of Subsection C.2. So finally,

$$\iint_{\Im z>0} |F(z)|\,d\mu(z) \;\leq\; CC_1\,\|F\|_1$$

for $F \in H_1$ of the special form given, hence for all $F \in H_1$.

We are done.

In like manner one can prove an analogue of the above result for the *unit circle*:

Theorem *Let μ be a positive Radon measure on $\{|z| < 1\}$. Then*

$$\iint_{|z|<1} |f(z)|\, d\mu(z) \leq k\,\|f\|_1$$

for all $f \in H_1(|z| < 1)$ iff

$$\mu(B_h) \leq Ch$$

for every curvilinear box B_h of the form shown:

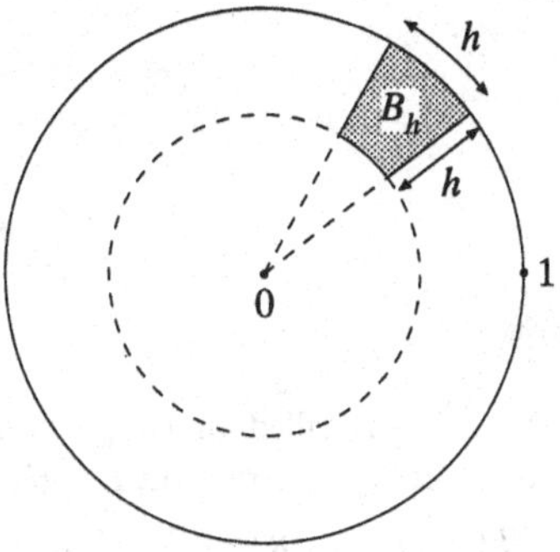

Problem 8 Let $0 = x_0 < x_1 < x_2 < \cdots < x_n = 1$ be any partition whatsoever of $[0, 1]$ into non-overlapping intervals $J_k = [x_{k-1}, x_k]$; put $|J_k| = 2\Delta_k$, $\overline{x_k} = \frac{1}{2}(x_{k-1} + x_k)$.

If

$$\Delta(x) = \sum_{k=1}^{n} \frac{\Delta_k^2}{(x - \overline{x}_k)^2 + \Delta_k^2},$$

show that, for $\lambda > 0$,

$$|\{x \in [0, 1];\ \Delta(x) > \lambda\}| \leq 2e^{-C\lambda}$$

with a numerical constant C independent of λ or of the particular choice of the x_k. (I *think* $C = 1/6\pi e$ works here.)

Hint: If $q = 1, 2, 3, \cdots$, $(1/q) + (1/p) = 1$, and $f \in L_p(0, 1)$, estimate

$$\int_0^1 \Delta(x) f(x)\, dx$$

in terms of $\|f^*\|_p$, and thence obtain an estimate for

$$\int_0^1 (\Delta(x))^q\, dx.$$

Note that

$$e^{\rho\Delta(x)} = \sum_0^\infty \frac{(\rho\Delta(x))^q}{q!}.$$

IX

Interpolation

Definition If $\Im z_n > 0$, $n = 1, 2, \ldots$, $\{z_n\}$ is called an *interpolating sequence for the upper half plane* iff, given *any* bounded sequence $\{c_n\}$, there is an $F \in H_\infty(\Im z > 0)$ such that

$$F(z_n) = c_n, \quad n = 1, 2, 3, \ldots .$$

It is possible to describe all interpolating sequences in terms of a simple geometric condition.

A. Necessary conditions

1. Uniformity lemma

Lemma *$\{z_n\}$ is an interpolating sequence iff there is a constant K such that, for any N, if $|c_n| \le 1$ for $n = 1, 2, \ldots, N$, there is an $F \in H_\infty$ with $\|F\|_\infty \le K$ such that $F(z_n) = c_n$, $n = 1, 2, \ldots, N$ (K being* independent *of N).*

Proof

If Given such a K, let $\{c_n\}$, $n = 1, 2, \ldots$ be an arbitrary sequence with $|c_n| \le 1$, and, for each N, let $F_N \in H_\infty$ with $\|F_N\|_\infty \le K$ be such that

$$F_N(z_n) = c_n \text{ for } n = 1, 2, \ldots, N .$$

Since $\|F_N\|_\infty \le K$, a subsequence of $\{F_N\}$ converges u.c.c. in $\Im z > 0$ to a function $F \in H_\infty$ with $\|F\|_\infty \le K$. Clearly $F(z_n) = c_n$ for all n.

Only if If $\{z_n\}$ is an interpolating sequence, let, for each L, S_L be the subset of $l_\infty(\mathbb{N})$ consisting of all the sequences $\{F(z_n)\}$ formed from the $F \in H_\infty$ with $\|F\|_\infty \le L$. (As *usual*, we denote the set of *positive integers* by $\mathbb{N}$.)

By assumption $\cup_{L=1}^{\infty} S_L = l_\infty(\mathbb{N})$. Also, each S_L is *closed* in $l_\infty(\mathbb{N})$, for if $F_k \in H_\infty$ and $\|F_k\|_\infty \le L$, and if $\{c_n\} \in l_\infty(\mathbb{N})$ is such that $\|\{F_k(z_n) - c_n\}\| = \sup_{n \in \mathbb{N}} |F_k(z_n) - c_n|$ *goes to zero* with k, then, extracting a *subsequence* of $\{F_k\}$ converging u.c.c. in $\Im z > 0$, the *limit*, F, of that subsequence belongs to H_∞ and has $\|F\|_\infty \le L$, with $F(z_n) = c_n$, $n \in \mathbb{N}$. So $\{c_n\} \in S_L$, and S_L *is closed* in $l_\infty(\mathbb{N})$.

The *Baire category theorem* now says that *some* S_L contains a *sphere*, say of radius $\rho > 0$, about one of its *points*, say the sequence $\{F_0(z_n)\}$, $F_0 \in H_\infty$, $||F_0||_\infty \leq L$. That is, if $|c_n - F_0(z_n)| \leq \rho$ for all n, then there is an $F \in H_\infty$, $||F||_\infty \leq L$, with $F(z_n) = c_n$, $n \in \mathbb{N}$. Therefore, if $|d_n| \leq \rho$ for all n, we can find a $G \in H_\infty$, $||G||_\infty \leq 2L$, with $G(z_n) = d_n$, $n \in \mathbb{N}$. (Just put $c_n = d_n + F_0(z_n)$ and then put $G = F - F_0$ with F from the preceding statement.)

It follows that if $|a_n| \leq 1$, $n \in \mathbb{N}$, we can find an $H \in H_\infty$ with $||H||_\infty \leq 2L/\rho$ and $H(z_n) = a_n$ for all n. The lemma holds with $K = 2L/\rho$.

2. $\prod_{k \neq n} |(z_n - z_k)/(z_n - \bar{z}_k)|$ must be bounded below

Lemma $\{z_n\}$ *is an interpolating sequence for* $\Im z > 0$ only if *there is a* $\delta > 0$ *such that, for* each n,

$$(*) \qquad \prod_{k \neq n} \left| \frac{z_n - z_k}{z_n - \bar{z}_k} \right| \geq \delta > 0.$$

Proof If $\{z_n\}$ is an interpolating sequence there is, by the previous lemma, a K such that, for *each* n we can find an $F \in H_\infty$ with $||F||_\infty \leq K$ and

$$F(z_k) = \begin{cases} 1, & k = n \\ 0, & k \neq n \end{cases}.$$

According to Chapter VI, Section C, we can form a Blaschke product $B(z)$ from the zeros z_k, $k \neq n$ of $F(z)$, and we'll have $F(z) = B(z)G(z)$ with a function $G \in H_\infty$ and $||G||_\infty = ||F||_\infty \leq K$. We have

$$|B(z_n)| = \prod_{k \neq n} \left| \frac{z_n - z_k}{z_n - \bar{z}_k} \right|,$$

so $1 = |F(z_n)| = |B(z_n)||G(z_n)| \leq K|B(z_n)|$ immediately yields $(*)$ with $\delta = 1/K$.

B. Carleson's theorem

The *necessary condition* $(*)$ for $\{z_n\}$ to be an interpolating sequence is also *sufficient*. This remarkable result is due to L. Carleson, and goes back to 1958 or thereabouts. Later, Shapiro and Shields found an *easier* proof of Carleson's theorem by making successive reductions with the help of the duality theory presented in Chapter VII, Subsections A.1 and A.2. This proof is given in the books by Duren and Hoffman.

Here, we give a proof which is more like Carleson's original one. It is simplified by applying directly the theorem on Carleson measures given in Chapter VIII, Section F. Thanks to an idea of Hörmander, it was possible to establish that result rather easily.

1. Computational lemma

Lemma *If, for every* n,

$$\prod_{k \neq n} \left| \frac{z_n - z_k}{z_n - \bar{z}_k} \right| \geq \delta > 0,$$

we have, for each n,

$$\sum_{k \neq n} \frac{\Im z_n \Im z_k}{|z_n - \bar{z}_k|^2} \leq \frac{1}{2} \log \frac{1}{\delta}.$$

Proof Write $z_k = x_k + iy_k$, $z_n = x_n + iy_n$; we have

$$0 < \left|\frac{z_n - z_k}{z_n - \bar{z}_k}\right| < 1,$$

and

$$\begin{aligned} 1 - \left|\frac{z_n - z_k}{z_n - \bar{z}_k}\right|^2 &= \frac{|z_n - \bar{z}_k|^2 - |z_n - z_k|^2}{|z_n - \bar{z}_k|^2} \\ &= \frac{(x_n - x_k)^2 + (y_n + y_k)^2 - (x_n - x_k)^2 - (y_n - y_k)^2}{|z_n - \bar{z}_k|^2} = \frac{4y_n y_k}{|z_n - \bar{z}_k|^2}. \end{aligned}$$

Writing $p_k = |(z_n - z_k)/(z_n - \bar{z}_k)|$, we have

$$2\log\frac{1}{p_k} = \log\frac{1}{1-(1-p_k^2)} = \sum_{m=1}^{\infty}\frac{(1-p_k^2)^m}{m} \geq 1 - p_k^2 = \frac{4\Im z_n \Im z_k}{|z_n - \bar{z}_k|^2},$$

by the calculation just made.

Since $\sum_k 2\log(1/p_k) \leq 2\log(1/\delta)$, the lemma follows.

2. A Carleson measure

Lemma *If, for all n,*

$$(*) \qquad \prod_{k \neq n}\left|\frac{z_n - z_k}{z_n - \bar{z}_k}\right| \geq \delta > 0,$$

the measure μ on $\Im z > 0$ given by

$$d\mu(z) = \sum_n \Im z_n d\delta_{z_n}(z),$$

(in other words, assigning mass $\Im z_n$ *to each* point z_n*) is a Carleson measure.*

Proof We will show that for any $x_0 \in \mathbb{R}$ and any $h > 0$

$$\mu([x_0, x_0 + h] \times (0, h)) \leq \left(1 + 5\log\frac{1}{\delta}\right)h;$$

this makes μ a Carleson measure by the theorem of Chapter VIII, Section F.

Pick any $x_0 \in \mathbb{R}$ and any $h > 0$, and denote the square $[x_0, x_0 + h] \times (0, h)$ by S:

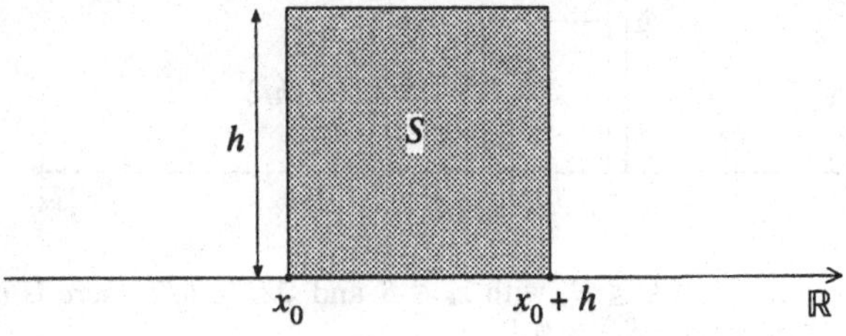

We have to show that

$$\sum_{z_n \in S} \Im z_n \leq \left(1 + 5\log\frac{1}{\delta}\right) h ;$$

this will evidently follow if, for *every* N we have

$$\sum_{z_n \in S \; n \leq N} \Im z_n \leq \left(1 + 5\log\frac{1}{\delta}\right) h .$$

Fix a large integer N. If there are *no* z_k, $1 \leq k \leq N$, with $z_k \in S$, *the desired inequality is surely true.*

Suppose now that there *are* some $z_k \in S$, $1 \leq k \leq N$. *If, for one of them, say* z_n, *we have* $\Im z_n \geq h/2$, *then the desired inequality is true*. Indeed, take the result of the preceding lemma:

$$\sum_{k \neq n} \Im z_k \cdot \frac{\Im z_n}{|z_n - \bar{z}_k|^2} \leq \frac{1}{2}\log\frac{1}{\delta} .$$

Observe that if, for $k \neq n$, $z_k \in S$, we *certainly have*

$$\frac{\Im z_n}{|z_n - \bar{z}_k|^2} \geq \frac{1}{10h},$$

because $|z_n - \bar{z}_k|^2 = (x_n - x_k)^2 + (y_n + y_k)^2 \leq h^2 + 4h^2 = 5h^2$ whilst $\Im z_n \geq h/2$. Therefore, by the previous relation,

$$\frac{1}{10h} \sum_{z_k \in S \; k \neq n} \Im z_k \leq \frac{1}{2}\log\frac{1}{\delta},$$

and, since $\Im z_n \leq h$,

$$\sum_{z_k \in S} \Im z_k \leq h + 5h\log\frac{1}{\delta},$$

implying the desired inequality (for *every* N).

It may be, however, that there are $z_k \in S$ *for* $1 \leq k \leq N$, *but that none of them has* $\Im z_k \geq h/2$. In that case, *we decompose the square* S into an *upper rectangle* R and *two lower quarter squares*, $S_{1,1}$, $S_{1,2}$, in the manner shown:

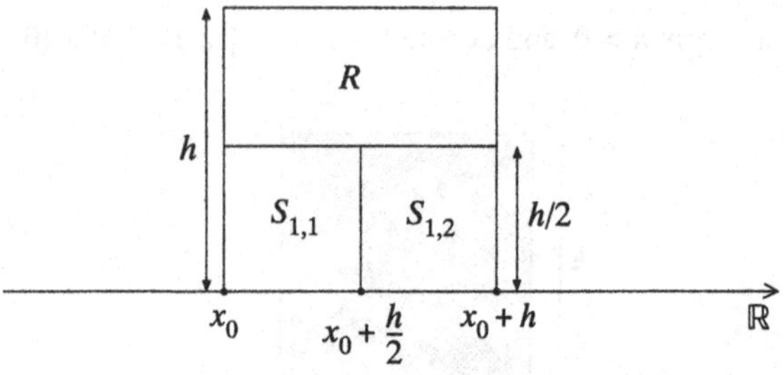

Because there is *no* z_k, $1 \le k \le N$, with $z_k \in S$ and $\Im z_k \ge h/2$, *there is no such* z_k *in* R, and therefore (note that $S_{1,1} \cap S_{1,2} \neq \emptyset$!)

$$\sum_{\substack{z_k \in S \\ 1 \le k \le N}} \Im z_k \le \textstyle\sum(S_{1,1}) + \sum(S_{1,2}),$$

where we write, during the rest of this discussion,

$$\textstyle\sum(A) = \displaystyle\sum_{\substack{z_k \in A \\ 1 \le k \le N}} \Im z_k .$$

If, now, there *is* a $z_n \in S_{1,1}$, $1 \le k \le N$, with $\Im z_n \ge h/4$, the above discussion shows that $\sum(S_{1,1}) \le \left(1 + 5\log(1/\delta)\right)(h/2)$. And if $S_{1,1}$ *has no* z_k *in it at all for* $1 \le k \le N$, $\sum(S_{1,1}) = 0$.

Similarly, $\sum(S_{1,2}) \le \left(1 + 5\log(1/\delta)\right)(h/2)$ if $S_{1,2}$ has in it a z_n, $1 \le n \le N$, with $\Im z_n \ge h/4$. If $S_{1,2}$ has *no* z_k *at all in it for* $1 \le k \le N$, then $\sum(S_{1,2}) = 0$.

If, now, there *are* z_k, $1 \le k \le N$ in $S_{1,1}$ but *all have* $\Im z_k < h/4$, we look at the *two lower quarter squares* $S_{1,1,1}$, $S_{1,1,2}$ in $S_{1,1}$, each of side $h/4$, and *see whether either has in it a* z_n, $1 \le n \le N$, *with* $\Im z_n \ge h/8$, *or else has no* z_k *in it at all for* $1 \le k \le N$.

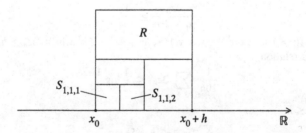

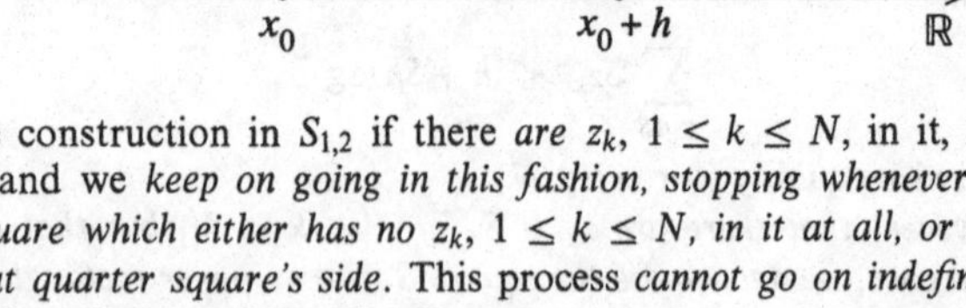

We do the same construction in $S_{1,2}$ if there *are* z_k, $1 \le k \le N$, in it, but *none of them has* $\Im z_k \ge h/4$, and we *keep on going in this fashion, stopping whenever we first get to a lower quarter square which either has no* z_k, $1 \le k \le N$, *in it at all, or else has one with* $\Im z_k \ge 1/2$ *of that quarter square's side*. This process *cannot go on indefinitely*, because we are *looking at only a finite number* (N) of z_k.

Here is an example of how the process could work out. In the figure, each *shaded* lower quarter square has in it a z_k, $1 \le k \le N$, with $\Im z_k \ge 1/2$ of that quarter square's side. The *unshaded* squares and rectangles have *no* z_k in them for $1 \le k \le N$.

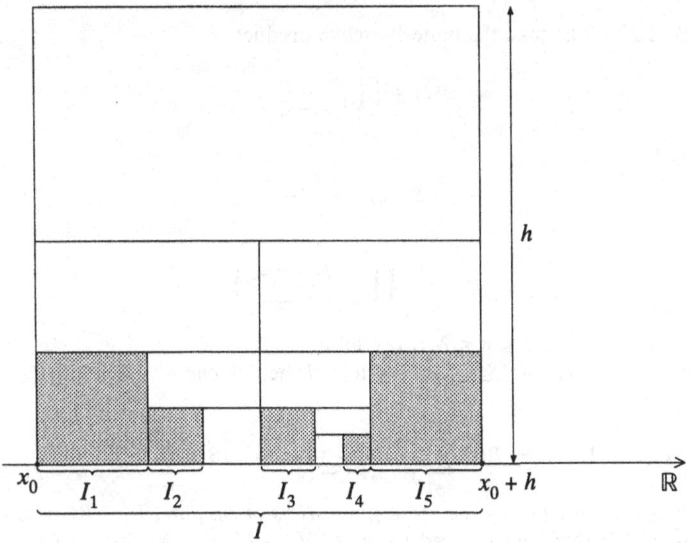

The construction just described leads to a finite number of *non-overlapping intervals*, say $I_1, I_2, \ldots, I_p$ inside the interval $I = [x_0, x_0 + h]$, each having length equal to that of I divided by some power of 2. Each I_k is the *base* of a *square*, say $S^{(k)}$, lying in S, and *all the* z_k, $1 \le k \le N$, belonging to S lie in the *union* of the $S^{(k)}$. Each $S^{(k)}$ has in it a z_k, $1 \le k \le N$, with $\Im z_k \ge \frac{1}{2}|I_k|$.

By the discussion at the beginning of this proof, we now have

$$\sum(S^{(k)}) \le \left(1 + 5\log\frac{1}{\delta}\right)|I_k|, \qquad k = 1, 2, \ldots, p.$$

Therefore

$$\sum(S) \le \sum_{k=1}^{p} \sum(S^{(k)}) \le \left(1 + 5\log\frac{1}{\delta}\right)\sum_{k=1}^{p}|I_k| \le \left(1 + 5\log\frac{1}{\delta}\right)|I|,$$

that is,

$$\sum_{z_k \in S \ 1 \le k \le N} \Im z_k \le \left(1 + 5\log\frac{1}{\delta}\right)h,$$

our desired inequality. The lemma is completely proved.

3. Proof of sufficiency of the condition (*)

Carleson's theorem *A sequence $\{z_n\}$ in the upper half plane with*

$$(*) \qquad \prod_{k \ne n}\left|\frac{z_n - z_k}{z_n - \bar{z}_k}\right| \ge \delta > 0$$

for every n is an interpolating sequence.

Proof By Subsection A.1 it is enough to show that there is a fixed constant K such that, for any positive integer N and any numbers c_n, $n = 1, 2, \ldots, N$ with $|c_n| \le 1$, there is an $F \in H_\infty$ with $\|F\|_\infty \le K$ and

$$F(z_n) = c_n, \qquad n = 1, 2, \ldots, N.$$

Having fixed N, let us take the finite Blaschke product

$$B(z) = \prod_{k=1}^{N} \left(\frac{z - z_k}{z - \bar{z}_k} \right).$$

For $1 \leq n \leq N$,

$$B'(z_n) = \frac{\beta_n}{2i\Im z_n},$$

where

$$\beta_n = \prod_{1 \leq k \leq N \ k \neq n} \left(\frac{z_n - z_k}{z_n - \bar{z}_k} \right),$$

so that surely $|\beta_n| \geq \delta$ for $1 \leq n \leq N$ if (*) holds.

Given the numbers c_n, $|c_n| \leq 1$, for $1 \leq n \leq N$, here is *one* $F \in H_\infty$ with $F(z_n) = c_n$ for $1 \leq n \leq N$:

$$F(z) = F_0(z) = B(z) \sum_{n=1}^{N} \frac{c_n}{B'(z_n)(z - z_n)} = 2iB(z) \sum_{n=1}^{N} \frac{\Im z_n}{\beta_n} \cdot \frac{c_n}{z - z_n}.$$

Any other $F \in H_\infty$ *with* $F(z_n) = c_n$ *for* $1 \leq n \leq N$ *is of the form* $F_0 + BG$ *where* $G \in H_\infty$, *and conversely*. We therefore proceed to see *if* $\|F_0 - BH_\infty\|_\infty$ *has a bound independent of* N; if it has one, say K, then *there will be* an $F \in H_\infty$ with $\|F\|_\infty \leq K$ and $F(z_n) = c_n$, $n = 1, 2, \ldots, N$, so *we will be done*, K being *independent* of N.

The idea now (due to D.J. Newman) is to use the duality theory of Chapter VII, Subsection A.2 to compute $\|F_0 - BH_\infty\|_\infty$. Since $|B(x)| \equiv 1$, $x \in \mathbb{R}$, (that's just why we *used* B !), we have

$$\|F_0 - BH_\infty\|_\infty = \inf\{\|(F_0/B) - G\|_\infty\,;\ G \in H_\infty\}.$$

By the analogue of a theorem in Chapter VII, Subsection A.2 for the *upper half plane* (see the table at end of Chapter VII, Subsection A.1), the last inf is equal to

$$\sup\left\{ \left| \int_{-\infty}^{\infty} \frac{F_0(x)}{B(x)} f(x)\, dx \right| ; \quad f \in H_1,\ \|f\|_1 \leq 1 \right\}.$$

For each $f \in H_1$, we have

$$\int_{-\infty}^{\infty} \frac{F_0(x)}{B(x)} f(x)\, dx = 2i \sum_{n=1}^{N} \int_{-\infty}^{\infty} \frac{\Im z_n}{\beta_n} \frac{c_n}{x - z_n} f(x)\, dx = -4\pi \sum_{n=1}^{N} \frac{c_n}{\beta_n} \Im z_n f(z_n).$$

Since $|\beta_n| \geq \delta$ and $|c_n| \leq 1$, the last expression is in absolute value $\leq 4\pi\delta^{-1} \sum_{n=1}^{N} \Im z_n |f(z_n)|$. But according to the lemma of Subsection 2 above, *the measure assigning mass* $\Im z_n$ *to each point* z_n *is Carleson*. Therefore, for $f \in H_1$,

$$4\pi\delta^{-1} \sum_{n=1}^{\infty} \Im z_n |f(z_n)| \leq 4\pi\delta^{-1} C\, \|f\|_1,$$

so, substituting in the previous relation, we see that $\|F_0 - BH_\infty\|_\infty \leq 4\pi\delta^{-1}C$, *a constant independent of* N.

The proof is complete.

Remark The proofs of the lemma in Subsection 2 and of the theorem on Carleson measures in Chapter VIII, Section F show that we can take the constant C figuring at the end of the above demonstration equal to $\tilde{C}\,(1 + 5\log(1/\delta))$, where $\tilde{C}$ is a purely *numerical* constant, independent of δ, hence of the sequence $\{z_n\}$. This means that whenever $|c_n| \le 1$ for all n we can find an $F \in H_\infty$ with $\|F\|_\infty \le K$ and $F(z_n) = c_n$, $n = 1,2,\ldots$, *where the constant K depends on the δ figuring in* (*) *like a numerical multiple of* $(1/\delta)\log(1/\delta)$.

Carleson's theorem can be carried over directly to the unit circle by conformal transformation. We just state the result:

Theorem *Let $|z_n| < 1$, $n = 1,2,\ldots$. A necessary and sufficient condition that there exist, for any bounded sequence $\{c_n\}$, an $f \in H_\infty(|z|<1)$ with $f(z_n) = c_n$, $n = 1,2,\ldots$, is that*

$$\prod_{k\neq n}\left|\frac{z_n - z_k}{1-\bar{z}_k z_n}\right| \ge \text{some } \delta > 0$$

for all n with δ independent of n.

It was indeed in *this* form that Carleson first published his result.

C. Weighted interpolation by functions in other H_p spaces

Theorem (Shapiro and Shields) *Let $1 \le p \le \infty$ and $\Im z_n > 0$. The sequences $\{(\Im z_n)^{1/p}F(z_n)\}$ fill out $l_p(\mathbb{N})$ as F ranges over $H_p(\Im z > 0)$ iff $\{z_n\}$ is an interpolating sequence.*

*Proof** By arguing precisely as in the proof of the lemma in Subsection A.1, we see that the sequences $\{(\Im z_n)^{1/p}F(z_n)\}$ fill out $l_p(\mathbb{N})$ as F ranges over H_p iff there is a $K < \infty$ such that, for *any* positive integer N and *any* c_n, $1 \le n \le N$, with $\sum_n |c_n|^p \le 1$, we can find an $F \in H_p$ with $\|F\|_p \le K$ and

$$(\Im z_n)^{1/p}F(z_n) = c_n\,, \quad n = 1,2,\ldots,N\,.$$

I claim first of all that *if this property holds, then $\{z_n\}$ is an interpolating sequence.* Indeed, fix any n and put

$$c_k = \begin{cases} 0, & k \neq n,\\ 1, & k = n\,. \end{cases}$$

For arbitrarily large N we can find an $F \in H_p$, $\|F\|_p \le K$, with

$$(\Im z_k)^{1/p}F(z_k) = c_k\,, \quad 1 \le k \le N\,.$$

Clearly, then, $F = BG$ where $G \in H_p$, $\|G\|_p \le K$, and $B(z)$ is the partial Blaschke product

$$B(z) = \prod_{k\neq n\ 1\le k\le N}\left(\frac{z-z_k}{z-\bar{z}_k}\right).$$

Since $1 = (\Im z_n)^{1/p}|F(z_n)| = |B(z_n)|(\Im z_n)^{1/p}|G(z_n)|$, we must have

$$\prod_{k\neq n\ 1\le k\le N}\left|\frac{z_n-z_k}{z_n-\bar{z}_k}\right| \ge \frac{1}{(\Im z_n)^{1/p}|G(z_n)|}.$$

Because $G \in H_p$ we can use Poisson's formula (Chapter VI) and we find, with

* We write out the details of the proof for $1 < p < \infty$. For $p = \infty$, the result has already been given, and for $p = 1$, the arguments below based on Hölder's inequality can be replaced by simpler more direct ones.

$1/q = 1 - (1/p)$,

$$(\Im z_n)^{1/p}|G(z_n)| = \left|\frac{1}{\pi}\int_{-\infty}^{\infty}\frac{(\Im z_n)^{1+1/p}G(t)\,dt}{|z_n - t|^2}\right| \le \frac{1}{\pi}\|G\|_p\left\{\int_{-\infty}^{\infty}\frac{y_n^{q+q/p}}{[(x_n-t)^2+y_n^2]^q}\,dt\right\}^{1/q}.$$

Now $\|G\|_p \le K$, and the integral in { } equals

$$\int_{-\infty}^{\infty}\frac{y_n^{2q-1}}{(y_n^2+\tau^2)^q}\,d\tau = \int_{-\infty}^{\infty}\frac{ds}{(s^2+1)^q},$$

a finite quantity, say C_p^q, *depending only on p*. Hence $(\Im z_n)^{1/p}|G(z_n)| \le \pi^{-1}KC_p$ for $\|G\|_p \le K$, and from the previous paragraph we get

$$\prod_{k\neq n\ 1\le k\le N}\left|\frac{z_n - z_k}{z_n - \bar{z}_k}\right| \ge \frac{\pi}{KC_p},$$

independently of N. Making now $N \to \infty$, we see that condition (*) of Carleson's theorem (Subsection B.3) *is* fulfilled, so $\{z_n\}$ *is* an *interpolating sequence.*

Conversely, suppose $\{z_n\}$ *is an interpolating sequence*; say

$$\prod_{k\neq n}\left|\frac{z_n - z_k}{z_n - \bar{z}_k}\right| \ge \delta > 0$$

for all n. Let N be any finite positive integer – *fix* it – and let numbers c_n, $n = 1, 2, \ldots, N$ be given with $\sum_1^N |c_n|^p \le 1$. Write, as in Subsection B.3,

$$B(z) = \prod_{n=1}^{N}\left(\frac{z - z_n}{z - \bar{z}_n}\right),$$

$$\beta_n = \prod_{1\le k\le N\ k\neq n}\left(\frac{z_n - z_k}{z_n - \bar{z}_k}\right).$$

Put $\gamma_n = (\Im z_n)^{-1/p}c_n$, $n = 1, \ldots, N$. Then, as in the proof of Carleson's theorem*,

$$F_0(z) = 2iB(z)\sum_{n=1}^{N}\frac{\gamma_n}{\beta_n}\frac{\Im z_n}{z - z_n}$$

has $F_0(z_n) = \gamma_n$, i.e., $(\Im z_n)^{1/p}F_0(z_n) = c_n$, and *we will be finished* if we can show that there is a $K < \infty$ *independent* of N and the *particular numbers* c_n, $n = 1, 2, \ldots, N$, such that $\|F_0 - BH_p\|_p \le K$.

Since $|B(x)| \equiv 1$ for $x \in \mathbb{R}$, we have, by the duality theory of Chapter VII, Section A,

$$\|F_0 - BH_p\|_p = \|(F_0/B) - H_p\|_p = \sup\left\{\left|\int_{-\infty}^{\infty}\frac{F_0(t)}{B(t)}f(t)\,dt\right|;\ f \in H_q,\ \|f\|_q \le 1\right\}.$$

* When $p = 1$, one should take

$$F_0(z) = 2iB(z)\sum_{n=1}^{N}\frac{\gamma_n}{\beta_n}\left(\frac{\Im z_n}{z - z_n} - \frac{\Im z_n}{z - \bar{z}_n}\right).$$

As in Subsection B.3, for $f \in H_q$, $\int_{-\infty}^{\infty}(F_0(t)f(t)/B(t))\,dt$ works out to

$$-4\pi \sum_{n=1}^{N} \Im z_n \gamma_n f(z_n)/\beta_n,$$

which, in the present case, is seen *by Hölder's inequality to be in absolute value*

$$\leq 4\pi \left[\sum_{n=1}^{N} \Im z_n |\gamma_n|^p/|\beta_n|^p\right]^{1/p} \left[\sum_{n=1}^{N} \Im z_n |f(z_n)|^q\right]^{1/q}.$$

Here, $\Im z_n |\gamma_n|^p = |c_n|^p$ and $|\beta_n|^p \geq \delta^p$, so, since $\sum_n |c_n|^p \leq 1$, the last expression is seen to be

$$\leq 4\pi\delta^{-1} \left[\sum_{n=1}^{\infty} \Im z_n |f(z_n)|^q\right]^{1/q}.$$

For $f \in H_q$ and $\|f\|_q \leq 1$ we can certainly write $f(z) = b(z)g(z)$ where $b(z)$ is a *Blaschke product* (hence in modulus ≤ 1 in $\Im z > 0$) and $g \in H_q$, $\|g\|_q = \|f\|_q$, is *free of zeros in the upper half plane*. So $g^q \in H_1$ with $\|g^q\|_1 \leq 1$, therefore, *by the lemma* of Subsection B.2,

$$\sum_{n=1}^{\infty} \Im z_n |f(z_n)|^q \leq \sum_{n=1}^{\infty} \Im z_n |g(z_n)|^q \leq C\,\|g^q\|_1 \leq C,$$

the measure assigning mass $\Im z_n$ to each point z_n being *Carleson*.

We see finally that

$$\|F_0 - BH_p\|_p \leq 4\pi\delta^{-1}C^{1/q},$$

a number independent of N and of the particular numbers c_n chosen for $1 \leq n \leq N$ with $\sum_n |c_n|^p \leq 1$. The theorem is completely proved.

D. Relations between some conditions on sequences $\{z_n\}$

In proving the lemma of Subsection B.2 we *really* showed that if

$$(**) \qquad \sum_k \frac{\Im z_n \Im z_k}{|z_n - \bar{z}_k|^2} \leq K,$$

a constant independent of n, for all n, *then* the measure assigning mass $\Im z_n$ to each point z_n is *Carleson*.

The converse is true.

Lemma *If the measure assigning mass $\Im z_n$ to each point z_n is Carleson, then there is a constant K with*

$$(**) \qquad \sum_k \frac{\Im z_n \Im z_k}{|z_n - \bar{z}_k|^2} \leq K$$

for every n.

Proof (Garnett) The test functions $F_n(z) = \Im z_n/(z - \bar{z}_n)^2$ belong to H_1 and $\|F_n\|_1 = \pi$ for every n. So by the Carleson measure property,

$$\sum_k \Im z_k |F_n(z_k)| = \sum_k \frac{\Im z_k \Im z_n}{|z_n - \bar{z}_k|^2} \leq K,$$

a constant independent of n. Q.E.D.

Lemma (Garnett) *If there is an $\eta > 0$ such that $|(z_n - z_m)/(z_n - \bar{z}_m)| \geq \eta$ for $n \neq m$ (in other words, if the* hyperbolic distance *between different points of the sequence $\{z_n\}$ is* bounded below*), and if*

$$(**) \qquad \sum_k \frac{\Im z_n \Im z_k}{|z_n - \bar{z}_k|^2} \leq K$$

for all n, then

$$(*) \qquad \prod_{k \neq n} \left| \frac{z_n - z_k}{z_n - \bar{z}_k} \right| \geq \delta > 0$$

with a suitable δ for all n, and $\{z_n\}$ is an interpolating sequence.

Proof Take any n and k with $k \neq n$, and put for the moment $r = |(z_n - z_k)/(z_n - \bar{z}_k)|$; then $0 < r < 1$, so

$$\log \frac{1}{r^2} = \log \frac{1}{1-(1-r^2)} = \sum_{l=1}^{\infty} \frac{(1-r^2)^l}{l} = (1-r^2) \sum_{m=0}^{\infty} \frac{(1-r^2)^m}{m+1}.$$

If $r \geq \eta$, the last expression is

$$\leq (1-r^2) \sum_{m=0}^{\infty} (1-\eta^2)^m = \frac{1-r^2}{\eta^2}.$$

By the computation in Subsection B.1,

$$1 - r^2 = \frac{4\Im z_n \Im z_k}{|z_n - \bar{z}_k|^2},$$

so we get by the previous calculation

$$\left| \frac{z_n - z_k}{z_n - \bar{z}_k} \right|^2 \geq \exp\left(-\frac{1-r^2}{\eta^2} \right) = \exp\left(-\frac{4\Im z_n \Im z_k}{\eta^2 |z_n - \bar{z}_k|^2} \right).$$

Using (**), we now find

$$\prod_{k \neq n} \left| \frac{z_n - z_k}{z_n - \bar{z}_k} \right|^2 \geq e^{-4K/\eta^2},$$

proving (*) with $\delta = e^{-4K/\eta^2}$. We're done.

The last result can now be combined with the lemma of Subsection B.1 to yield the

Theorem *For sequences $\{z_n\}$ in the upper half plane, the condition that*

$$(*) \qquad \prod_{k \neq n} \left| \frac{z_n - z_k}{z_n - \bar{z}_k} \right| \geq \text{some } \delta > 0$$

for all n is equivalent *to*

$$(**) \qquad \sum_k \frac{\Im z_n \Im z_k}{|z_n - \bar{z}_k|^2} \leq K$$

independent of n for all n plus *the* uniform hyperbolic separation condition

$$\left| \frac{z_n - z_m}{z_n - \bar{z}_m} \right| \geq \eta > 0, \quad n \neq m.$$

It is good to summarize the relations between various properties of sequences $\{z_n\}$ in $\Im z > 0$ by means of the following table:

$$\prod_k \frac{\Im z_k \Im z_n}{|z_n - \bar{z}_k|^2} \le K \text{ independent of } n$$

EQUIVALENT TO

$$\sum_{z_k \in S_h} \Im z_k \le Ch \text{ for any square}$$

$$S_h = [x_0, x_0 + h] \times (0, h)$$

EQUIVALENT TO

$$\sum_k \Im z_k |F(z_k)| \le \tilde{C}\, \|F\|_1 \text{ for any}$$

$F \in H_1$ (Carleson measure property)

+

$$\left| \frac{z_n - z_m}{z_n - \bar{z}_m} \right| \ge \eta > 0$$

if $n \ne m$.

TOGETHER EQUIVALENT TO

$$\prod_{k \ne n} \left| \frac{z_n - z_k}{z_n - \bar{z}_k} \right| \ge \delta > 0 \text{ independent of } n$$

EQUIVALENT TO

$\{z_n\}$ is an interpolating sequence for $\Im z > 0$.

E. Interpolation by bounded harmonic functions. Garnett's theorem

In establishing the following result, we have to consider the *average* of an expression $(\pm a_1 \pm a_2 \pm a_3 \pm \ldots \pm a_N)^2$ *for all possible choices of the* + *and* − *signs* (there are 2^N such choices).

This average is simply

$$a_1^2 + a_2^2 + a_3^2 + \ldots + a_N^2,$$

because the *cross terms* $2(\pm a_k)(\pm a_l)$ obtained in the squaring operation *average out to zero.*

Garnett's theorem (ca. 1970) *Let* $\Im z_n > 0$. *Suppose that for any bounded sequence* $\{c_n\}$ *there is a* $U(z)$ bounded *and* harmonic *(*not *necessarily analytic!) in* $\Im z > 0$ *with* $U(z_n) = c_n$. *Then* $\{z_n\}$ *is an interpolating sequence for* $\Im z > 0$.

Proof (Varopoulos, ca. 1972 or 1973) If the sequences $\{U(z_n)\}$ fill out $l_\infty(\mathbb{N})$ as U ranges over the bounded harmonic functions (on the upper half plane), the argument used to prove the lemma of Subsection A.1 shows that there is a K such that, for any sequence $\{c_n\}$ with $|c_n| \le 1$, there is a harmonic function $U(z)$ with $|U(z)| \le K$ for $\Im z > 0$ and $U(z_n) = c_n$. If $\{c_n\}$ is *real*, this holds for a *real valued* U, for if $U(z)$ is (complex valued) harmonic, so is $\Re U(z)$.

Given any n, let

$$c_k = \begin{cases} 0, & k \neq n \\ 1, & k = n, \end{cases}$$

and take a real valued harmonic U with $|U(z)| \leq K$ in $\Im z > 0$ and $U(z_k) = c_k$. Then $U(z) + K$ is harmonic and ≥ 0 in $\Im z > 0$, $U(z_n) + K = K + 1$, and $U(z_k) + K = K$ for $k \neq n$. *Harnack's theorem** now gives us

$$\frac{K+1}{K} = \frac{U(z_n)+K}{U(z_k)+K} \leq \frac{1 + |(z_n - z_k)/(z_n - \bar{z}_k)|}{1 - |(z_n - z_k)/(z_n - \bar{z}_k)|},$$

from which we see that

(†) $$\left| \frac{z_n - z_k}{z_n - \bar{z}_k} \right| \geq \frac{1}{2K+1}, \quad k \neq n.$$

The equivalences established in Section D show that *if we can establish*

$$\sum_{z_k \in S_h} \Im z_k \leq Ch$$

for *all* S_h of the form $[x_0, x_0 + h] \times (0, h)$, *we will be done*, thanks to (†).

Observe first of all that

(§) $$\sum_n |a_n| \leq \frac{K}{\pi} \int \left| \sum_n \frac{a_n \Im z_n}{|z_n - t|^2} \right| dt$$

for any sequence $\{a_n\}$ in $l_1(\mathbb{N})$. *Indeed*, we can find numbers c_n, $|c_n| = 1$, with $\sum_n |a_n| = \sum_n a_n c_n$, and then we can get $U(z)$ harmonic in $\Im z > 0$, $|U(z)| \leq K$ there, with $U(z_n) = c_n$. By Chapter VI,

$$U(z_n) = \frac{1}{\pi} \int_{-\infty}^{\infty} \frac{\Im z_n}{|z_n - t|^2} U(t)\, dt$$

with $|U(t)| \leq K$ a.e., so

$$\sum_n |a_n| = \sum_n a_n c_n = \sum_n a_n U(z_n) = \frac{1}{\pi} \int_{-\infty}^{\infty} \left(\sum_n \frac{a_n \Im z_n}{|z_n - t|^2} \right) U(t)\, dt,$$

which is in absolute value

$$\leq \frac{K}{\pi} \int_{-\infty}^{\infty} \left| \sum_n \frac{a_n \Im z_n}{|z_n - t|^2} \right| dt.$$

Writing $z_n = x_n + iy_n$, let S_h be any *square* of side h whose *base*, I, lies on $\mathbb{R}$. We are to show that

$$\sum_{z_n \in S_h} y_n \leq Ch$$

with a constant C independent of h or the position of I on $\mathbb{R}$.

Let the *numerical* constant M be *so large* that

$$\frac{K}{\pi} \int_{(M-1)/2}^{\infty} \frac{ds}{s^2+1} \leq \frac{1}{4};$$

* To obtain the following relation from the lemma at the beginning of the scholium to Subsection D.3, Chapter I, make a preliminary conformal mapping of $\{\Im z > 0\}$ onto the unit circle which sends z_n to 0.

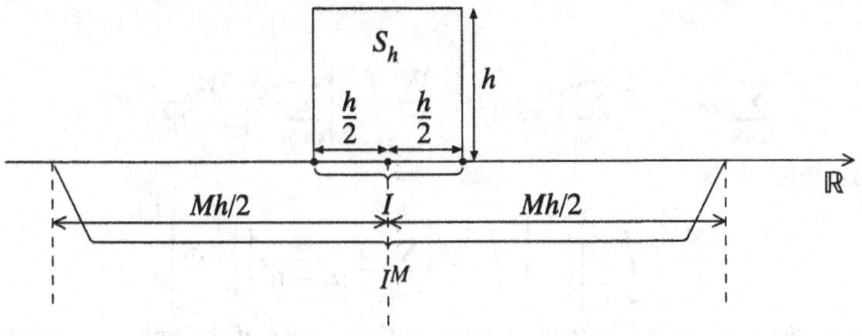

fix such an M *once and for all*. Whatever, I, the base of S_h, may be, let I^M be an interval on $\mathbb{R}$ having *the same midpoint as* I, *but* M *times the length of* I.

If $z_n \in S_h$, our choice of M gives

$$\frac{K}{\pi}\int_{\sim I^M} \frac{\Im z_n}{|z_n - t|^2}\,dt \le \frac{1}{4} + \frac{1}{4} = \frac{1}{2}$$

as is easily seen by making the change of variable $(t - x_n)/y_n = s$.

> In order to estimate $\sum_{z_n \in S_h} y_n$, *Varopoulos' idea is to use* (§) *with* $a_n = \pm y_n$ *if* $z_n \in S_h$ *and* $a_n = 0$ *otherwise, then to average over all possible choices of plus and minus signs!*

For technical reasons, we also *limit* n *to the range* $1, 2, \ldots, N$ where N is some arbitrary large integer. Our estimates *will not depend* on N, so at the end we can make $N \to \infty$. *Thus, for the time being, in all sums,* n *is restricted to the range* $1 \le n \le N$, *but this restriction is not explicitly stated.* From (§), we have

$$\sum_{z_n \in S_h} y_n = \sum_{z_n \in S_h} |\pm y_n| \le \frac{K}{\pi}\int_{-\infty}^{\infty} \left| \sum_{z_n \in S_h} \frac{\pm y_n^2}{|z_n - t|^2} \right| dt.$$

As we have just seen, our choice of M makes the last integral

$$\le \frac{K}{\pi}\int_{I^M} \left| \sum_{z_n \in S_h} \frac{\pm y_n^2}{|z_n - t|^2} \right| dt + \sum_{z_n \in S_h} \frac{K y_n}{\pi} \int_{\sim I^M} \frac{y_n}{|z_n - t|^2}\,dt$$

$$\le \frac{K}{\pi}\int_{I^M} \left| \sum_{z_n \in S_h} \frac{\pm y_n^2}{|z_n - t|^2} \right| dt + \frac{1}{2}\sum_{z_n \in S_h} y_n.$$

Subtracting $\frac{1}{2}\sum_{z_n \in S_h} y_n$ and multiplying by 2, we find

$$\sum_{z_n \in S_h} y_n \le \frac{2K}{\pi}\int_{I^M} \left| \sum_{z_n \in S_h} \pm \frac{y_n^2}{|z_n - t|^2} \right| dt.$$

> Let $\mathscr{E}$ denote the operation of averaging over all possible choices of plus and minus signs.

Then

$$\sum_{y_n\in S_h} y_n = \mathscr{E}\sum_{y_n\in S_h} |\pm y_n| \le \frac{2K}{\pi}\int_{I^M}\mathscr{E}\left|\sum_{z_n\in S_h}\pm\frac{y_n^2}{|z_n-t|^2}\right| dt.$$

By Schwarz, this last is

$$\le \frac{2K}{\pi}\left[\int_{I^M} dt\right]^{1/2}\left[\int_{I^M}\left\{\mathscr{E}\left|\sum_{z_n\in S_h}\pm\frac{y_n^2}{|z_n-t|^2}\right|\right\}^2 dt\right]^{1/2}.$$

The square of the average is less than or equal to the average of the square, so the preceding expression is

$$\le \frac{2K}{\pi}\sqrt{Mh}\left[\int_{I^M}\mathscr{E}\left(\sum_{z_n\in S_h}\pm\frac{y_n^2}{|z_n-t|^2}\right)^2 dt\right]^{1/2} = \frac{2K}{\pi}\sqrt{Mh}\left[\int_{I^M}\sum_{z_n\in S_h}\frac{y_n^4}{|z_n-t|^4}dt\right]^{1/2},$$

using the observation made at the beginning of this Section.

But

$$\int_{I^M}\frac{y_n^4}{|z_n-t|^4}dt \le \int_{-\infty}^{\infty}\frac{y_n^4\,dt}{|z_n-t|^4} = y_n\int_{-\infty}^{\infty}\frac{ds}{(s^2+1)^2} = cy_n,$$

say, where c is a *numerical constant* whose *value* need not concern us here.

Substituting into the previous expression and going back to the chain of inequalities from which it came, we see that

$$\sum_{z_n\in S_h} y_n \le \frac{2K}{\pi}\sqrt{Mh\cdot c\sum_{z_n\in S_h} y_n}.$$

As stipulated above, the sum here is really for $1\le n\le N$, hence it is certainly finite, so, after squaring and cancelling,

$$\sum_{z_n\in S_h\ 1\le n\le N} y_n \le \frac{4K^2Mc}{\pi^2}h.$$

The coefficient on the right does not depend on N, so, *now* making $N\to\infty$, we have

$$\sum_{z_n\in S_h} y_n \le \frac{4K^2Mc}{\pi^2}h.$$

Since (†) also holds, the theorem is completely proved.

Problem 9 In this problem $\mathscr{A}_0$ denotes the subspace of H_∞ consisting of functions *continuous* in $\Im z\ge 0$ and *tending to zero* for $z\to\infty$ in the closed upper half plane. c_0 denotes the subspace of l_∞ consisting of sequences *tending to zero*.

Let $\Im z_n>0$ and $z_n \underset{n}{\longrightarrow} \infty$.

(a) *If*, for *each* $\{y_n\}\in c_0$ there *is* a $\Phi\in\mathscr{A}_0$ with $\Phi(z_n)=y_n$, *then* $\{z_n\}$ *is* an interpolating sequence.

(b)* If $\{z_n\}$ *is* an interpolating sequence, then, *given* $\{y_n\}\in c_0$ there *is* a $\Phi\in\mathscr{A}_0$ with $\Phi(z_n)=y_n$.

(Hint: *First show* how to get a $\phi\in\mathscr{A}_0$ with $\|\phi\|_\infty\le K\sup_k|y_k|$ such that $|\phi(z_n)-y_n|\le\frac{1}{2}\sup_k|y_k|$. Pay attention to the continuity properties of the Blaschke product having the z_k as its zeros.)

X

Functions of Bounded Mean Oscillation

During this whole chapter we do the work for the circle $\{|z| < 1\}$. Analogous results (with similar proofs) hold for the half-plane $\Im z > 0$.

In Chapter VII we saw that $H_1(0) = \{zf(z);\ f \in H_1\}$ has the dual L_∞/H_∞. Towards the end of the 1960s, *C. Fefferman* saw that the dual of $\Re H_1(0)$ could be represented as an *actual space of functions*, rather than as a quotient space. The functions in this space are characterized by a *simple geometric property*, namely, that of having *bounded mean oscillation*. This property was discovered some years ago by Nirenberg and John, in connection with rather unrelated work on differential equations — it was somewhat of a surprise to see that it is relevant to the study of H_p spaces.

A. Dual of $\Re H_1(0)$

1. An identity

If $\psi(\theta) \in L_\infty$ is *periodic* of period 2π, we know from Chapter I, Section E that its *harmonic conjugate*

$$\tilde{\psi}(\theta) = \frac{1}{2\pi} \fint_{-\pi}^{\pi} \frac{\psi(\theta - t)}{\tan(t/2)}\, dt$$

exists a.e. and belongs to $L_2(-\pi, \pi)$. In fact, we saw in Chapter V that $\tilde{\psi} \in L_p(-\pi, \pi)$ for *all* $p < \infty$ (and more!).

Lemma *Let $\psi \in L_\infty$ be periodic of period 2π. Then for each $f \in H_1(0)$,*

$$\lim_{r \to 1} \int_{-\pi}^{\pi} \tilde{\psi}(\theta) \Re f(re^{i\theta})\, d\theta$$

exists and equals $-\int_{-\pi}^{\pi} \psi(\theta) \Im f(e^{i\theta})\, d\theta$.

Proof We may suppose ψ *real* — the general case follows from this one by superposition.

Since $\psi + i\tilde{\psi} \in H_2$ for instance (by Chapter I, Section E!), if $f \in H_1(0)$, for each $r < 1$, the function

$$(\psi(\theta) + i\tilde{\psi}(\theta)) f(re^{i\theta})$$

certainly belongs to $H_1(0)$, so by Cauchy's theorem,

$$\int_{-\pi}^{\pi} (\psi(\theta) + i\tilde{\psi}(\theta)) f(re^{i\theta}) \, d\theta = 0 .$$

Taking imaginary parts, we find

$$\int_{-\pi}^{\pi} \tilde{\psi}(\theta) \Re f(re^{i\theta}) \, d\theta = -\int_{-\pi}^{\pi} \psi(\theta) \Im f(re^{i\theta}) \, d\theta .$$

Since $f \in H_1$, we have

$$\int_{-\pi}^{\pi} |f(re^{i\theta}) - f(e^{i\theta})| \, d\theta \longrightarrow 0 \text{ as } r \longrightarrow 1$$

by Chapter II, Section B. (*Once again* we are using the theorem of the brothers Riesz!) So, since $\psi \in L_\infty$,

$$\int_{-\pi}^{\pi} \psi(\theta) \Im f(re^{i\theta}) \, d\theta \longrightarrow \int_{-\pi}^{\pi} \psi(\theta) \Im f(e^{i\theta}) \, d\theta$$

as $r \to 1$. The lemma is proved.

2. Real linear functionals on $H_1(0)$

Theorem *Every real valued linear functional L on $H_1(0)$ can be written in the form*

$$Lf = \lim_{r \to 1} \int_{-\pi}^{\pi} (\phi(\theta) + \tilde{\psi}(\theta)) \cdot \Re f(re^{i\theta}) \, d\theta, \quad f \in H_1(0),$$

with real valued ϕ and ψ belonging to L_∞ and of period 2π.

Conversely, *if ϕ and ψ* are *real, of period 2π, and belong to L_∞, the limit on the right side of the above formula* exists *for every $f \in H_1(0)$* and *defines a real linear functional on $H_1(0)$.*

Proof The *converse* statement follows directly from the above lemma and the theorem of Chapter II, Section B used in its proof.

To get the *direct* statement, let L be a real linear functional on $H_1(0)$. Then, for $f \in H_1(0)$,

$$Lf = \Re \Lambda f$$

where Λ is some *complex valued* linear functional on $H_1(0)$. (Bohnenblust-Sobczyk theorem.) By Chapter VII, Λ is identified with an element of L_∞/H_∞ so there is certainly a 2π-periodic function in L_∞ — call it $\phi + i\psi$, with ϕ and ψ *real* — such that

$$\Lambda f = \int_{-\pi}^{\pi} (\phi(\theta) + i\psi(\theta)) f(e^{i\theta}) \, d\theta, \quad f \in H_1(0).$$

Taking real parts, we get

$$Lf = \int_{-\pi}^{\pi} (\phi(\theta) \Re f(e^{i\theta}) - \psi(\theta) \Im f(e^{i\theta})) \, d\theta .$$

According to the lemma and the theorem of Chapter II, Section B already used, this equals

$$\lim_{r\to 1}\int_{-\pi}^{\pi}(\phi(\theta)+\tilde{\psi}(\theta))\,\Re f(re^{i\theta})\,d\theta\,.$$

We are done.

Scholium There is *only one* $f \in H_1(0)$ *having a given real part* $\Re f$. Therefore we may look on $\Re H_1(0)$ as a *real Banach space with the norm*

$$\|\Re f\| = \|f\|_1\,,\quad f \in H_1(0)\,.$$

A *linear functional* on $\Re H_1(0)$ corresponds to a *real linear functional* on $H_1(0)$ and conversely.

By the theorem, the (real) *dual* of $\Re H_1(0)$ is thus *identified with the set of sums* $\Re L_\infty + \widetilde{\Re L_\infty}$. When does such a *sum* $\phi+\tilde{\psi}$ correspond to the *zero functional* on $\Re H_1(0)$? In other words, when is

$$\lim_{r\to 1}\int_{-\pi}^{\pi}(\phi(\theta)+\tilde{\psi}(\theta))\,\Re f(re^{i\theta})\,d\theta \;=\; 0$$

for *all* $f \in H_1(0)$? Taking, *first* $f(z) = z^n$ and *then* $f(z) = iz^n$ with $n = 1,2,3,\ldots$ we see that

$$\int_{-\pi}^{\pi}(\phi(\theta)+\tilde{\psi}(\theta))\begin{Bmatrix}\sin n\theta\\ \cos n\theta\end{Bmatrix}d\theta \;=\; 0,\quad n = 1,2,3,\ldots.$$

so $\phi(\theta)+\tilde{\psi}(\theta) = \text{const}$. The *converse* statement is *also clear*. Thus:

> *The dual of* $\Re H_1(0)$ *is* $\Re L_\infty + \widetilde{\Re L_\infty}$
> *modulo the constant functions.*

B. Introduction of *BMO*

A function which can be written as $\phi(\theta)+\tilde{\psi}(\theta)$ with ϕ and ψ *real*, 2π*-periodic*, and *bounded* can of course be thus written in *more than one way*. That's why *intrinsic* characterizations of such sums are important. Fefferman *found* such a characterization.

1. Definition of *BMO*

Notation If $G(\theta)$ is locally integrable and I is *any interval*, we henceforth write

$$\boxed{G_I = \frac{1}{|I|}\int_I G(\theta)\,d\theta\,.}$$

G_I is the *average value* of G on I.

Definition A locally integrable 2π-periodic function $G(\theta)$ is said to be of *bounded mean oscillation*, in symbols, $G \in BMO$, provided that

$$\frac{1}{|I|}\int_I |G(\theta)-G_I|\,d\theta \;\le\; \text{some finite constant}$$

for *all* INTERVALS $I \subseteq \mathbb{R}$.

We write, for $G \in BMO$,

$$\|G\|_* \;=\; \sup_I \frac{1}{|I|}\int_I |G(\theta)-G_I|\,d\theta\,,$$

the sup being taken over all *intervals* I.

Comment Although it's *horrible* to represent a *space of functions by three letters* (*BMO*), this designation has now become standard usage!

Remark 1 If C is a *constant*, $\|C\|_* = 0$ and $\|G - C\|_* = \|G\|_*$ for $G \in BMO$. This is evident from the definition.

Remark 2 For 2π-*periodic* locally integrable functions G (the only ones considered systematically in this chapter), in order

to check that $\|G\|_* < \infty$, *it is enough to verify that*

$$\frac{1}{|I|}\int_I |G(\theta) - G_I|\, d\theta$$

is bounded above for all intervals I *of length* $\leq$ *an arbitrary given* $l > 0$.

The proof of this fact, *important in applications*, is left as an exercise.

Remark 3 In studying the dual of $\Re H_1(\Im z > 0)$, one deals with *non-periodic functions of bounded mean oscillation*. Their definition is *formally* the same as that of the *periodic* ones given above. *Of course, in their case, Remark 2 does not apply.*

2. If ϕ and ψ are bounded, $\phi + \tilde{\psi}$ is in *BMO*

Lemma *Let I be any interval. Then if C is any constant,*

$$\frac{1}{|I|}\int_I |G(t) - G_I|\, dt \;\leq\; \frac{2}{|I|}\int_I |G(t) - C|\, dt.$$

Proof $G(t) - G_I = [G(t) - C] - [G_I - C]$. But $|G_I - C| \leq (1/|I|)\int_I |G(t) - C|\, dt$, so

$$\frac{1}{|I|}\int_I |G(t) - G_I|\, dt \;\leq\; \frac{1}{|I|}\int_I |G(t) - C|\, dt + |G_I - C| \;\leq\; \frac{2}{|I|}\int_I |G(t) - C|\, dt.$$

Q.E.D.

Our first connection between the result of Section A and *BMO* is provided by the

Theorem *Let* $G = \phi + \tilde{\psi}$, *with* ϕ, ψ periodic *of period* 2π *and* bounded. *Then* $G \in BMO$.

Proof It is obvious that $\|\phi\|_* \leq 2\|\phi\|_\infty$, so it is enough to verify that $\|\tilde{\psi}\|_* < \infty$. By Remark 2 of Subsection 1 it is enough to show that

$$\frac{1}{|J|}\int_J |\tilde{\psi}(t) - \tilde{\psi}_J|\, dt \;\leq\; \text{some } C < \infty$$

for *any interval* J of *length* $\leq 2\pi/3$.

We do this by the rudimentary application of a *typical BMO technique. Given* J of length $\leq 2\pi/3$, let J' be an *interval* having *the same midpoint* as J, *but three times its length*:

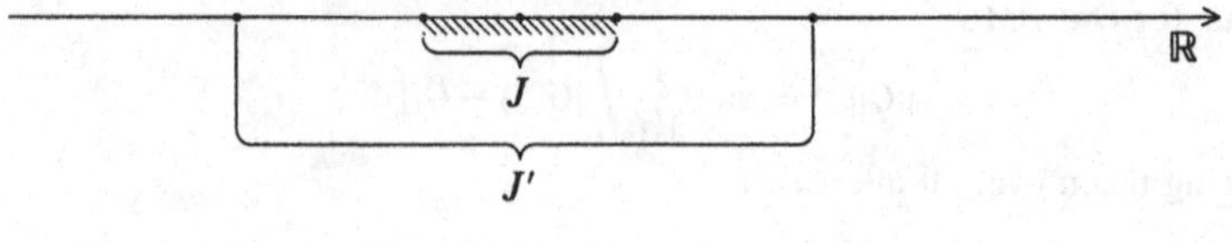

We then take

$$\psi_1(t) = \begin{cases} \psi(t), & t \in \bigcup_{n=-\infty}^{\infty} (J' + 2\pi n) \\ 0, & \text{elsewhere;} \end{cases}$$

$$\psi_2(t) = \psi(t) - \psi_1(t).$$

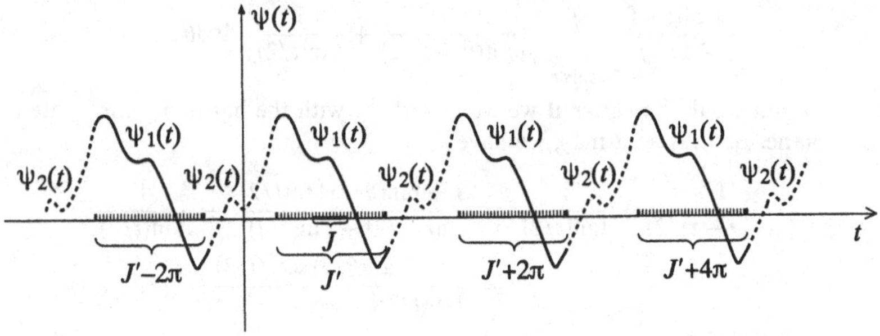

Since $\tilde{\psi} = \tilde{\psi}_1 + \tilde{\psi}_2$ and $\tilde{\psi}_J = (\tilde{\psi}_1)_J + (\tilde{\psi}_2)_J$,

$$\frac{1}{|J|}\int_J |\tilde{\psi}(t) - \tilde{\psi}_J|\,dt \;\le\; \frac{1}{|J|}\int_J |\tilde{\psi}_1(t) - (\tilde{\psi}_1)_J|\,dt + \frac{1}{|J|}\int_J |\tilde{\psi}_2(t) - (\tilde{\psi}_2)_J|\,dt.$$

By the lemma,

$$\frac{1}{|J|}\int_J |\tilde{\psi}_1(t) - (\tilde{\psi}_1)_J|\,dt \;\le\; \frac{2}{|J|}\int_J |\tilde{\psi}_1(t)|\,dt,$$

and we use *Hilbert's inequality* to estimate the right hand side. We have:

$$\int_J |\tilde{\psi}_1(t)|\,dt \;\le\; \left[|J| \int_{-\pi}^{\pi} |\tilde{\psi}_1(t)|^2\,dt\right]^{1/2}$$

which by Hilbert's inequality (part of theorem in Chapter I, Subsection E.4) is $\le \left[|J| \int_0^{2\pi} |\psi_1(t)|^2\,dt\right]^{1/2}$. Since, on $[0, 2\pi]$, $\psi_1(t)$ *vanishes* outside a set of measure $|J'| = 3|J|$, the last expression is $\le (|J| \cdot 3|J|\,\|\psi\|_\infty^2)^{1/2} = \sqrt{3}|J|\,\|\psi\|_\infty$, and therefore

$$\frac{1}{|J|}\int_J |\tilde{\psi}_1(t) - (\tilde{\psi}_1)_J|\,dt \;\le\; 2\sqrt{3}\,\|\psi\|_\infty\,.$$

We turn now to $(1/|J|)\int_J |\tilde{\psi}_2(t) - (\tilde{\psi}_2)_J|\,dt$ which, by the lemma, is

$$\le\; \frac{2}{|J|}\int_J |\tilde{\psi}_2(t) - c|\,dt,$$

where, *here*, we use a constant c equal to $\tilde{\psi}_2(m)$, where m is the *midpoint* of J. Without loss of generality, take J to be $(-\alpha, \alpha)$, $0 < \alpha \le \pi/3$; then $m = 0$, so $c = \tilde{\psi}_2(0)$ and, on

$[-\pi, \pi]$, $\psi_2(t)$ *vanishes* on $J' = [-3\alpha, 3\alpha]$. Thus, we are using

$$c = \psi_2(0) = -\frac{1}{2\pi} \int\limits_{3\alpha\leq|t|\leq\pi} \frac{\psi(t)}{\tan(t/2)}\,dt.$$

We hence have to estimate

$$\frac{1}{2\alpha}\int_{-\alpha}^{\alpha}\left|\frac{1}{2\pi}\int_{3\alpha\leq|t|\leq\pi}\left\{\frac{1}{\tan((\theta-t)/2)}+\frac{1}{\tan(t/2)}\right\}\psi(t)\,dt\right|\,d\theta$$

which we simply majorize by

$$\frac{\|\psi\|_\infty}{4\pi\alpha}\int\limits_{-\alpha}^{\alpha}\int\limits_{2\alpha\leq|t|\leq\pi}\left|\frac{1}{\tan((\theta-t)/2)}+\frac{1}{\tan(t/2)}\right|\,dt\,d\theta.$$

This computation would be *easier* if we were working with the harmonic conjugate for the upper half plane. Be that as it may, we have

$$\frac{1}{\tan((\theta-t)/2)}+\frac{1}{\tan(t/2)} = \frac{1+\tan(\theta/2)\tan(t/2)}{\tan(\theta/2)-\tan(t/2)}+\frac{1}{\tan(t/2)}$$

$$= \frac{\tan(\theta/2)\sec^2(t/2)}{\tan(t/2)[\tan(\theta/2)-\tan(t/2)]},$$

whence, for $-\alpha \leq \theta \leq \alpha$,

$$\int\limits_{2\alpha\leq|t|\leq\pi}\left|\frac{1}{\tan((\theta-t)/2)}+\frac{1}{\tan(t/2)}\right|\,dt$$

$$= 2\left|\tan\frac{\theta}{2}\right|\int\limits_{2\alpha\leq|t|\leq\pi}\frac{d\tan(t/2)}{\tan^2(t/2)-\tan(t/2)\tan(\theta/2)} \leq 8\left|\tan\frac{\theta}{2}\right|\int\limits_{\tan\alpha}^{\infty}\frac{d\tau}{\tau^2} = \frac{8|\tan(\theta/2)|}{\tan\alpha},$$

so finally,

$$\frac{1}{|J|}\int\limits_J|\tilde{\psi}_2(\theta)-(\tilde{\psi}_2)_J|\,d\theta \leq \frac{2}{|J|}\int\limits_J|\tilde{\psi}_2(\theta)-c|\,d\theta$$

$$= \frac{1}{\alpha}\int\limits_{-\alpha}^{\alpha}|\tilde{\psi}_2(\theta)-\tilde{\psi}_2(0)|\,d\theta \leq \frac{4\|\psi\|_\infty}{\pi\alpha\tan\alpha}\int\limits_{-\alpha}^{\alpha}\left|\tan\frac{\theta}{2}\right|\,d\theta \leq \frac{4}{\pi}\|\psi\|_\infty.$$

Putting this together with the preceding estimate for $(1/|J|)\int_J|\tilde{\psi}_1(\theta)-(\tilde{\psi}_1)_J|\,d\theta$, we obtain

$$\frac{1}{|J|}\int\limits_J|\tilde{\psi}(t)-\tilde{\psi}_J|\,dt \leq \left(2\sqrt{3}+(4/\pi)\right)\|\psi\|_\infty,$$

valid whenever $|J| \leq 2\pi/3$.

We are done.

Remark The above calculation, in conjunction with the one behind Remark 2 of Subsection 1 (*that* one was left to the reader!) shows that

$$\|\phi+\tilde{\psi}\|_* \leq K(\|\phi\|_\infty+\|\psi\|_\infty)$$

for 2π-periodic functions ϕ and ψ, with a *strictly numerical* constant K.

C. Garsia's norm

Garsia has observed that there is *another* norm for functions in *BMO* which is easier to use.

Notation If $\phi(\theta)$ is locally integrable and 2π-periodic, we put, for $|z| < 1$,

$$U_\phi(z) = \frac{1}{2\pi}\int_{-\pi}^{\pi} \frac{1-|z|^2}{|e^{it}-z|^2}\phi(t)\,dt\,;$$

U_ϕ is a *harmonic extension* of $\phi(t)$ to $\{|z| < 1\}$.

1. The norm $\mathfrak{N}(\phi)$

Lemma *For* $|z| < 1$,

$$U_{\phi^2}(z) - \left(U_\phi(z)\right)^2 = \frac{1}{8\pi^2}\int_{-\pi}^{\pi}\int_{-\pi}^{\pi} (\phi(t)-\phi(s))^2 \cdot \frac{1-|z|^2}{|z-e^{it}|^2}\cdot\frac{1-|z|^2}{|z-e^{is}|^2}\,dt\,ds\,.$$

Proof Multiply out in the integrand.

Definition If ϕ is *real valued* and periodic of period 2π,

$$\mathfrak{N}(\phi) = \sup_{|z|<1}\left\{U_{\phi^2}(z) - \left(U_\phi(z)\right)^2\right\}^{1/2}.$$

We call $\mathfrak{N}(\phi)$ the *Garsia norm* of ϕ. It is indeed a *norm*.

Lemma *If a is a real constant,*

$$\mathfrak{N}(a\phi) = |a|\mathfrak{N}(\phi) \quad\text{and}\quad \mathfrak{N}(\phi+\psi) \le \mathfrak{N}(\phi) + \mathfrak{N}(\psi)\,.$$

Proof The first relation is obvious. The *second* follows from the previous lemma and the triangle inequality for the Hilbert space norm. For, by that lemma, $\{U_{\phi^2}(z)-[U_\phi(z)]^2\}^{1/2}$ can, for each fixed z, be looked upon as a Hilbert space norm of $\phi(s)-\phi(t)$ on $[-\pi,\pi]\times[-\pi,\pi]$, using a certain positive measure on that square.

2. Two simple inequalities for $\mathfrak{N}(\phi)$

It will turn out that $\mathfrak{N}(\phi)$ is *equivalent* to the *BMO* norm $\|\phi\|_*$ introduced in Section B.

For the *moment*, let us just observe how much *easier* $\mathfrak{N}(\theta)$ is to *work* with than $\|\phi\|_*$.

Theorem *If ϕ and ψ are real, 2π-periodic, and bounded,*

$$\mathfrak{N}(\phi+\tilde{\psi}) \le \sqrt{2}(\|\phi\|_\infty + \|\psi\|_\infty)\,.$$

Proof We see easily (from the first lemma of Subsection 1, for instance), that

$$\mathfrak{N}(\phi) \le \sqrt{2}\,\|\phi\|_\infty\,.$$

Observe now the elementary relation $(\psi + i\tilde{\psi})^2 = \psi^2 - \tilde{\psi}^2 + 2i\psi\tilde{\psi}$, which makes $\left(U_\psi(z)\right)^2 - \left(U_{\tilde{\psi}}(z)\right)^2$ *harmonic* in $|z| < 1$. Since $\psi + i\tilde{\psi} \in H_p$ for every $p < \infty$ (Chapter V), the harmonic function $\left(U_\psi(z)\right)^2 - \left(U_{\tilde{\psi}}(z)\right)^2$ can be *recovered* from its *boundary values* $\psi^2 - \tilde{\psi}^2$ by *Poisson's formula*, i.e.,

$$\left(U_\psi(z)\right)^2 - \left(U_{\tilde{\psi}}(z)\right)^2 = U_{\psi^2-\tilde{\psi}^2}(z) = U_{\psi^2}(z) - U_{\tilde{\psi}^2}(z),$$

and we thus have the identity

$$\boxed{U_{\tilde{\psi}^2}(z) - \left(U_{\tilde{\psi}}(z)\right)^2 = U_{\psi^2}(z) - \left(U_\psi(z)\right)^2}\,.$$

From this we see that $\mathfrak{N}(\tilde{\psi}) = \mathfrak{N}(\psi)$; so $\mathfrak{N}(\tilde{\psi}) = \mathfrak{N}(\psi) \le \sqrt{2}\,\|\psi\|_\infty$ by the statement at the beginning of this proof. Combining the last inequality with that statement, we are done.

The inequality $\mathfrak{N}(\phi) \le \text{const.}\,\|\phi\|_*$ will be proved in Section F — it is a consequence of some rather delicate work of Nirenberg and John, the discoverers of *BMO* functions. The *reverse* inequality is, however, completely elementary.

Theorem *For ϕ real and 2π-periodic,*

$$\|\phi\|_* \le K\mathfrak{N}(\phi),$$

where K is a constant independent of ϕ.

Proof If I is any interval, by Schwarz' inequality,

$$\frac{1}{|I|}\int_I |\phi(t) - \phi_I|\,dt \le \left\{\frac{1}{|I|}\int_I (\phi(t) - \phi_I)^2\,dt\right\}^{1/2}$$
$$= \left\{\frac{1}{2|I|^2}\int_I\int_I (\phi(t) - \phi(s))^2\,dt\,ds\right\}^{1/2},$$

as is seen on multiplying out the integrands in the two last integrals.

Suppose $|I| \le 2\pi$, then without loss of generality put $I = (-\alpha, \alpha)$, $0 < \alpha \le \pi$. Use the formula

$$U_{\phi^2}(r) - (U_\phi(r))^2 = \frac{1}{8\pi^2}\int_{-\pi}^{\pi}\int_{-\pi}^{\pi} (\phi(s) - \phi(t))^2 \cdot \frac{1 - r^2}{1 + r^2 - 2r\cos s} \cdot \frac{1 - r^2}{1 + r^2 - 2r\cos t}\,ds\,dt,$$

taking for r the special value

$$r = 1 - \sin\frac{\alpha}{2}.$$

For this value of r, and $-\alpha \le t \le \alpha$,

$$\frac{1 - r^2}{1 + r^2 - 2r\cos t} = \frac{(1 + r)(1 - r)}{(1 - r)^2 + 4r\sin^2(t/2)} \ge \frac{1}{5\sin(\alpha/2)} \ge \frac{2}{5\alpha},$$

so

$$U_{\phi^2}(r) - (U_\phi(r))^2 \ge \frac{1}{50\pi^2\alpha^2}\int_{-\alpha}^{\alpha}\int_{-\alpha}^{\alpha} (\phi(s) - \phi(t))^2\,ds\,dt$$
$$= \left(\frac{2}{5\pi}\right)^2 \cdot \frac{1}{2|I|^2}\int_I\int_I (\phi(s) - \phi(t))^2\,ds\,dt.$$

Combining this with the inequality given at the beginning of the proof, we see that

$$\frac{1}{|I|}\int_I |\phi(t) - \phi_I|\,dt \le \frac{5\pi}{2}\mathfrak{N}(\phi)$$

if $|I| \le 2\pi$.

Suppose finally that n is a *positive integer* and $2\pi n < |I| \le 2\pi(n + 1)$. Let $J \supseteq I$ be an interval of length $2\pi(n + 1)$; then

$$\frac{1}{|I|}\int_I |\phi(t) - \phi_J|\,dt \le \frac{n + 1}{n} \cdot \frac{1}{|J|}\int_J |\phi(t) - \phi_J|\,dt \le \frac{2}{|J|}\int_J |\phi(t) - \phi_J|\,dt.$$

But *by the 2π-peridocity of ϕ, $\phi_J = (1/2\pi)\int_0^{2\pi}\phi(s)\,ds = c$*, say, and

$$\frac{1}{|J|}\int_J |\phi(t)-\phi_J|\,dt \;=\; \frac{1}{2\pi}\int_{-\pi}^{\pi} |\phi(t)-c|\,dt,$$

which is $\leq (5\pi/2)\mathfrak{N}(\phi)$, as we have just proved.

Thus, by a lemma of Subsection B.2,

$$\frac{1}{|I|}\int_I |\phi(t)-\phi_I|\,dt \leq \frac{2}{|I|}\int_I |\phi(t)-\phi_J|\,dt \leq \frac{4}{|J|}\int_J |\phi(t)-\phi_J|\,dt \leq 10\pi\mathfrak{N}(\phi).$$

We have proved $\|\phi\|_* \leq 10\pi\mathfrak{N}(\phi)$ (and, incidently, *almost completely worked out* the *exercise* that was 'left to the reader' in Subsection B.1 (Remark 2), if he or she has not already *done* it by now!). We are done.

Remark Taken together, the above two theorems provide us with a new proof of the fact that $\|\phi+\tilde{\psi}\|_* \leq C(\|\phi\|_\infty + \|\psi\|_\infty)$, already shown in Section B. The splitting used in the proof of Section B, is, however, an important technique which is applied frequently in studying *BMO*. That is why we gave that proof.

3. Where we are heading

In the scholium at the end of Subsection A.2 it was observed that the dual of $\mathfrak{R}H_1(0)$ is $\mathfrak{R}L_\infty + \widetilde{\mathfrak{R}L_\infty}$ modulo the constant functions. And as we have now (twice!) seen, $\|F\|_* < \infty$ for any F in $\mathfrak{R}L_\infty + \widetilde{\mathfrak{R}L_\infty}$. What Fefferman discovered is that this relation *characterizes* the functions F therein; those are, in other words, *precisely* the ones belonging to *BMO*.

This will follow if, for any $F \in BMO$, we can prove that

$$\lim_{r\to 1}\int_{-\pi}^{\pi} \mathfrak{R}f(re^{i\theta})F(\theta)\,d\theta$$

exists and has modulus bounded by a constant multiple of $\|F\|_*\,\|f\|_1$ for all $f \in H_1(0)$. Establishment of that property turns out to be the *main difficulty* in the present study.

We will explain *two different procedures* for arriving at the result in question. The *first*, close to Fefferman's original one, is taken up in the next three sections; use of the Garsia norm $\mathfrak{N}(\phi)$ helps us to face the considerable technical complications arising in the work. *That norm alone is involved in the next two of those sections, and the original BMO norm* $\|\phi\|_*$ *is not brought back in until* Section F. *Another* much quicker approach is presented in Section G. It is based on the *atomic decomposition* of functions belonging to $\mathfrak{R}H_1$ deduced in Chapter VIII, Section E, and avoids use of most of the machinery developed below as well as of Garsia's norm. The reader looking for an easy road to Fefferman's theorem may thus skip over Subsections D.1 and D.2, and then pass directly on to Section G after stopping for the first lemma in Subsection D.3. But by doing so he or she will miss out on the elaboration of some very useful technical material involving Green's theorem and Carleson measures. That material has been and continues to be very important for the investigation of matters taken up in the present book or related to them; first of all in Wolff's proof of the corona theorem (Chapter XI), then in the study of H_p spaces of functions of several variables, of Lipschitz domains, and so forth. For this reason, the reader who *has* the time is urged to go through Sections D, E and F.

D. Computations based on Green's theorem

1. Some identities

Lemma *Let $W(z)$ be $\mathscr{C}_\infty$ on $\{|z| \le 1\}$ and let $W(0) = 0$. Then*

$$\int_{-\pi}^{\pi} W(e^{i\theta})\,d\theta = \iint_{|z|<1} \left(\log\frac{1}{|z|}\right) \nabla^2 W(z)\,dx\,dy.$$

Notation As usual, $\nabla^2 W = W_{xx} + W_{yy}$.

Proof of lemma Writing $z = re^{i\theta}$ and taking ρ, $0 < \rho < 1$, we have by Green's theorem

$$\iint_{\rho<r<1} \left(\log\frac{1}{r}\right) \nabla^2 W\,dx\,dy = \iint_{\rho<r<1} \left(\left(\log\frac{1}{r}\right)\nabla^2 W - W\nabla^2\log\frac{1}{r}\right) dx\,dy$$

$$= \int_0^{2\pi} \left(\log\frac{1}{r}\frac{\partial W}{\partial r} - W\frac{\partial \log(1/r)}{\partial r}\right)_{r=1} d\theta - \int_0^{2\pi} \left(\log\frac{1}{r}\frac{\partial W}{\partial r} - W\frac{\partial \log(1/r)}{\partial r}\right)_{r=\rho} \cdot \rho\,d\theta.$$

Since $W(0) = 0$, $W(\rho e^{i\theta}) = O(\rho)$, and the *second integral* on the *right* goes to *zero* as $\rho \to 0$. The *first* integral on the right is just $\int_0^{2\pi} W(e^{i\theta})\,d\theta$. The lemma follows on making $\rho \to 0$.

Lemma *Let $W(z) = |z|W_1(z)$ with a function W_1* twice continuously differentiable *on $\{|z| \le 1\}$. Then*

$$\int_{-\pi}^{\pi} W(e^{i\theta})\,d\theta = \iint_{|z|<1} \left(\log\frac{1}{|z|}\right) \nabla^2 W(z)\,dx\,dy.$$

Proof Just like that of the preceding lemma.

Notation *At this point, we introduce the (physicists') vector operator*

$$\underset{\to}{\nabla} = \underset{\to}{i}\frac{\partial}{\partial x} + \underset{\to}{j}\frac{\partial}{\partial y},$$

$\underset{\to}{i}$ and $\underset{\to}{j}$ being unit vectors in the directions of the x and y axes respectively. We use '$\cdot$' to denote the dot product of two-dimensional vectors.

Theorem *Let $u(z)$ and $V(z)$ be harmonic in $\{|z| < R\}$ where $R > 1$, and let $u(0) = 0$. Then*

$$\boxed{\int_{-\pi}^{\pi} u(e^{i\theta})V(e^{i\theta})\,d\theta = 2\iint_{|z|<1} \left(\log\frac{1}{|z|}\right) (\underset{\to}{\nabla} u\cdot\underset{\to}{\nabla} V)\,dx\,dy.}$$

Proof $W(z) = u(z)V(z)$ satisfies the hypothesis of the *first* lemma above. Therefore the integral on the right equals

$$\iint_{|z|<1} \left(\log\frac{1}{|z|}\right) \nabla^2(uV)\,dx\,dy.$$

However,

$$\nabla^2(uV) = (\nabla^2 u)V + u\nabla^2 V + 2\underset{\to}{\nabla} u\cdot\underset{\to}{\nabla} V = 2\underset{\to}{\nabla} u\cdot\underset{\to}{\nabla} V,$$

since $\nabla^2 u = \nabla^2 V = 0$ (harmonicity).

That does it.

2. Expression of H_1 norm as a double integral

Lemma *Let $f(z)$ be* analytic *in $\{|z| < R\}$ where $R > 1$, and* vanish only at the origin *in that circle. Then $|f(z)|$ is $\mathscr{C}_\infty$ in $\{0 < |z| < R\}$ and*

$$\nabla^2 |f| \;=\; \frac{\nabla \Re f \cdot \nabla \Re f}{|f|}, \quad 0 < |z| < R.$$

(Due to *P.* Stein (*not* E.M. Stein!), ca. 1933.)

Proof Writing $f = u + iv$ with u and v real and harmonic in $\{|z| < R\}$, we have $|f| = (u^2+v^2)^{1/2}$ and this is $\mathscr{C}_\infty$ away from the origin because $u^2+v^2 \neq 0$ if we are not there.

We see that

$$\frac{\partial |f|}{\partial x} \;=\; \frac{uu_x + vv_x}{(u^2+v^2)^{1/2}},$$

$$\frac{\partial^2 |f|}{\partial x^2} \;=\; \frac{u_x^2+v_x^2}{(u^2+v^2)^{1/2}} + \frac{uu_{xx}+vv_{xx}}{(u^2+v^2)^{1/2}} - \frac{(uu_x+vv_x)^2}{(u^2+v^2)^{3/2}};$$

similarly,

$$\frac{\partial^2 |f|}{\partial y^2} \;=\; \frac{u_y^2+v_y^2}{(u^2+v^2)^{1/2}} + \frac{uu_{yy}+vv_{yy}}{(u^2+v^2)^{1/2}} - \frac{(uu_y+vv_y)^2}{(u^2+v^2)^{3/2}}.$$

Since $u_x = v_y,\;\; u_y = -v_x$, we have

$$(uu_x+vv_x)^2 + (uu_y+vv_y)^2 = (u^2+v^2)(u_x^2+u_y^2).$$

This relation, together with $u_{xx}+u_{yy} = 0,\;\; v_{xx}+v_{yy} = 0$, gives us

$$\frac{\partial^2 |f|}{\partial x^2} + \frac{\partial^2 |f|}{\partial y^2} = \frac{u_x^2+u_y^2}{(u^2+v^2)^{1/2}}$$

away from the origin, proving the lemma.

Theorem *Let $f(z)$ be* analytic *for $|z| < R$ where $R > 1$, and suppose that f has a* simple zero *at the* origin *and* no other zeros *in $|z| < R$. Then*

$$\int_{-\pi}^{\pi} |f(e^{i\theta})|\, d\theta \;=\; \iint_{|z|<1} \left(\log \frac{1}{|z|}\right) \frac{\nabla \Re f \cdot \nabla \Re f}{|f(z)|}\, dx\, dy.$$

Proof Apply the *second lemma* of Subsection 1 with $W(z) = |f(z)| = |z|\;|f(z)/z|$. Here $f(z)/z$ never vanishes in $|z| < R$ and is analytic there, so, by work used in proving the above lemma, $|f(z)/z|$ is $\mathscr{C}_\infty$ for $\{|z| < R\}$. The identity established by that lemma then gives us what we want.

3. Expressing the property that F generates a linear functional on $\Re H_1(0)$

We return to the problem of finding an *intrinsic characterization of the functions of the form $\phi + \tilde{\psi}$, with ϕ* and *ψ real, 2π-periodic, and bounded*. We already *know* from Subsection C.2 that if $F = \phi + \tilde{\psi}$, then $\mathfrak{N}(F) < \infty$, and *now* set out along the road of *proving* that *if $\mathfrak{N}(F) < \infty$, then F can be written as a sum $\phi + \tilde{\psi}$ with ϕ, ψ real, bounded and of* *period* 2π. According to the theorem in Section A, *this will follow if we can show that*

$$\lim_{R\to 1}\int_{-\pi}^{\pi} \Re f(Re^{i\theta})F(\theta)\,d\theta$$

exists for every $f \in H_1(0)$, *and represents a linear functional on* $\Re H_1(0)$, *whenever* $\mathfrak{N}(F) < \infty$.

We perform a series of successive reductions.

Lemma

$$\lim_{R\to 1}\int_{-\pi}^{\pi} \Re f(Re^{i\theta})F(\theta)\,d\theta$$

exists and represents a linear functional on $\Re H_1(0)$ *if there exists a constant C such that, for any* $f \in H_1(0)$ *and any* R, $0 < R < 1$,

$$\left|\int_{-\pi}^{\pi} \Re f(Re^{i\theta})F(\theta)\,d\theta\right| \le C\,\|f\|_1 .$$

Proof Let $f \in H_1(0)$ and let $\epsilon > 0$. By what is essentially the theorem of the brothers Riesz (Chapter II, Section B), there is an $R < 1$ such that $\int_{-\pi}^{\pi} |f(e^{i\theta}) - f(R'e^{i\theta})|\,d\theta < \epsilon$ whenever $R < R' < 1$. Let now $R < R_1 < R_2 < 1$; then, if $R' = R_1/R_2$, $R < R' < 1$, so if $g(z) = f(z) - f(R'z)$, $g \in H_1(0)$ and $\|g\|_1 < \epsilon$. Therefore, if the inequality in the hypothesis holds,

$$\left|\int_{-\pi}^{\pi} \Re g(R_2 e^{i\theta})F(\theta)\,d\theta\right| < C\epsilon,$$

i.e.,

$$\left|\int_{-\pi}^{\pi} \Re f(R_2 e^{i\theta})F(\theta)\,d\theta - \int_{-\pi}^{\pi} \Re f(R_1 e^{i\theta})F(\theta)\,d\theta\right| < C\epsilon$$

whenever $R < R_1 < R_2 < 1$.

It follows that the desired limit exists; clearly it is in absolute value $\le C\,\|f\|_1$. Q.E.D.

Lemma *Let* $F \in L_1(-\pi, \pi)$. *Then*

$$\lim_{R\to 1}\int_{-\pi}^{\pi} \Re f(Re^{i\theta})F(\theta)\,d\theta$$

exists for all $f \in H_1(0)$ *and represents a linear functional on* $\Re H_1(0)$ *if there is a constant C such that for all R and* $\rho < 1$ *and all* $f \in H_1(0)$,

$$\left|\int_{-\pi}^{\pi} \Re f(Re^{i\theta})U_F(\rho e^{i\theta})\right| \le C\,\|f\|_1 .$$

(For meaning of the symbol U_F, *see the beginning of Section C.)*

Proof If $R < 1$, $f(Re^{i\theta})$ *is continuous*, so, since (Chapter I !)

$$\int_{-\pi}^{\pi} |U_F(\rho e^{i\theta}) - F(\theta)|\,d\theta \longrightarrow 0 \text{ as } \rho \to 1,$$

we have, for any $R < 1$,

$$\int_{-\pi}^{\pi} \Re f(Re^{i\theta})F(\theta)\,d\theta = \lim_{\rho\to 1}\int_{-\pi}^{\pi} \Re f(Re^{i\theta})U_F(\rho e^{i\theta})\,d\theta .$$

The result now follows by the preceding lemma.

Lemma *Let* $F \in L_1(-\pi, \pi)$. *In order to prove that*

$$\lim_{R\to 1}\int_{-\pi}^{\pi} \Re f(Re^{i\theta})F(\theta)\,d\theta$$

exists for all $f \in H_1(0)$ *and represents a linear functional on* $\Re H_1(0)$, *it is enough to show that there exists a constant* C *such that, for all* R *and* $\rho < 1$,

$$\left| \int_{-\pi}^{\pi} \Re f(Re^{i\theta}) U_F(\rho e^{i\theta})\, d\theta \right| < C \|f\|_1$$

whenever $f \in H_1(0)$ *has* precisely one simple zero at the origin, and no other zeros *in* $\{|z| < 1\}$.

Proof By a trick already mentioned in Chapter IV. Let $f \in H_1(0)$, then we can write $f(z) = zB(z)g(z)$, where $B(z)$ is a Blaschke product, $g(z)$ *has no zeros in* $|z| < 1$, and $\|g\|_1 = \|f\|_1$. We can now put

$$f_1(z) = \tfrac{1}{2} z(B(z) - 1)g(z)$$
$$f_2(z) = \tfrac{1}{2} z(B(z) + 1)g(z)\,;$$

then f_1 and f_2 each *have one simple zero at the origin, and no others in the unit circle*. We have $f = f_1 + f_2$, so, if the inequality holds under the conditions stated in the hypothesis,

$$\left| \int_{-\pi}^{\pi} \Re f(Re^{i\theta}) U_F(\rho e^{i\theta})\, d\theta \right| \leq \sum_{k=1}^{2} \left| \int_{-\pi}^{\pi} \Re f_k(Re^{i\theta}) U_F(\rho e^{i\theta})\, d\theta \right|$$
$$\leq C(\|f_1\|_1 + \|f_2\|_1) \leq 2C \|f\|_1 ,$$

because $\|f_1\|_1 \leq \|f\|_1$, $\|f_2\|_1 \leq \|f\|_1$.

Our result now follows by the preceding lemma.

We now combine the last of the above lemmas with the theorem of Subsection 1 to obtain:

Theorem *Let* $F \in L_1(-\pi, \pi)$. *In order to prove that*

$$\lim_{R \to 1} \int_{-\pi}^{\pi} \Re f(Re^{i\theta}) F(\theta)\, d\theta$$

exists for all $f \in H_1(0)$ *and represents a linear functional on* $\Re H_1(0)$, *it is enough to show that there exists a constant* C *with*

$$\left| \iint_{|z|<1} \left(\log \frac{1}{|z|} \right) (\nabla u \cdot \nabla V)\, dx\, dy \right| \leq C \|f\|_1$$

for $u(z) = \Re f(Rz)$, $V(z) = U_F(\rho z)$ *and* all R, $\rho < 1$, *whenever* $f \in H_1(0)$ *has* a simple zero at 0 and no other zeros *in* $|z| < 1$.

E. Fefferman's theorem with the Garsia norm

We continue, basing our work on the somewhat ungainly theorem in Subsection D.3. According to that result and the theorem of Section A, given a real 2π-periodic F with $\mathfrak{N}(F) < \infty$, we will *have* $F \in \Re L_\infty + \widetilde{\Re L_\infty}$ as soon as we prove, with a numerical constant

K, that

$$\left|\iint_{|z|<1} \left(\log\frac{1}{|z|}\right)(\nabla u\cdot\nabla V)\,dx\,dy\right| \leq K\,\|f\|_1\,\mathfrak{N}(F) \text{ for}$$
$$u(z) = \Re f(Rz),\ V(z) = U_F(\rho z),\ R \text{ and } \rho \text{ arbitrary } < 1,$$

whenever $f \in H_1(0)$ has a simple zero at 0 and no other zeros in $\{|z| < 1\}$.

The rest of this section, really the heart of the whole chapter, is devoted to the proof of the boxed inequality.

We continue to let $u(z)$ stand for $\Re f(Rz)$ and $V(z)$ for $U_F(\rho z)$, where $f \in H_1(0)$ and R, ρ are arbitrary positive numbers < 1.

1. Use of Schwarz' inequality

We are interested in the case where $f \in H_1(0)$ has a simple zero at 0, and no others in the unit circle. Fefferman had the idea of applying Schwarz' inequality to the left hand member of the above boxed inequality, in such a way as to take advantage of the theorem in Subsection D.2. We have, namely,

$$\left|\iint_{|z|<1} \left(\log\frac{1}{|z|}\right)(\nabla u\cdot\nabla V)\,dx\,dy\right|$$
$$\leq \left[\iint_{|z|<1} \left(\log\frac{1}{|z|}\right)\frac{\nabla u\cdot\nabla u}{|f(Rz)|}\,dx\,dy\right]^{1/2} \left[\iint_{|z|<1} \left(\log\frac{1}{|z|}\right)(\nabla V\cdot\nabla V)|f(Rz)|\,dx\,dy\right]^{1/2}.$$

The factor $|f(Rz)|$, which was not even present, has thus been brought in!

Since $u(z) = \Re f(Rz)$ with $f \in H_1(0)$ having *a simple zero at* 0, *and no others in the unit circle*, the theorem of Subsection D.2 says that

$$\iint_{|z|<1} \left(\log\frac{1}{|z|}\right)\frac{\nabla u\cdot\nabla u}{|f(Rz)|}\,dx\,dy = \int_{-\pi}^{\pi} |f(Re^{i\theta})|\,d\theta < \|f\|_1,$$

so our boxed inequality will be proved as soon as we show that

$$(*) \qquad \iint_{|z|<1} \left(\log\frac{1}{|z|}\right)(\nabla V\cdot\nabla V)|f(Rz)|\,dx\,dy \leq K\,\|f\|_1\,(\mathfrak{N}(F))^2$$

for all $f \in H_1(0)$ of the aforementioned special form, with a purely numerical constant K.

2. A measure to be proved Carleson

In Subsection 1, restriction to $f \in H_1(0)$ with *a simple zero at* 0, *and no others in the unit circle*, is *on account* of the *requirement* of the *theorem* in Subsection D.2. That is also why the *third* lemma of Subsection D.3 was needed.

Now, however, there is no obstacle to our proving $(*)$ for *all* $f \in H_1(0)$. Observe that for

$f \in H_1(0)$, $g(z) = f(Rz)/z$ is in H_1, and clearly $\|g\|_1 \le \|f\|_1$. It is therefore enough to show that

$$\iint_{|z|<1} \left(|z| \log \frac{1}{|z|}\right) (\nabla V \cdot \nabla V) |g(z)| \, dx \, dy \;\le\; K(\mathfrak{N}(F))^2 \|g\|_1$$

for *all* $g \in H_1$, with a numerical constant K.

Recall now the definition of *Carleson measures* (for the unit circle) in Chapter VIII, Section F. We see that

we have to prove that $\left(|z| \log \frac{1}{|z|}\right) (\nabla V \cdot \nabla V) \, dx \, dy$
is a Carleson measure with 'Carleson constant'
equal to a numerical multiple of $(\mathfrak{N}(F))^2$.

Part of this program is carried through rather easily.

Lemma $\iint_{|z|<1/2} (|z| \log(1/|z|)) (\nabla V \cdot \nabla V) |g(z)| \, dx \, dy \;\le\; K'(\mathfrak{N}(F))^2 \|g\|_1$ *for all* $g \in H_1$, *with a numerical constant* K'.

Proof For $|z| \le \frac{1}{2}$, $|z| \log(1/|z|) \le e^{-1}$ and $|g(z)| \le (1/\pi) \|g\|_1$.

Again, $V(z) = U_F(\rho z)$ where $0 \le \rho < 1$, so $(\nabla V)(z) = \nabla U_F(\rho z) = \nabla U_{F_1}(\rho z)$, where

$$F_1(\theta) = F(\theta) - \frac{1}{2\pi} \int_{-\pi}^{\pi} F(t) \, dt.$$

Now by direct differentiation of Poission's formula, we have, for $|z| \le \frac{1}{2}$,

$$|\nabla U_{F_1}(\rho z)| \;\le\; \frac{4}{\pi} \int_{-\pi}^{\pi} |F_1(t)| \, dt,$$

i.e.,

$$|\nabla U_{F_1}(\rho z)| \;\le\; \frac{8}{2\pi} \int_{-\pi}^{\pi} \left| F(\theta) - \frac{1}{2\pi} \int_{-\pi}^{\pi} F(t) \, dt \right| d\theta$$

for $|z| \le \frac{1}{2}$.

From the proof of the second theorem in Subsection C.2, we see that the *quantity on the right* in the last expression is $\le 20\pi \mathfrak{N}(F)$. Using this, together with the other estimates, we find

$$\iint_{|z|<1/2} \left(|z| \log \frac{1}{|z|}\right) (\nabla V \cdot \nabla V) |g(z)| \, dx \, dy \;\le\; \frac{100\pi^2}{e} (\mathfrak{N}(F))^2 \|g\|_1,$$

Q.E.D.

3. Main lemma

We now come to the *main work* of this whole chapter, which is to prove that

$$\iint_{1/2<|z|<1} \left(|z| \log \frac{1}{|z|}\right) (\nabla V \cdot \nabla V) |g(z)| \, dx \, dy \;\le\; K''(\mathfrak{N}(F))^2 \|g\|_1$$

with a numerical constant K'' for all $g \in H_1$. This proof depends on the fundamental *theorem on Carleson measures* given in Chapter VIII, Section F.

Lemma *Let* $0 < h \le \frac{1}{2}$ *and let* S_h *be* any curvilinear box *of the form* $1 - h \le r < 1$, $\theta_0 < \theta < \theta_0 + h$.

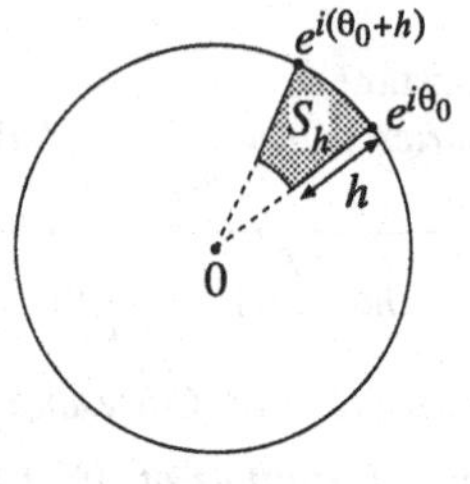

Then, with a numerical constant C,

$$\iint_{S_h} \left(|z| \log \frac{1}{|z|}\right) (\nabla V \cdot \nabla V)\, dx\, dy \;\le\; C(\mathfrak{N}(F))^2 \cdot h.$$

Proof Without loss of generality $\theta_0 = -h/2$, so that

$$S_h \;=\; \{re^{i\theta}\,;\; 1 - h \le r < 1,\; -h/2 \le \theta \le h/2\}.$$

In that case we estimate the integral in question by *first making a change of variable* $z \mapsto \zeta$, where $(i - \zeta)/(i + \zeta) = z$. This is a *conformal mapping* of $\{|z| < 1\}$ onto $\Im\zeta > 0$. As usual, we write $\zeta = \xi + i\eta$.

Put $V(z) = w(\zeta)$ for z corresponding to ζ in the above fashion.

We have, for $\frac{1}{2} \le |z| < 1$,

$$|z| \log \frac{1}{|z|} \;\le\; \frac{1}{2} \log \frac{1}{|z|^2} \;=\; \frac{(1 - |z|^2)}{2} \left(1 + \tfrac{1}{2}(1 - |z|^2)^2 + \tfrac{1}{3}(1 - |z|^2)^3 + \cdots\right)$$

$$\le\; \frac{(1 - |z|^2)}{2|z|^2} \;\le\; 2(1 - |z|^2).$$

In terms of $\zeta = \xi + i\eta$, $|z|^2 = (\xi^2 + (\eta - 1)^2)/(\xi^2 + (\eta + 1)^2)$, so $1 - |z|^2 = 4\eta/(\xi^2 + (1 + \eta)^2) \le 4\eta$, and finally

$$\boxed{|z| \log \frac{1}{|z|} \;\le\; 8\eta \text{ for } \tfrac{1}{2} \le |z| < 1.}$$

Let now S_h correspond to the set Σ_h in the ζ-plane under the mapping $z \longmapsto \zeta$:

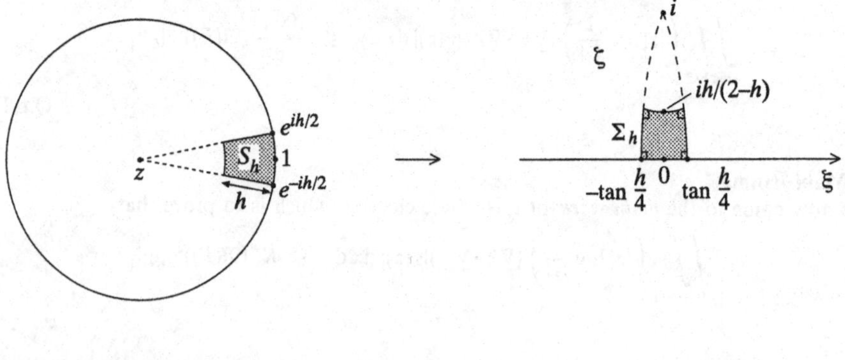

Then, since the mapping is *conformal*, the quantity we are trying to estimate,

$$\iint_{S_h} \left(|z| \log \frac{1}{|z|}\right) \left\{ \left(\frac{\partial V}{\partial x}\right)^2 + \left(\frac{\partial V}{\partial y}\right)^2 \right\} dx\, dy,$$

equals

$$\iint_{\Sigma_h} \left(|z| \log \frac{1}{|z|}\right) \left(\left(\frac{\partial w}{\partial \xi}\right)^2 + \left(\frac{\partial w}{\partial \eta}\right)^2 \right) d\xi\, d\eta$$

which is

$$\leq \iint_{\Sigma_h} 8\eta \left(\left(\frac{\partial w}{\partial \xi}\right)^2 + \left(\frac{\partial w}{\partial \eta}\right)^2 \right) d\xi\, d\eta$$

by the above boxed inequality.

Since $0 \leq h \leq \frac{1}{2}$, Σ_h *is included in the square*

$$B_h = \{(\xi,\eta);\ -h/2 < \xi < h/2,\ 0 < \eta < h\},$$

as some simple calculations show. (Actually, it is manifest *without any calculation that for* $0 \leq h \leq \frac{1}{2}$, Σ_h is *always included* in B_{Ch} for *some numerical constant* C, and the reader *may content himself* with this *evident fact*. We take here the *permissible value* $C = 1$ in order to *combat* the *proliferation of numerical constants in our formulas*!) Here is the situation:

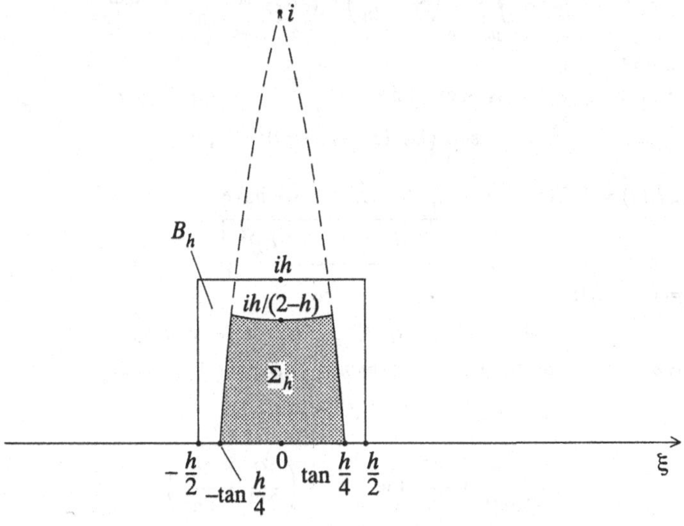

We therefore have

$$\iint_{S_h} \left(|z| \log \frac{1}{|z|}\right) |\nabla V|^2\, dx\, dy \;\leq\; 8 \iint_{B_h} \eta \left(w_\xi^2 + w_\eta^2\right) d\xi\, d\eta\,,$$

and our lemma will be proved when we show that the quantity on the right is $\leq K''(\mathfrak{N}(F))^2 \cdot h$ with a numerical constant K''.

Now we use a trick. B_h is entirely contained in the semi-circle $|\zeta| < 2h$, $\eta > 0$, and on

B_h, $1-(|\zeta|/2h) > 1/4$. Therefore

$$\iint_{B_h} \eta(w_\xi^2 + w_\eta^2)\,d\xi\,d\eta \;\le\; 4 \iint_{|\zeta|<2h,\ \eta>0} \left(1-\frac{|\zeta|}{2h}\right)\eta(w_\xi^2+w_\eta^2)\,d\xi\,d\eta .$$

The factor $(1-(|\zeta|/2h))\,\eta$ in the right-hand integrand *vanishes on the boundary of the region of integration*, and *that will help us*, as we shall see in a moment.

We have $w(\zeta) = V(z)$ with $z = (i-\zeta)/(i+\zeta)$, so $w(\zeta)$ is *harmonic* for $\Im\zeta > 0$. (*Recall that we are using $V(z)$ to denote $U_F(\rho z)$ throughout this whole section, with ρ a fixed but arbitrary positive number < 1. Recall also that*

$$U_F(z) = \frac{1}{2\pi}\int_{-\pi}^{\pi} \frac{1-|z|^2}{|z-e^{it}|^2} F(t)\,dt ,$$

the *Poisson integral* of F.) *On account of the* HARMONICITY *of $w(\zeta)$, we have the identity* (TRICK!)

$$\boxed{\left(\frac{\partial^2}{\partial\xi^2}+\frac{\partial^2}{\partial\eta^2}\right) w^2 = 2(w_\xi^2+w_\eta^2)}$$

(The second edition of Zygmund's book already contains an application of this identity to the study of a related question.) Our problem thus reduces to the estimate of

$$J = 2\iint_{|\zeta|<2h,\ \eta>0} \left(1-\frac{|\zeta|}{2h}\right)\eta\left(\frac{\partial^2 w^2}{\partial\xi^2}+\frac{\partial^2 w^2}{\partial\eta^2}\right) d\xi\,d\eta$$

in terms of h and $\mathfrak{N}(F)$.

In Subsection C.1, $\mathfrak{N}(F)$ was defined as

$$\sup_{|z|<1} \left(U_{F^2}(z) - (U_F(z))^2\right)^{1/2} ,$$

so if we put $P(z) = U_{F^2}(z) - [U_F(z)]^2$ for $|z|<1$, we have

$$\boxed{0 \le P(z) \le (\mathfrak{N}(F))^2 .}$$

Now for $z = (i-\zeta)/(i+\zeta)$,

$$(w(\zeta))^2 = (U_F(\rho z))^2 = U_{F^2}(\rho z) - P(\rho z),$$

where the term $U_{F^2}(\rho z)$ on the right is harmonic. Therefore, if we put

$$b(\zeta) = P(\rho z),$$

we have

$$\left(\frac{\partial^2}{\partial\xi^2}+\frac{\partial^2}{\partial\eta^2}\right) w^2 = -\left(\frac{\partial^2 b}{\partial\xi^2}+\frac{\partial^2 b}{\partial\eta^2}\right),$$

and the integral we have to estimate boils down to

$$J = -2\iint_{|\zeta|<2h,\ \eta>0} \left(1-\frac{|\zeta|}{2h}\right)\eta\left(\frac{\partial^2 b}{\partial\xi^2}+\frac{\partial^2 b}{\partial\eta^2}\right) d\xi\,d\eta ,$$

where

$$\boxed{0 \le b(\zeta) \le (\mathfrak{N}(F))^2 .}$$

Write $\nabla_\zeta^2 = \dfrac{\partial^2}{\partial\xi^2} + \dfrac{\partial^2}{\partial\eta^2}$. Applying *Green's theorem* to the integral just written, we find

$$J = -2 \iint\limits_{|\zeta|<2h,\ \eta>0} b(\zeta)\nabla_\zeta^2\left(\eta\left(1 - \frac{|\zeta|}{2h}\right)\right) d\xi\, d\eta$$
$$+ 2\int_\Gamma \left(b(\zeta)\frac{\partial\{\eta\,(1-(|\zeta|/2h))\}}{\partial n_\zeta} - \eta\left(1 - \frac{|\zeta|}{2h}\right)\frac{\partial b(\zeta)}{\partial n_\zeta}\right)|d\zeta|,$$

where Γ is this contour,

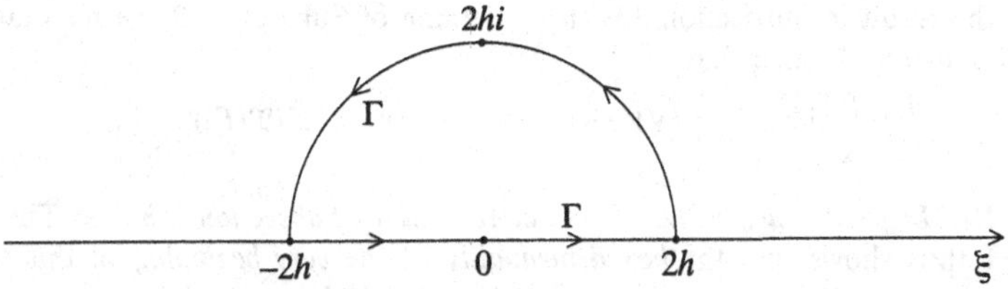

and $\partial/\partial n_\zeta$ denotes *differentiation with respect to distance* in the *direction* of the *outward normal* to Γ at ζ.

The line integral around Γ *is negative*. Indeed, $\eta\,(1-(|\zeta|/2h)) \equiv 0$ on Γ (*now we see why* the extra factor $1-(|\zeta|/2h)$ was brought in!); also, $\eta\,(1-(|\zeta|/2h)) > 0$ *inside* Γ, so clearly $\partial(\eta\,(1-(|\zeta|/2h)))/\partial n_\zeta \le 0$ on Γ. Therefore $b(\zeta)\,\partial(\eta\,(1-(|\zeta|/2h)))/\partial n_\zeta \le 0$ on Γ in view of the above boxed inequality, and the line integral around Γ *does* come out to be ≤ 0. (If the reader has trouble *keeping track of signs here*, he or she may simply *compute*

$$\int_\Gamma |b(\zeta)| \left|\frac{\partial\,(\eta\,(1-(|\zeta|/2h)))}{\partial n_\zeta}\right| |d\zeta|$$

using *polar coordinates*. A value $\le$ const. h $\sup_\zeta |b(\zeta)| \le$ const.$(\mathfrak{N}(F))^2 h$ will be found.)

We thus have

$$J \le -2 \iint\limits_{|\zeta|<2h,\ \eta>0} b(\zeta)\nabla_\zeta^2\left(\eta\left(1 - \frac{|\zeta|}{2h}\right)\right) d\xi\, d\eta\,.$$

To evaluate this, use the polar coordinates $\zeta = \sigma e^{i\phi}$; then

$$\nabla_\zeta^2 = \frac{1}{\sigma}\frac{\partial}{\partial\sigma}\left(\sigma\frac{\partial}{\partial\sigma}(\quad)\right) + \frac{1}{\sigma^2}\frac{\partial^2}{\partial\phi^2}$$

and $\eta\,(1-(|\zeta|/2h)) = (1-(\sigma/2h))\,\sigma\sin\phi$, from which we easily find

$$\nabla_\zeta^2\,(\eta\,(1-(|\zeta|/2h))) = -\frac{3}{2h}\sin\phi,$$

which, substituted into the above integral, yields

$$J \le \frac{3}{h}\int_0^\pi\int_0^{2h} b(\sigma e^{i\phi})\sin\phi\;\sigma\, d\sigma\, d\phi \;\le\; \frac{3}{h}(\mathfrak{N}(F))^2\int_0^\pi\int_0^{2h}\sin\phi\;\sigma\, d\sigma\, d\phi \;=\; 12\,(\mathfrak{N}(F))^2\, h.$$

We are at the end of the calculation. Going back to what we started with, we find

$$\iint\limits_{S_h}\left(|z|\log\frac{1}{|z|}\right)|\nabla V|^2\, dx\, dy \;\le\; 32J \;\le\; 384\,(\mathfrak{N}(F))^2\, h,$$

and the lemma is proved.

Combining this last lemma with the *theorem on Carleson measures for the unit circle*, given at the end of Chapter VIII, Section F, we find:

Theorem *There is a numerical constant K'' such that*

$$\iint_{1/2<|z|<1} \left(|z| \log \frac{1}{|z|}\right) (\underset{\rightarrow}{\nabla} V \cdot \underset{\rightarrow}{\nabla} V) |g(z)| \, dx \, dy \;\leq\; K''(\mathfrak{N}(F))^2 \, \|g\|_1$$

for all $g \in H_1$.

4. Fefferman's theorem with $\mathfrak{N}(F)$

Combining the result in Subsection 3 with the lemma of Subsection 2, we see that there is a numerical constant K such that

$$\iint_{|z|<1} \left(|z| \log \frac{1}{|z|}\right) (\underset{\rightarrow}{\nabla} V \cdot \underset{\rightarrow}{\nabla} V) \, |g(z)| \, dx \, dy \;\leq\; K(\mathfrak{N}(F))^2 \, \|g\|_1$$

for all $g \in H_1$. *In particular, inequality* (*) *at the end of Subsection* 1 *holds*. The work in Subsection 1 thus shows that the *boxed inequality* at the *very beginning* of this section is *valid*, whence, by the theorem at the end of Subsection D.3, we obtain:

Theorem *Let F be real valued and periodic, of period 2π. If $\mathfrak{N}(F) < \infty$,*

$$\lim_{R \to 1} \int_{-\pi}^{\pi} \Re f(Re^{i\theta}) F(\theta) \, d\theta$$

exists for every $f \in H_1(0)$, and represents a linear functional on $\Re H_1(0)$.

Scholium The *norm* of the linear functional on $\Re H_1(0)$ furnished by this theorem is $\leq \text{const} \cdot \mathfrak{N}(F)$.

Indeed, the proofs of the lemmas in Subsection D.3 show that the *norm* of the functional in question is $\leq 2C$, where C is the *constant figuring in the statement of the theorem* in Subsection D.3. The *boxed inequality* at the very beginning of this section is *the same as the one in the theorem*, Subsection D.3, with $C = K\mathfrak{N}(F)$, K being a *numerical constant*. And we *proved* the boxed inequality in Subsections 1–3.

We can, however, say more than this, namely, $\mathfrak{N}(F)$ is also $\leq$ a constant times the norm of the above linear functional!

Indeed, call

$$Lf = \lim_{R \to 1} \int_{-\pi}^{\pi} \Re f(Re^{i\theta}) F(\theta) \, d\theta$$

for $f \in H_1(0)$, and denote the *norm* of L by $|||L|||$. The Hahn-Banach argument in Section A shows that we can find $\phi + i\psi$ periodic of period 2π in L_∞ with ϕ *and* ψ *real*, and

$$\|\phi + i\psi\|_\infty \leq |||L|||,$$

such that

$$Lf = \Re \int_{-\pi}^{\pi} (\phi(t) + i\psi(t)) f(e^{it}) \, dt$$

for $f \in H_1(0)$. By the lemma in Section A,

$$Lf = \lim_{R \to 1} \int_{-\pi}^{\pi} \Re f(Re^{it}) (\phi(t) + \tilde{\psi}(t)) \, dt$$

so that

$$F(t) = \phi(t) + \tilde{\psi}(t) + c$$

with a constant c, according to the scholium at the very end of Section A.

With the first lemma of Subsection C.1, we see that

$$\mathfrak{N}(F) = \mathfrak{N}(\phi + \tilde{\psi}),$$

but by Subsection C.2, the quantity on the right is

$$\leq \sqrt{2}(\|\phi\|_\infty + \|\psi\|_\infty) \leq 2\sqrt{2}\,\|\phi + i\psi\|_\infty \leq 2\sqrt{2}\,|||L|||.$$

Let us combine the above theorem with its scholium, the material in Subsection A.2, and the first result of Subsection C.2. We get

Garsia's version of Fefferman's theorem *If $F(\theta)$ is real and periodic of period 2π, the following three conditions on F are equivalent:*

(i) $L\Re f = \lim_{R\to 1}\int_{-\pi}^{\pi}(\Re f Re^{i\theta})F(\theta)\,\mathrm{d}\theta$ *exists for every $f \in H_1(0)$, and L is a linear functional on $\Re H_1(0)$.*

(ii) $F(\theta) = \phi(\theta) + \tilde{\psi}(\theta)$, *with ϕ, ψ real, 2π-periodic, and bounded.*

(iii) $\mathfrak{N}(F) < \infty$.

Moreover, the norms $|||L|||$ (of L as a linear functional on $\Re H_1(0)$) and $\mathfrak{N}(F)$ are equivalent, $\mathfrak{N}(F) \leq \sqrt{2}(\|\phi\|_\infty + \|\psi\|_\infty)$ with ϕ and ψ the bounded functions in (ii), *and we can find ϕ_1, ψ_1, periodic of period 2π and real valued, such that*

$$F(\theta) = \phi_1(\theta) + \tilde{\psi}_1(\theta) + c,$$

c a constant, with

$$\|\phi_1\|_\infty + \|\psi_1\|_\infty \leq M \cdot \mathfrak{N}(F),$$

M being a numerical constant.

F. Fefferman's theorem with the *BMO* norm

We now show that in the theorem of Subsection E.4, the condition $\mathfrak{N}(F) < \infty$ can be replaced by $\|F\|_* < \infty$, i.e., by the fact that $F \in BMO$. Since we already saw in Subsection C.2 that $\|F\|_* \leq \text{const.}\,\mathfrak{N}(F)$, we have here to establish the reverse inequality, $\mathfrak{N}(F) \leq \text{const.}\,\|F\|_*$, proving *equivalence* of the norms $\mathfrak{N}(\ )$ and $\|\ \|_*$. Proof of the reverse inequality depends on some deep work of Nirenberg and John.

1. Theorem of Nirenberg and John

Recall first of all the definition from Section B:

$$\|F\|_* = \sup_I \frac{1}{|I|}\int_I |F(t) - F_I|\,\mathrm{d}t$$

with I ranging over the set of *intervals*.

Lemma *If $F \in BMO$ and if I and J are intervals with the same midpoint and $I \subset J$, then*

$$|F_I - F_J| \leq 2\left(\log_2\frac{|J|}{|I|} + 1\right)\|F\|_*.$$

(Here, $\log_2$ *means logarithm to the base 2.)*

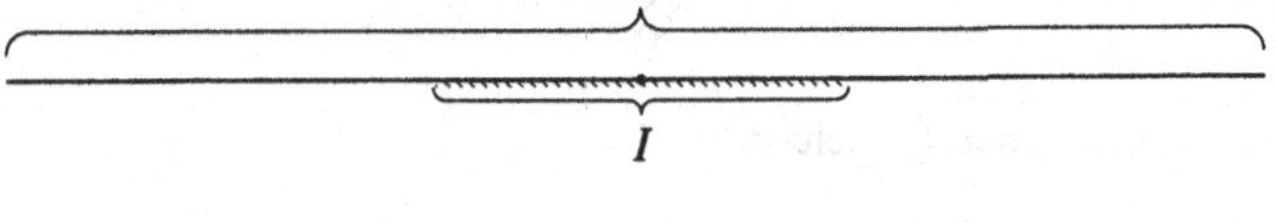

Proof First of all, if $|J| \le 2|I|$,

$$|F_I - F_J| = \frac{1}{|I|}\left|\int_I (F(t) - F_J)\,dt\right| \le \frac{1}{|I|}\int_I |F(t) - F_J|\,dt \le \frac{2}{|J|}\int_J |F(t) - F_J|\,dt \le 2\|F\|_* ,$$

and the result is true in this case. (*Here*, we *only* used the relations $I \subset J$, $|J| \le 2|I|$.)

In the general case, argue by induction. If $2^n|I| < |J| \le 2^{n+1}|I|$ and the result is proved for intervals J' with $|J'| \le 2^n|I|$, *take* an interval J' having the *same midpoint* as I, but *half the length* of J, so that $I \subset J' \subset J$. By the induction hypothesis,

$$|F_I - F'_J| \le 2\|F\|_* \left(\log_2 \frac{|J'|}{|I|} + 1\right).$$

Since $|J| = 2|J'|$, by the argument given above, $|F_J - F'_J| \le 2\|F\|_*$, so

$$|F_J - F_I| \le |F_J - F'_J| + |F'_J - F_I| \le 2\|F\|_* + 2\|F\|_* \left(\log_2 \frac{|J'|}{|I|} + 1\right)$$

$$= 2\|F\|_* \left(\log_2 \frac{|J|}{|I|} + 1\right),$$

and the result holds for intervals J with $2^n|I| < |J| \le 2^{n+1}|I|$.

We are done.

BMO actually first appeared when the following result was published. In it, the function F is assumed to be *real valued*.

Theorem (Nirenberg and John) *For any interval I and positive integer n,*

$$|\{x \in I;\ F(x) - F_I > 4n\|F\|_*\}| \le 2^{-n}|I|.$$

Proof (Garnett) By making a *change of scale*, we may reduce our situation to one where $\|F\|_* \le \frac{1}{2}$, and, under this assumption, it is *enough* to prove that

$$|\{x \in I;\ F(x) - F_I > 2n\}| \le 2^{-n}|I|.$$

A *change of variable* now allows us to take $I = [0,1]$; there is also no loss of generality in taking F_I to be *zero*.

These reductions made, we shall work with the *dyadic subintervals* of $[0,1] = I$. *These are those of the very special form* $[k/2^l,\ (k+1)/2^l]$, with k and l non-negative integers.

> *Dyadic intervals have the very nice property that if two of them overlap, one must be contained in the other.*

Let $E = \{x \in I;\ F(x) > 2n\}$. By Lebesgue's theorem (differentiation of indefinite integrals), almost every $x \in E$ is contained in some *small dyadic interval* $J \subset I$ with $F_J > 2n$. Since $F_I = 0$, there *must*, for almost every $x \in E$, be a *dyadic interval $J \subset I$ and containing x, of greatest possible length**, such that $F_J > 1$. Let such a J of greatest possible length containing x be called $J(x)$. After throwing away from E a set of measure zero, we will have

$$E \subseteq \bigcup_{x \in E} J(x).$$

* being, we emphasize, *properly* included in I

Let $\mathscr{E} = \bigcup_{x \in E} J(x)$. $\mathscr{E}$ can actually be written as a *non-overlapping union* of $J(x)$'s. Indeed, if $x, y \in E$ and $J(x)$ and $J(y)$ *overlap*, then $J(x) = J(y)$. For, by the above boxed remark, either $J(x) \subseteq J(y)$ or $J(y) \subseteq J(x)$. If, say, $J(y) \supseteq J(x)$ and $|J(y)| > |J(x)|$, then $x \in J(y)$ and $F_{J(y)} > 1$ together *contradict* the fact that $J = J(x)$ is a dyadic interval $\subset I$ of *maximum length containing* x such that $F_J > 1$. So if $J(y) \supseteq J(x)$ then $|J(y)| = |J(x)|$, and $J(y)$ must in fact *equal* $J(x)$.

We see that we can extract a *sequence of points* $x_i \in E$ such that the $J(x_i)$ *do not overlap* and

$$\bigcup_{i=1}^{\infty} J(x_i) = \mathscr{E}.$$

(This sequence, and the union in question, *may* in fact be *finite* or even *void*. A *countable* set of the $J(x)$ already serves to cover $\mathscr{E}$ because each $J(x)$ *contains an open set*.) The above dissection of $\mathscr{E}$ into a *countable number of non-overlapping dyadic intervals* is called a *Calderón–Zygmund decomposition*†.

Now $|\mathscr{E}| < \frac{1}{2}|I|$. Indeed, since $F_I = 0$,

$$\frac{1}{|I|}\int_{\mathscr{E}} F(x)\,dx \;\le\; \frac{1}{|I|}\int_{\mathscr{E}} |F(x) - F_I|\,dx \;\le\; \frac{1}{|I|}\int_I |F(x) - F_I|\,dx \;\le\; \|F\|_* \;\le\; \frac{1}{2},$$

whilst

$$\frac{1}{|I|}\int_{\mathscr{E}} F(x)\,dx \;=\; \frac{1}{|I|}\sum_i \int_{J(x_i)} F(t)\,dt$$

because the $J(x_i)$ don't overlap, and this equals

$$\frac{1}{|I|}\sum_i |J(x_i)|\,F_{J(x_i)} \;>\; \frac{1}{|I|}\sum_i |J(x_i)| \;=\; \frac{|\mathscr{E}|}{|I|}$$

(except when $|\mathscr{E}| = 0$), so in any event $|\mathscr{E}|/|I| < \frac{1}{2}$.

Since $E \subseteq \mathscr{E}$, our theorem is now proved for the case $n = 1$.

To prove the theorem for $n > 1$, we proceed by induction. I say that for each i, $F_{J(x_i)} \le 2$. Indeed, *given* $J(x_i)$, let J' be the dyadic interval $\subseteq I$ *containing* $J(x_i)$ *and having twice its length*. Note that there *is one*‡, *and only one* such:

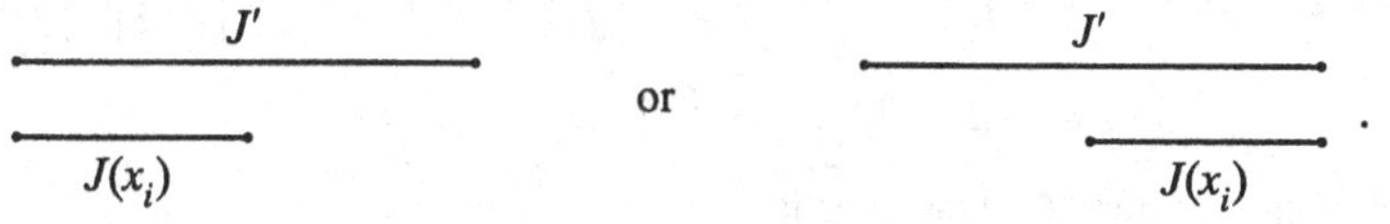

Then, by the maximal property according to which $J(x_i)$ was chosen (and because $F_I = 0$!),

$$F_{J'} \le 1.$$

By the *first part of the proof of the above lemma*, with $J = J'$ and $I = J(x_i)$, we now have $|F_{J'} - F_{J(x_i)}| \le 2\|F\|_* \le 1$, so $F_{J(x_i)} \le 2$, as claimed.

Let now $E \cap J(x_i) = E_i$. Since the $J(x_i)$ don't overlap, and cover $\mathscr{E} \supseteq E$,

$$E = \bigcup_i E_i, \quad \text{and} \quad |E| = \sum_i |E_i|.$$

† Cf. in Section E of Chapter VIII

‡ recall that $J(x_i)$ is *properly* included in I

If $x \in E_i$, $F(x) > 2n$, so, *since* $F_{J(x_i)} \le 2$,

$$F(x) - F_{J(x_i)} > 2(n-1).$$

We see that we can make an induction step at this point. For, as just seen,

$$E_i \subseteq \{x \in J(x_i);\ F(x) - F_{J(x_i)} > 2(n-1)\},$$

so, *if the theorem is true with* $n-1$ *in place of* n,

$$|E_i| \le \left(\tfrac{1}{2}\right)^{n-1} |J(x_i)|.$$

Therefore, since the $J(x_i)$ are non-overlapping and add up to $\mathscr{E}$,

$$|E| = \sum_i |E_i| \le \left(\tfrac{1}{2}\right)^{n-1} \sum_i |J(x_i)| = \left(\tfrac{1}{2}\right)^{n-1} |\mathscr{E}| < \left(\tfrac{1}{2}\right)^{n-1} \cdot \tfrac{1}{2}|I| = \left(\tfrac{1}{2}\right)^{n} |I|,$$

and the theorem holds with n.

We are done.

Corollary *If* $F \in BMO$ *and* I *is an interval,*

$$\frac{1}{|I|}\int_I [F(x) - F_I]^2\, dx \le C\, \|F\|_*^2,$$

with C a numerical constant.

Remark This is like a *reversed Schwarz inequality*!

Proof Let $m(\lambda) = |\{x \in I;\ |F(x) - F_I| > \lambda\}|$; by the theorem of Nirenberg and John just proven, for every non-negative integer n,

$$m(\lambda) \le 2 \times \left(\tfrac{1}{2}\right)^n |I| \quad \text{if } \lambda > 4n\,\|F\|_*$$

(the extra factor of 2 coming from the fact that we are looking at the measure of the set of x where $F(x) - F_I < -\lambda$ *or* $F(x) - F_I > \lambda$). Therefore

$$m(\lambda) \le 2|I| \exp\left\{-\left[\frac{\lambda}{4\,\|F\|_*}\right]\log 2\right\},$$

where $[t]$ denotes the *greatest integer* $\le t$.

By Chapter VIII, Section A, we now have

$$\int_I |F(x) - F_I|^2\, dx = 2\int_0^\infty \lambda m(\lambda)\, d\lambda \le 4|I| \int_0^\infty \lambda \exp\left(-(\log 2)\left[\frac{\lambda}{4\,\|F\|_*}\right]\right) d\lambda$$

$$= 64|I|\,\|F\|_*^2 \int_0^\infty s e^{-[s]\log 2}\, ds = C\,\|F\|_*^2\, |I|$$

with a numerical constant C. This does it.

Remark *Clearly much stronger inequalities for integrals taken over I and involving* $|F(x) - F_I|$ *are valid.*

2. Equivalence of *BMO* norm and Garsia norm

Lemma *Let* $f(t) \ge 0$. *Then there is a numerical constant C such that*

$$\int_{-\pi}^{\pi} \frac{1-r^2}{1+r^2-2r\cos t} f(t)\, dt \le C \int_0^\infty \frac{(1-r)s^2}{\left((1-r)^2+s^2\right)^2}\left(\frac{1}{2s}\int_{-s}^{s} f(t)\, dt\right) ds.$$

Proof $(1-r^2)/(1+r^2-2r\cos t) = (1+r)(1-r)/((1-r)^2 + 4r\sin^2(t/2))$, and the right hand side is clearly $\le k(1-r)/((1-r)^2+t^2)$ for $0 \le r < 1$ and $-\pi \le t \le \pi$, where k is a suitable numerical constant.

Write $1-r=u$. Then, if $f(t)\geq 0$,

$$\int_0^\pi \frac{1-r^2}{1+r^2-2r\cos t}(f(t)+f(-t))\,dt \;\leq\; k\int_0^\pi \frac{u}{u^2+t^2}(f(t)+f(-t))\,dt$$
$$\leq k\int_0^\infty \frac{u}{u^2+t^2}(f(t)+f(-t))\,dt\,.$$

The last integral equals

$$k\int_0^\infty (f(t)+f(-t))\int_t^\infty \frac{2us}{(u^2+s^2)^2}\,ds\,dt \;=\; k\int_0^\infty \frac{2us}{(u^2+s^2)^2}\int_0^s (f(t)+f(-t))\,dt\,ds$$
$$=\; 4k\int_0^\infty \frac{us^2}{(u^2+s^2)^2}\left(\frac{1}{2s}\int_{-s}^s f(t)\,dt\right)ds,$$

an expression of the required form.

Theorem *If F is real and periodic, of period 2π, we have*

$$\mathfrak{N}(F)\leq \tilde{K}\,\|F\|_*$$

with a numerical constant $\tilde{K}$.

Proof Using once more the notation introduced at the beginning of Section C, we have

$$(\mathfrak{N}(F))^2 \;=\; \sup_{|z|<1}\,\{U_{F^2}(z)-(U_F(z))^2\}\,;$$

it suffices to estimate $U_{F^2}(z)-(U_F(z))^2$ in terms of $\|F\|_*$. Since $F(x)$ and its translates $F_h(x)=F(x-h)$ *clearly have the same BMO norm* $\|\ \|_*$, there is no loss of generality in taking $z=r$ with $0<r<1$. By the first lemma of Subsection C.1,

$$U_{F^2}(r)-(U_F(r))^2 \;=\; \frac{1}{8\pi^2}\int_{-\pi}^{\pi}\int_{-\pi}^{\pi}\frac{(1-r^2)^2\,(F(s)-F(t))^2\,ds\,dt}{(1+r^2-2r\cos s)(1+r^2-2r\cos t)}\,.$$

Since $[F(s)-F(t)]^2\geq 0$, we can apply the above lemma *twice* to estimate the above right-hand double integral, and find it to be

$$\leq\; C^2\int_0^\infty\int_0^\infty \frac{us^2}{(u^2+s^2)^2}\cdot\frac{ut^2}{(u^2+t^2)^2}\left(\frac{1}{4st}\int_{-s}^s\int_{-t}^t (F(\sigma)-F(\tau))^2\,d\sigma\,d\tau\right)ds\,dt,$$

where $u=1-r$ and C is a numerical constant.

Now, if I and J are intervals with *the same* midpoint, we have:

$$\left(\frac{1}{|J|}\cdot\frac{1}{|I|}\int_J\int_I [F(\sigma)-F(\tau)]^2\,d\sigma\,d\tau\right)^{1/2}$$
$$\leq\; \left(\frac{1}{|J|}\cdot\frac{1}{|I|}\int_J\int_I [F(\sigma)-F_I]^2\,d\sigma\,d\tau\right)^{1/2} + \left(\frac{1}{|I|}\cdot\frac{1}{|J|}\int_I\int_J [F(\tau)-F_J]^2\,d\tau\,d\sigma\right)^{1/2}$$
$$+\;\left(\frac{1}{|I|}\cdot\frac{1}{|J|}\int_I\int_J [F_I-F_J]^2\,d\tau\,d\sigma\right)^{1/2}$$
$$=\; \left(\frac{1}{|I|}\int_I [F(\sigma)-F_I]^2\,d\sigma\right)^{1/2} + \left(\frac{1}{|J|}\int_J [F(\tau)-F_J]^2\,d\tau\right)^{1/2} + |F_I-F_J|\,.$$

By the Corollary to the Nirenberg-John theorem in Subsection 1, this is

$$\leq\; 2\sqrt{C}\,\|F\|_* + |F_I-F_J|,$$

and by the lemma at the beginning of Subsection 1, the expression just found is in turn

$$\leq\; 2\sqrt{C}\,\|F\|_* + 2\,\|F\|_*\left(1+\left|\,\log_2\frac{|J|}{|I|}\,\right|\right).$$

We see that for s and $t >, 0$,

$$\boxed{\left(\frac{1}{4st}\int_{-s}^{s}\int_{-t}^{t}(F(\sigma)-F(\tau))^2\,d\sigma\,d\tau\right)^{1/2} \le 2c\,\|F\|_* + 2\,\|F\|_*\,(1+|\log_2(s/t)|).}$$

Plugging this expression into the quadruple integral obtained at the end of the preceding paragraph, we see that

$$U_{F^2}(r)-(U_F(r))^2 \le K\int_0^\infty\int_0^\infty \frac{u^2s^2t^2\,\|F\|_*^2\,(1+(\log(s/t))^2)}{(u^2+s^2)^2(u^2+t^2)^2}\,ds\,dt,$$

where K is a suitable numerical constant and $u = 1-r$. Making the changes of variable $s/u = x$, $t/u = y$, the integral on the right becomes

$$K\,\|F\|_*^2\int_0^\infty\int_0^\infty \frac{x^2y^2\left(1+(\log x-\log y)^2\right)\,dx\,dy}{(1+x^2)^2(1+y^2)^2}$$

$$\le K\,\|F\|_*^2\int_0^\infty\int_0^\infty \frac{x^2y^2\left(1+2(\log x)^2+2(\log y)^2\right)\,dx\,dy}{(1+x^2)^2(1+y^2)^2}.$$

But the double integral in this last expression is *clearly finite*, and has a *numerical value independent* of u, hence of r. Therefore, with a numerical constant $\tilde{K}$,

$$\left(U_{F^2}(r)-(U_F(r))^2\right)^{1/2} \le \tilde{K}\,\|F\|_*,$$

and finally, $\mathfrak{N}(F) \le \tilde{K}\,\|F\|_*$. Q.E.D.

Corollary *The norms* $\|\ \|_*$ *and* $\mathfrak{N}(\ )$ *are equivalent.*

Proof By the *above* theorem and the *second* one of Subsection C.2.

3. Fefferman's theorem

Combining the equivalence of $\mathfrak{N}$ and $\|\ \|_*$ established in Subsection 2 with Garsia's version of the Fefferman theorem (Subsection E.4), we obtain the fundamental theorem about BMO, namely:

Theorem (Fefferman) *The dual of* $\Re H_1(0)$ *is* BMO/(constant functions). *If* $F(\theta)$ *is real, of period* 2π, *and locally integrable, the following three conditions on* F *are equivalent:*

(i) $L\Re f = \lim_{R\to 1}\int_{-\pi}^{\pi}\Re f(Re^{i\theta})F(\theta)\,d\theta$ *exists for every* $f \in H_1(0)$ *and* L *is a linear functional on* $\Re H_1(0)$.

(ii) $F(\theta) = \phi(\theta)+\tilde{\psi}(\theta)$ *with* ϕ *and* ψ *real,* 2π*-periodic, and bounded.*

(iii) $F \in BMO$.

In case $F \in BMO$, *the* norm *of the linear functional* L *given by* (i) is equivalent *to* $\|F\|_*$, *the* BMO *norm of* F, *and we can find bounded functions* $\phi_1(\theta)$, $\psi_1(\theta)$, *periodic of period* 2π, *such that*

$$F(\theta) = \phi_1(\theta)+\tilde{\psi}_1(\theta)+\text{ const.},$$

with

$$\|\phi_1\|_\infty+\|\psi_1\|_\infty \le A\,\|F\|_* \le B(\|\phi_1\|_\infty+\|\psi_1\|_\infty),$$

A and B being numerical constants. Thus, $BMO = \Re L_\infty + \widetilde{\Re L_\infty}$.

Remark At the time the lectures preceding the first edition of this book were being given (in 1978) one could still say that, for an arbitrary $F \in BMO$, no method was known for *actually constructing* ϕ and $\psi \in L_\infty$ such that $\phi + \tilde{\psi} = F$, and that the only known proof for this decomposition was *indirect*, being based on combination of the Hahn-Banach duality argument of Section A with the inequalities established in Sections D and E and the present section. Of course, the Hahn-Banach theorem *does* have a *constructive side* to it, at least insofar as linear functionals over *separable Banach spaces* are considered. Provided the dual we are looking at *is* identified with a space of functions, careful examination of what the *steps used in proving* the Hahn-Banach theorem *signify concretely* in a given situation can sometimes lead to an *explicit construction procedure* for the function corresponding to the linear functional in question. Be that as it may, the existence proof *given here is certainly not constructive as it stands.*

A constructive procedure for obtaining the decomposition has, however, been found; it is due to *Peter Jones.* He was in fact *working* on it at the *same time that this chapter was being covered* in the lectures (late Spring of 1978), and had obtained his result by the time those lectures were finished!

G. Alternative proof, based on the atomic decomposition in $\Re H_1$

As promised in Subsection C.3, we give now an easier proof of the Fefferman theorem based on the results in Chapter VIII, Subsection E.2. The reader should refer to the definition of an *atom* for $\Re H_1(|z| < 1)$ appearing near the beginning of that discussion.

Lemma *Let $f(e^{i\theta})$ be an atom for $\Re H_1(|z| < 1)$ and $F(\theta) \in BMO$. Then*

$$\left| \int_{-\pi}^{\pi} f(e^{i\theta})F(\theta)\,d\theta \right| \le \|F\|_* .$$

Proof Let J be the arc of the unit circumference associated with f. Then, since $f(e^{i\theta})$ vanishes for $e^{i\theta} \notin J$ and $\int_J f(e^{i\theta})\,d\theta = 0$, we have

$$\int_{-\pi}^{\pi} f(e^{i\theta})F(\theta)\,d\theta = \int_J f(e^{i\theta})F(\theta)\,d\theta = \int_J f(e^{i\theta})(F(\theta) - F_J)\,d\theta .$$

But $|f(e^{i\theta})| \le 1/|J|$ a.e. The right-hand integral therefore has modulus

$$\le \frac{1}{|J|}\int_J |F(\theta) - F_J|\,d\theta \le \|F\|_* .$$

Done.

Corollary *Under the conditions of the lemma,*

$$\left| \int_{-\pi}^{\pi} f(e^{i(\theta-\tau)})F(\theta)\,d\theta \right| \le \|F\|_*$$

for each $\tau \in \mathbb{R}$.

Proof When $f(e^{i\theta})$ is an atom associated to the arc J of the unit circumference, $f_\tau(e^{i\theta}) = f(e^{i(\theta-\tau)})$ is also an atom, associated to the rotated arc $e^{i\tau}J$. Refer to the lemma.

Adapting the notation introduced at the beginning of Section C, we write, for integrable functions $f(e^{i\theta})$,

$$U_f(z) = \frac{1}{2\pi}\int_{-\pi}^{\pi} \frac{1-|z|^2}{|z-e^{it}|^2} f(e^{it})\,dt, \quad |z|<1.$$

Lemma *If $f(e^{i\theta})$ is an atom for $\Re H_1(|z|<1)$ and $F \in BMO$, we have*

$$\left|\int_{-\pi}^{\pi} U_f(Re^{i\theta})F(\theta)\,d\theta\right| \leq \|F\|_{*}$$

for $0 \leq R < 1$.

Proof After making the change of variable $t = \theta - \tau$ in the integral used to get $U_f(Re^{i\theta})$, we find that

$$\int_{-\pi}^{\pi} U_f(Re^{i\theta})F(\theta)\,d\theta = \frac{1}{2\pi}\int_{-\pi}^{\pi}\int_{-\pi}^{\pi} \frac{(1-R^2)f(e^{i(\theta-\tau)})}{|R-e^{-i\tau}|^2} F(\theta)\,d\tau\,d\theta$$

$$= \frac{1}{2\pi}\int_{-\pi}^{\pi}\left(\int_{-\pi}^{\pi} f(e^{i(\theta-\tau)})F(\theta)\,d\theta\right)\frac{1-R^2}{|R-e^{i\tau}|^2}\,d\tau,$$

use of Fubini's theorem being justified by absolute convergence (f is in L_∞ and F in $L_1(-\pi,\pi)$!). Apply the preceding corollary.

We can now use the atomic decomposition obtained in Chapter VIII, Subsection E.2 to establish the fundamental

Theorem *If $f \in H_1(0)$ and $F \in BMO$,*

$$\lim_{R\to 1}\int_{-\pi}^{\pi} \Re f(Re^{i\theta})F(\theta)\,d\theta$$

exists and is in modulus $\leq B\,\|f\|_1\,\|F\|_{*}$, where B is an absolute constant.

Proof According to the first lemma of Subsection D.3 (with the C figuring therein replaced by $B\,\|F\|_{*}$), it is enough to show that for every $R<1$,

$$\left|\int_{-\pi}^{\pi} \Re f(Re^{i\theta})F(\theta)\,d\theta\right| \leq B\,\|F\|_{*}\,.$$

From the theorem at the end of Chapter VIII, Subsection E.2 we have, since $f(0)=0$,

$$\Re f(e^{i\theta}) = \sum_n a_n f_n(e^{i\theta}) \text{ a.e.}$$

with the series on the right convergent in $L_1(-\pi,\pi)$, where the f_n are atoms for $\Re H_1(|z|<1)$, and the a_n constants with

$$\sum_n |a_n| \leq B\,\|f\|_1\,.$$

For each $R < 1$, $\Re f(Re^{i\theta}) = U_{\Re f}(Re^{i\theta})$ (Chapter II, Subsection B.1!), so by the previous relation,

$$\Re f(Re^{i\theta}) = \sum_n a_n U_{f_n}(Re^{i\theta}), \quad R < 1,$$

with the series on the right now *uniformly convergent* in θ. Thence, since $F \in L_1(-\pi, \pi)$,

$$\int_{-\pi}^{\pi} \Re f(Re^{i\theta}) F(\theta)\, d\theta = \sum_n a_n \int_{-\pi}^{\pi} U_{f_n}(Re^{i\theta}) F(\theta)\, d\theta,$$

and the sum on the right is in absolute value

$$\leq \sum_n |a_n| \, \|F\|_* \leq B \|f\|_1 \, \|F\|_*$$

by the last lemma. This does it.

On combining the result just found with the statement in Subsection A.2 and the remark at the end of Subsection B.2, we arrive again at the Fefferman theorem of Subsection F.3. The reader who has any doubts about how everything is put together may refer to the discussion in the scholium of Subsection E.4.

H. Representation in terms of radially bounded measures

Fefferman's theorem, given in Subsection F.3, can be combined with the radial maximal function characterization of $\Re H_1$ (Chapter VIII, Subsection D.2) to yield, *by duality*, a curious representation of functions in BMO. This representation has some resemblance to the decomposition $BMO = \Re L_\infty + \widetilde{\Re L_\infty}$, and was important in the development of the subject because *Carleson* published a *constructive procedure for getting it* in 1976, when a *constructive method* for decomposing arbitrary BMO functions into a sum of something in $\Re L_\infty$ and something else in $\widetilde{\Re L_\infty}$ was *not yet known* (see remark at end of Subsection F.3). It is still interesting in its own right, and shows a connection between the material of Chapter VIII, Section D and the present chapter.

1. Radially bounded measures

We will be interested in signed measures ν on $\{|z| \leq 1\}$ which, roughly speaking, can be put in the form

$$d\nu(re^{i\theta}) = d\mu_\theta(r)\, d\theta,$$

with $\int_0^1 |d\mu_\theta(r)|$ *bounded* for $0 \leq \theta < 2\pi$. In order to avoid measure-theoretic niceties and digressions, we simply take the

Definition A real measure ν on $\{|z| \leq 1\}$ is called *radially bounded* if there is a constant C such that, for any *sector* (nota bene!) S_h of the form

$$S_h = \{re^{i\theta};\ 0 \leq r \leq 1,\ \theta_0 \leq \theta \leq \theta_0 + h\},$$

we have

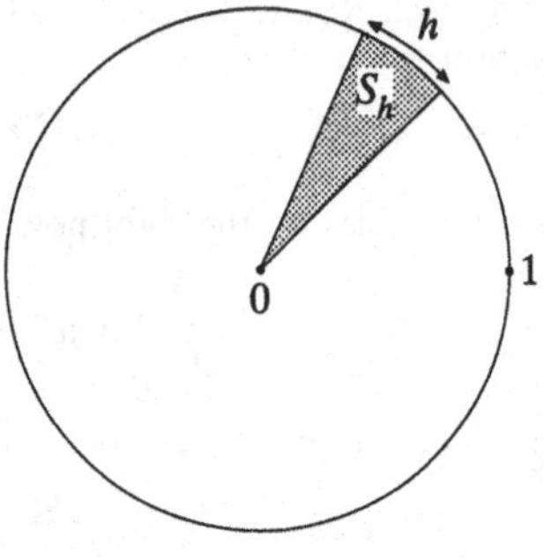

$$\iint_{S_h} |d\nu(z)| \leq Ch.$$

The smallest permissible value of C for which this relation holds (for all S_h) is denoted by $\|\nu\|^*$.

$\| \ \|^*$ is a *norm* for radially bounded measures, and the *ordinary measure norm* $\|\nu\| = \int\int_{|z|\leq 1} |d\nu(z)|$ is clearly $\leq 2\pi\|\nu\|^*$. By the last theorem in Chapter VIII, Section F *a radially bounded measure is surely Carleson.*

Lemma *A radially bounded measure ν can be written as a sum $\mu + \sigma$, where:*

(i) *μ is radially bounded, and carried on the* open circle $\{|z| < 1\}$;

(ii) *σ is carried on* $\{|z| = 1\}$, *and* $d\sigma(e^{i\theta}) = s(\theta)\,d\theta$ *with* $s \in L_\infty$.

Proof Let μ simply be the restriction of ν to $\{|z| < 1\}$, and σ the restriction of ν to $\{|z| = 1\}$. Then, if I is any arc of $\{|z| = 1\}$,

$$\int_I |d\sigma| \leq \|\nu\|^* \cdot \text{length } I,$$

so σ is *absolutely continuous* with respect to *linear Lebesgue measure* on $\{|z| = 1\}$ and has a *bounded density with respect to the latter*, i.e., $d\sigma(e^{i\theta}) = s(\theta)\,d\theta$ with $s \in L_\infty$.

Definition If ν is *radially bounded* and $\nu = \mu + \sigma$ with μ carried on $\{|z| < 1\}$ and σ on $\{|z| = 1\}$, and if $d\sigma(e^{i\theta}) = s(\theta)\,d\theta$, we write

$$P_\nu(\theta) = s(\theta) + \frac{1}{2\pi}\iint_{|z|<1} \frac{1-|z|^2}{|z-e^{i\theta}|^2}\,d\mu(z).$$

Remark The integral on the right converges absolutely for almost all θ, and yields a function in $L_1[\pi, \pi]$. *This is true for any finite measure μ on* $\{|z| < 1\}$.

Indeed,

$$\frac{1}{2\pi}\int_{-\pi}^{\pi}\iint_{|z|<1} \frac{1-|z|^2}{|z-e^{i\theta}|^2}|d\mu(z)|\,d\theta = \frac{1}{2\pi}\iint_{|z|<1}\int_{-\pi}^{\pi} \frac{1-|z|^2}{|z-e^{i\theta}|^2}\,d\theta\,|d\mu(z)| = \iint_{|z|<1} |d\mu(z)|.$$

We will eventually prove in this section that *BMO coincides* with the *set of P_ν for ν ranging over the family of radially bounded measures.* In the present subsection, let us just show that $P_\nu \in BMO$ whenever ν is radially bounded.

Lemma *If $U(z)$ is* continuous *on* $\{|z| \leq 1\}$ *and* harmonic *in* $\{|z| < 1\}$, *then*

$$\int_{-\pi}^{\pi} U(e^{i\theta})P_\nu(\theta)\,d\theta = \iint_{|z|\leq 1} U(z)\,d\nu(z),$$

whenever ν *is radially bounded.*

Proof Put $\nu = \mu+\sigma$ where μ is the restriction of ν to $\{|z|<1\}$ and $d\sigma(e^{i\theta}) = s(\theta)d\theta$, $s \in L_\infty$; we have by definition $P_\nu(\theta) = P_\mu(\theta)+s(\theta)$, with $\int_{-\pi}^{\pi} U(e^{i\theta})s(\theta)\,d\theta = \int_{-\pi}^{\pi} U(e^{i\theta})\,d\sigma(e^{i\theta})$. Again, by Poisson's formula (Chapter I!), for $|z|<1$,

$$U(z) = \frac{1}{2\pi}\int_{-\pi}^{\pi} \frac{1-|z|^2}{|z-e^{i\theta}|^2} U(e^{i\theta})\,d\theta,$$

from which, by Fubini's theorem,

$$\iint_{|z|<1} U(z)\,d\mu(z) = \int_{-\pi}^{\pi} U(e^{i\theta})P_\mu(\theta)\,d\theta.$$

We're done.

Theorem *If* ν *is radially bounded,* $P_\nu(\theta)$ *is in* BMO, *and* $\|P_\nu\|_* \leq C\|\nu\|^*$ *with a numerical constant* C.

Proof By the first lemma of Subsection D.3 and the scholium in Subsection E.4, it is enough to show that there is a numerical constant C such that

$$\int_{-\pi}^{\pi} \Re f(Re^{i\theta})P_\nu(\theta)\,d\theta \;\leq\; C\,\|f\|_1\,\|\nu\|^*$$

for all $f \in H_1(0)$ and all R, $0 \leq R < 1$.

By the above lemma,

$$\int_{-\pi}^{\pi} \Re f(Re^{i\theta})P_\nu(\theta)\,d\theta \;=\; \iint_{|z|\leq 1} \Re f(Rz)\,d\nu(z),$$

since $\Re f(Rz)$ is harmonic for $|z|<1$ and continuous for $|z|\leq 1$. Taking a large integer N, we put $h = 2\pi/N$ and break the closed unit disk up into the N non-overlapping sectors

$$S_h(n) \;=\; \{re^{i\theta};\; 0 \leq r \leq 1,\; nh \leq \theta \leq (n+1)h\},\quad n = 0,1,2,\ldots,N-1.$$

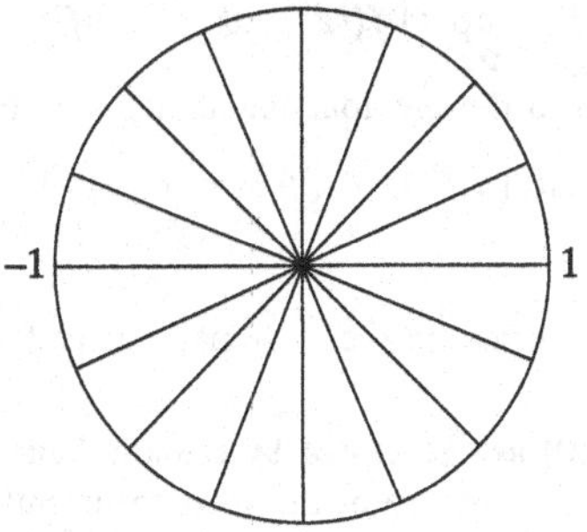

Fixing R, $0 \leq R < 1$, we have

$$\left|\iint_{|z|\leq 1} \Re f(Rz)\,d\nu(z)\right| \;\leq\; \sum_{n=0}^{N-1} \iint_{S_h(n)} |\Re f(Rz)|\,|d\nu(z)|.$$

For each n there is, by the continuity of $\Re f(Rz)$ on $\{|z| \leq 1\}$, a $z_n \in S_h(n)$ with

$$\iint_{S_h(n)} |\Re f(Rz)|\,|\mathrm{d}\nu(z)| \;=\; |\Re f(Rz_n)| \iint_{S_h(n)} |\mathrm{d}\nu(z)|\,,$$

and the right side is in turn $\leq |\Re f(Rz_n)| \|\nu\|^* h$ by definition of $\|\nu\|^*$. Writing $z_n = r_n e^{i\theta_n}$ where $nh \leq \theta_n \leq (n+1)h$ and $0 \leq r_n \leq 1$, we see that $|\Re f(Rz_n)| \leq \sup_{0\leq r\leq 1} |\Re f(Rre^{i\theta_n})|$, and the preceding relations make

$$\left| \iint_{|z|\leq 1} \Re f(Rz)\,\mathrm{d}\nu(z) \right| \;\leq\; \|\nu\|^* \sum_{n=0}^{N-1} \sup_{0\leq r\leq 1} |\Re f(Rre^{i\theta_n})| \cdot h.$$

On the right we have a Riemann sum for the integral

$$\int_0^{2\pi} \sup_{0\leq r\leq 1} |\Re f(Rre^{i\theta})|\,\mathrm{d}\theta$$

whose integrand is a continuous function of θ (since $R < 1$). For large N, the right-hand side of the last inequality is hence as close as we like to

$$\|\nu\|^* \int_0^{2\pi} \sup_{0\leq r\leq 1} |\Re f(Rre^{i\theta})|\,\mathrm{d}\theta\,,$$

and thus

$$\left| \iint_{|z|\leq 1} \Re f(Rz)\,\mathrm{d}\nu(z) \right| \;\leq\; \|\nu\|^* \int_0^{2\pi} \sup_{0\leq r\leq 1} |\Re f(Rre^{i\theta})|\,\mathrm{d}\theta.$$

We therefore have

$$\left| \int_{-\pi}^{\pi} \Re f(Re^{i\theta}) P_\nu(\theta)\,\mathrm{d}\theta \right| \;\leq\; \|\nu\|^* \int_{-\pi}^{\pi} \sup_{0\leq r\leq 1} |\Re f(re^{i\theta})|\,\mathrm{d}\theta$$

for every $f \in H_1(0)$ and all $R < 1$, using the formula at the beginning of this proof. By a corollary at the end of Subsection C.3 in Chapter VIII,

$$\int_{-\pi}^{\pi} \sup_{0\leq r\leq 1} |\Re f(re^{i\theta})|\,\mathrm{d}\theta \;\leq\; C\,\|f\|_1$$

for $f \in H_1$. Substituting this into the previous inequality, we obtain

$$\left| \int_{-\pi}^{\pi} \Re f(Re^{i\theta}) P_\nu(\theta)\,\mathrm{d}\theta \right| \;\leq\; C\|\nu\|^*\,\|f\|_1\,,$$

and the proof is complete.

Remark This result only depends on the 'easy part' of Fefferman's theorem.

2. Linear functionals on $\Re H_1(0)$ are generated by radially bounded measures

We now set out to establish the converse of the theorem in Subsection 1.

Lemma* *Let* $\|\nu_k\|^* \leq M$, $k = 1, 2, 3, \ldots$. *Then there is a subsequence* $\{\nu_{k_j}\}$ *of* $\{\nu_k\}$, *a*

* A mistake in the first edition's version of this lemma was pointed out to me by Jim Thomson.

constant c, $|c| \le 2\pi M$, *and a radially bounded* ν, $\|\nu\|^* \le M$, *with*

$$d\nu_{k_j} \underset{j}{\longrightarrow} c\,d\delta + d\nu \quad w^*$$

on $\{|z| \le 1\}$. *Here,* δ *is the unit point mass at* 0.

Proof By a remark at the beginning of Subsection 1, $\|\nu_k\| \le 2\pi\|\nu_k\|^* \le 2\pi M$, so some subsequence of the ν_k does converge w* to a measure μ on $\{|z| \le 1\}$ with $\|\mu\| \le 2\pi M$. Let ν be the restriction of μ to $\{0 < |z| \le 1\}$. Then it is easy to verify that $\|\nu\|^* \le M$. Since μ is a multiple of δ plus ν, the lemma follows.

Theorem *Let* $F(\theta)$, *real and* 2π*-periodic, belong to BMO. There exists a radially bounded real measure* ν *with*

$$F(\theta) = P_\nu(\theta) + \text{constant.}$$

We can take ν *so as to satisfy*

$$\|\nu\|^* \le K\,\|F\|_* \,,$$

where K is a numerical constant.

Remark Putting $\mu =$ restriction of ν to $\{|z| < 1\}$ we have, by a lemma of Subsection 1,

$$F(\theta) = \text{const} + s(\theta) + \frac{1}{2\pi}\iint_{|z|<1} \frac{1-|z|^2}{|z-e^{i\theta}|^2}\,d\mu(z)$$

with $s(\theta) \in L_\infty$. This looks a little like the basic decomposition

$$F(\theta) = \phi(\theta) + \tilde{\psi}(\theta)$$

with ϕ and $\psi \in L_\infty$.

Proof By the hard part of Fefferman's theorem (Subsection F.3 or Section G), the function $F \in BMO$ defines a functional L on $\Re H_1(0)$ according to the formula

$$L(\Re f) = \lim_{R\to 1}\int_{-\pi}^{\pi} \Re f(Re^{i\theta})F(\theta)\,d\theta\,, \quad f \in H_1(0)\,.$$

We have $|||L||| \le A\,\|F\|_*$ with a numerical constant A.

Let, for each $M > 0$,

$$K_M = \{P_\nu;\ \nu \text{ real},\ \|\nu\|^* \le M|||L|||\}\,.$$

K_M is a *convex set.*

I claim that, for sufficiently large M, L is in the w* *closure, over* $\Re H_1(0)$, *of the linear functionals corresponding to the* P_ν *in* K_M.

If ν is radially bounded, it is convenient to denote by L_ν the linear functional on $\Re H_1(0)$ corresponding to P_ν: it is given by the formula

$$L_\nu(\Re f) = \lim_{R\to 1}\int_{-\pi}^{\pi} \Re f(Re^{i\theta})P_\nu(\theta)\,d\theta\,, \quad f \in H_1(0)\,.$$

We are to prove that if M is *large enough*, L is in the w* closure of the L_ν, $\|\nu\|^* \le M|||L|||$.

For given M, suppose that L is *not* in the w* closure of the L_ν, $\|\nu\|^* \le M|||L|||$. Then there is a w*-closed *hyperplane* separating L from the convex set of these L_ν. That is, there is a $g \in H_1(0)$ with $L_\nu(\Re g) < L(\Re g)$ whenever $\|\nu\|^* \le M|||L|||$. Now the set of L_ν

under consideration contains 0 and is taken onto itself on multiplication by -1. Therefore $|L_v(\Re g)| < L(\Re g)$ for $\|v\|^* \leq M|||L|||$, and without loss of generality we can take $L(\Re g) = 1$; otherwise, use a suitable positive multiple of g in place of g.

Thus, for $\|v\|^* \leq M|||L|||$, $|L_v(\Re g)| < 1$ *whilst* $L(\Re g) = 1$.

For the moment, write $M|||L||| = B$. *We can find v with* $\|v\|^* = B$ *such that* $L_v(\Re g)$ *is as close as we please to*

$$B \cdot \int_{-\pi}^{\pi} \sup_{0\leq r<1} |\Re g(re^{i\theta})|\, d\theta .$$

Indeed, we can, by Lebesgue's monotone convergence theorem first find an $R < 1$ such that

$$B \cdot \int_{-\pi}^{\pi} \sup_{0\leq r\leq 1} |\Re g(Rre^{i\theta})|\, d\theta$$

is already within, say, ϵ of the preceding (finite!) expression. The function $g_R^+(\theta) = \sup_{0\leq r\leq 1} |\Re g(Rre^{i\theta})|$ now being *continuous*, we see that the sum

$$B \sum_{n=1}^{N} g_R^+\left(\frac{2\pi}{N}n\right) \cdot \frac{2\pi}{N}$$

is in turn within ϵ of the previous integral if N is *sufficiently large*. Let, for $\theta_n = (2\pi/N)n$, the *actual maximum*

$$\sup_{0\leq r\leq 1} |\Re g(Rre^{i\theta_n})| \; = \; g_R^+(\theta_n)$$

be attained* for $r = r_n$. Then the previous sum equals

$$B \sum_{n=1}^{N} |\Re g(Rr_n e^{i\theta_n})| \cdot \frac{2\pi}{N} .$$

$\Re g(Rz)$ is *uniformly continuous* for $|z| \leq 1$. *Therefore, if N is large,* $\Re g(Rr_n e^{i\theta})$ *oscillates by less than* $\epsilon/2\pi B$ *when* θ *runs from* $2\pi(n-1)/N$ *to* $2\pi n/N$, *no matter what the value of* $r_n \in [0,1]$ *may be*. Thence,

$$\left| \frac{2\pi}{N} \Re g(Rr_n e^{i\theta_n}) - \int_{\theta_{n-1}}^{\theta_n} \Re g(Rr_n e^{i\theta})\, d\theta \right| < \frac{\epsilon}{NB}$$

for $n = 1, 2, \ldots, N$, *no matter what the* r_n *are, as long as N is large*. Let $\epsilon_n = \operatorname{sgn} \Re g(Rr_n e^{i\theta_n})$. Then

$$B \sum_{n=1}^{N} \int_{\theta_{n-1}}^{\theta_n} \Re g(Rr_n e^{i\theta}) \cdot \epsilon_n \, d\theta$$

differs at most by ϵ from

$$B \sum_{n=1}^{N} |\Re g(Rr_n e^{i\theta_n})| \cdot \frac{2\pi}{N} ,$$

---

* r_n is surely *not* 0, because $g \in H_1(0)$ is zero at the origin but does not vanish identically

hence at most by 3ϵ from

$$B\int_{-\pi}^{\pi} \sup_{0\le r<1} |\Re g(re^{i\theta})|\, d\theta,$$

if N is large.

For such a large N, which we now *fix*, let us define a measure ν on $\{|z| \le R\}$ as follows: *in the sector $\theta_{n-1} < \theta < \theta_n$, ν is carried on the circular arc $|z| = Rr_n$, and on that arc,*

$$d\nu(Rr_n e^{i\theta}) = B\epsilon_n\, d\theta, \quad \theta_{n-1} < \theta < \theta_n.$$

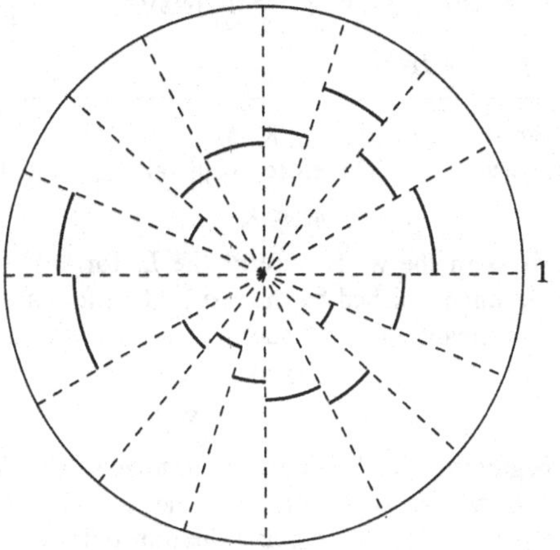

Clearly,

$$B\sum_{n=1}^{N} \int_{\theta_{n-1}}^{\theta_n} \Re g(Rr_n e^{i\theta})\epsilon_n\, d\theta = \iint_{|z|\le R} \Re g(z)\, d\nu(z),$$

and it is evident that $\|\nu\|^* = B$.

Since ν is supported on $\{|z| \le R\}$ and $R < 1$,

$$\iint_{|z|\le 1} \Re g(z)\, d\nu(z) = \lim_{r\to 1} \iint \Re g(rz)\, d\nu(z) = \lim_{r\to 1} \int_{-\pi}^{\pi} \Re g(re^{i\theta}) P_\nu(\theta)\, d\theta,$$

the last relation holding by a lemma in Subsection 1, because, when $0 \le r < 1$, $\Re g(rz)$ is *continuous* for $|z| \le 1$ and *harmonic* for $|z| < 1$. That is,

$$L_\nu(\Re g) = \iint_{|z|\le 1} \Re g(z)\, d\nu(z).$$

We have thus found a ν, $\|\nu\|^* = B$, such that $L_\nu(\Re g)$ is within 3ϵ of

$$B\int_{-\pi}^{\pi} \sup_{0\le r<1} |\Re g(re^{i\theta})|\, d\theta,$$

proving our assertion.

Because of this, $|L_\nu(\Re g)| < 1$ for $\|\nu\|^* \leq M|||L||| = B$ *implies that*

$$M|||L||| \int_{-\pi}^{\pi} \sup_{0\leq r<1} |\Re g(re^{i\theta})|\, d\theta \;\leq\; 1.$$

Now, by Chapter VIII, Subsection D.2, there is a numerical constant K such that $$\|g\|_1 \;\leq\; K \int_{-\pi}^{\pi} \sup_{0\leq r<1} |\Re g(re^{i\theta})|\, d\theta$$ *for* $g \in H_1(0)$.

On account of this, we see that $|||L|||\, \|g\|_1 \leq K/M$.

However, we *also* had $L(\Re g) = 1$. Therefore $|||L|||\, \|g\|_1 \geq 1$ and $1 \leq K/M$, that is

$$M \leq K.$$

We see that L certainly IS in the w* closure of the L_ν for $\|\nu\|^* \leq M|||L|||$ *provided that* $M > K$, the *numerical constant* furnished by the result of Fefferman and Stein in Subsection D.2 of Chapter VIII. That means that, for such M, we can find a sequence* of measures ν_k, $\|\nu_k\|^* \leq M|||L|||$, with

$$L_{\nu_k} \xrightarrow[k]{} L \qquad \text{w}^*.$$

The lemma at the beginning of this subsection now shows that the $d\nu_k$ belonging to a *subsequence converge* w*, *as measures*, to some $d\mu$ supported on $\{|z| \leq 1\}$ with $d\mu = c d\delta + d\nu$ and that $\|\nu\|^* \leq M|||L|||$, δ being the unit point mass at the origin. Without loss of generality,

$$d\nu_k \xrightarrow[k]{} c d\delta + d\nu \qquad \text{w}^*$$

AS MEASURES.

Recall that L was the linear functional on $\Re H_1(0)$ corresponding to the *BMO* function F:

$$L(\Re f) = \lim_{R\to 1} \int_{-\pi}^{\pi} \Re f(Re^{i\theta}) F(\theta)\, d\theta.$$

I claim that

$$F(\theta) = P_\nu(\theta) \;+\; \text{const.}$$

For each $n = 1, 2, 3, \ldots$ the function z^n is in $H_1(0)$, so each of the functions $r^n \begin{Bmatrix} \sin \\ \cos \end{Bmatrix} n\theta$ is in $\Re H_1(0)$. Take any such function and call it $U(z)$. Because $U(z)$ is *continuous* on $\{|z| \leq 1\}$, we here have simply

$$LU \;=\; \lim_{r\to 1} \int_{-\pi}^{\pi} U(re^{i\theta}) F(\theta)\, d\theta \;=\; \int_{-\pi}^{\pi} U(e^{i\theta}) F(\theta)\, d\theta\,;$$

* our set of L_ν is *bounded*, and $\Re H_1(0)$ *separable*

similarly, for each ν_k,

$$L_{\nu_k} U = \int_{-\pi}^{\pi} U(e^{i\theta}) P_{\nu_k}(\theta)\, d\theta .$$

By a lemma in Subsection 1, due to *continuity* of $U(z)$ on $\{|z| \le 1\}$ and its *harmonicity* in the *interior* thereof,

$$\int_{-\pi}^{\pi} U(e^{i\theta}) P_{\nu_k}(\theta)\, d\theta = \iint_{|z|\le 1} U(z)\, d\nu_k(z).$$

Since $U(z)$ is *continuous* on $|z| \le 1$ and $d\nu_k \underset{k}{\longrightarrow} cd\delta + d\nu$ w* *as measures while* $U(0) = 0$,

$$\iint_{|z|\le 1} U(z)\, d\nu_k(z) \underset{k}{\longrightarrow} cU(0) + \iint_{|z|\le 1} U(z)\, d\nu(z) = \iint_{|z|\le 1} U(z)\, d\nu(z);$$

the integral on the right *is*, however, $\int_{-\pi}^{\pi} U(e^{i\theta}) P_{\nu}(\theta)\, d\theta$ by the lemma already used. That is,

$$L_{\nu_k} U \underset{k}{\longrightarrow} \int_{-\pi}^{\pi} U(e^{i\theta}) P_{\nu}(\theta)\, d\theta .$$

At the same time, by w* convergence of the *functionals* L_{ν_k} to L,

$$L_{\nu_k} U \underset{k}{\longrightarrow} LU = \int_{-\pi}^{\pi} U(e^{i\theta}) F(\theta)\, d\theta .$$

In fine,

$$\int_{-\pi}^{\pi} U(e^{i\theta}) P_{\nu}(\theta)\, d\theta = \int_{-\pi}^{\pi} U(e^{i\theta}) F(\theta)\, d\theta$$

whenever $U(e^{i\theta})$ is of the form $\cos n\theta$ or $\sin n\theta$ for $n = 1, 2, 3, \ldots$. It follows that

$$F(\theta) - P_{\nu}(\theta) = \text{const.}$$

Here, $\|\nu\|^* \le M|||L||| \le MA\,\|F\|_*$ where A is a numerical constant and M is any number larger than K, the numerical constant furnished by the theorem of Fefferman and Stein given in Subsection D.2 of Chapter VIII.

The theorem is completely proved.

Remark Together with the theorem of Subsection 1, we see that BMO is *identical* with the set of transforms $P_{\nu}(\theta)$ for *radially bounded* ν, and that the norms $\|P_{\nu}\|_*$ and $\|\nu\|^*$ *are equivalent, if two functions in BMO which differ by a constant are considered to be the same.*

Problem 10 Let $\Phi(z)$ be $\mathscr{C}_\infty$ in $\{|z| < 1\}$, *not necessarily harmonic there*, and suppose that the radial boundary value $\Phi(e^{i\theta})$ exists a.e., and that $\int_{-\pi}^{\pi} |\Phi(e^{i\theta}) - \Phi(\rho e^{i\theta})|\, d\theta \to 0$ as $\rho \to 1$.

If $|\nabla\Phi|\, dx\,dy = \sqrt{\Phi_x^2 + \Phi_y^2}\, dx\,dy$ is a *Carleson measure* on $\{|z| < 1\}$, *show that* $\Phi(e^{i\theta})$ *is BMO*.

HINT With $F \in H_1(0)$, $R < 1$, $F(Rz) = u(z) + iv(z)$ (u, v real) and $\Phi_\rho(z) = \Phi(\rho z)$, $\rho < 1$, first show that

$$\int_{-\pi}^{\pi} u(e^{i\theta}) \Phi_\rho(e^{i\theta})\, d\theta = \iint_{|z|<1} \frac{1}{r}\left(u\frac{\partial \Phi_\rho}{\partial r} - \frac{v}{r}\frac{\partial \Phi_\rho}{\partial \theta} \right) dx\,dy .$$

XI

Wolff's Proof of the Corona Theorem

A. Homomorphisms of H_∞ and maximal ideals

H_∞ is *actually a Banach algebra over* $\mathbb{C}$, because if f and $g \in H_\infty$, $fg \in H_\infty$ and $\|fg\|_\infty \le \|f\|_\infty \|g\|_\infty$. Because H_∞ has this multiplicative structure, it is natural to consider the *algebraic multiplicative homomorphisms* of H_∞ onto $\mathbb{C}$.

Let $L : H_\infty \mapsto \mathbb{C}$ be such a homomorphism. Since H_∞ contains the multiplicative identity 1, we must have $L(1) = 1$. If $f \in H_\infty$ and λ is any complex number of modulus $> \|f\|_\infty$, the function $(\lambda - f(z))^{-1}$ belongs to H_∞, so, since $(\lambda - f(z))^{-1}(\lambda - f(z)) = 1$, taking L of both sides shows that $L(\lambda - f)$ *can't be zero*. Letting λ range over *all* complex numbers of modulus $> \|f\|_\infty$, we see that $|L(f)| \le \|f\|_\infty$; an *algebraic* multiplicative homomorphism L of H_∞ onto $\mathbb{C}$ is *necessarily continuous*, and of *norm* ≤ 1 as a *linear functional* on the Banach *space* H_∞. It is in fact of norm *equal* to 1 because $L(1) = 1$. The set of such L is obviously a w* closed subset of the *unit sphere* in H_∞'s *dual*; as such it is w* *compact*.

For a multiplicative homomorphism L of H_∞ onto $\mathbb{C}$, the set $\mathfrak{m}$ of elements of H_∞ taken onto 0 by L is a *maximal* (proper) *ideal* in H_∞ because $\mathbb{C}$ is a *field*. So to every *homomorphism* corresponds a *maximal ideal. Conversely, to every maximal ideal corresponds a homomorphism of* H_∞ *onto* $\mathbb{C}$.

Indeed, if $\mathfrak{m}$ is a *proper* ideal in H_∞, so is $\bar{\mathfrak{m}}$, its *norm closure*. For if $f \in H_\infty$ and $\|1 - f\|_\infty < 1$, then $f^{-1} \in H_\infty$, so $\mathfrak{m}$ cannot *contain* f *without* $1 = f^{-1}f$ *also* being in $\mathfrak{m}$. So, $\mathfrak{m}$ being *proper* makes $\|1 - \mathfrak{m}\|_\infty \ge 1$, and $1 \notin \bar{\mathfrak{m}}$. From this it is manifest that if $\mathfrak{m}$ is a (proper) *maximal ideal*, we must *already have* $\mathfrak{m} = \bar{\mathfrak{m}}$.

Take any maximal ideal $\mathfrak{m}$. Since it is norm-closed, the quotient ring $H_\infty/\mathfrak{m}$ is a (*complete*!) Banach algebra over $\mathbb{C}$. *It is a field* because $\mathfrak{m}$ is *maximal*. But now a celebrated theorem of Gelfand tells us that *the only complete normed field over* $\mathbb{C}$ *is* $\mathbb{C}$ *itself*! So $H_\infty/\mathfrak{m}$ is indeed isomorphic to $\mathbb{C}$, and the canonical homomorphism L of H_∞ onto $H_\infty/\mathfrak{m}$ is *in fact* one of H_∞ onto $\mathbb{C}$; we can define $L(f)$ as the unique complex number λ for which $\lambda - f \in \mathfrak{m}$ – there *must be one* because $H_\infty/\mathfrak{m}$ is isomorphic to $\mathbb{C}$.

In this way the set of multiplicative homomorphisms L of H_∞ onto $\mathbb{C}$ is in natural one-to-one correspondence with the set of maximal ideals $\mathfrak{m}$ in H_∞.

> *If* $\mathfrak{m}$ *is such a maximal ideal and* L *is the multiplicative homomorphism of* H_∞ *corresponding to it, it is customary to write*
>
> $$h(\mathfrak{m}) \text{ for } L(h)$$
>
> *when* $h \in H_\infty$. *This we do henceforth.*

The set of maximal ideals $\mathfrak{m}$ is denoted by $\mathfrak{M}$. We take for $\mathfrak{M}$ the topology of *pointwise convergence* of maximal ideals $\mathfrak{m}$ (*as multiplicative homomorphisms*) over H_∞ – $\mathfrak{M}$ is then *compact* for the reasons stated at the beginning of this section.

It now becomes natural to look at H_∞ as a *Banach algebra of functions* on its set $\mathfrak{M}$ of *maximal ideals, associating* to each $f \in H_\infty$ the *function* $f(\mathfrak{m})$, $\mathfrak{m} \in \mathfrak{M}$. $\mathfrak{M}$ is called the *maximal ideal space of* H_∞. This approach, a rather abstract one, has proven quite fruitful; it is in fact the main point of view adopted in books like Gamelin's. One complication is, however, that the space $\mathfrak{M}$ is very large – so vast is it, in fact, that it has many bizarre properties.

There is, however, a simple subset of $\mathfrak{M}$ ready at hand. If $|z| < 1$, the *point evaluation*

$$f \mapsto f(z)$$

is a homomorphism of H_∞ onto $\mathbb{C}$! So each point z in the open unit circle corresponds in obvious fashion to a certain *maximal ideal*, namely the ideal of functions $f \in H_\infty$ *vanishing at* z. *We denote that maximal ideal by* z, *also*. If we *do* this, then we can consider the open unit circle $\{|z| < 1\}$ as a *subset* of $\mathfrak{M}$.

The natural question *now arises*:

Is $\{|z| < 1\}$ w* *dense in* $\mathfrak{M}$? If the answer is *yes*, there is some *hope* of being able to arrive at a *more concrete description* of the very complicated space $\mathfrak{M}$. The conjecture that the response *is* positive was known as the *corona conjecture*. To *prove* it or *disprove* it was the celebrated corona problem.

Carleson solved the corona problem in 1962, in the *positive sense*. His proof of what has come to be known as the *corona theorem* was based on an intricate geometrical construction, of combinatorial character. The construction itself has proved useful for the study of *other problems*, especially in the hands of Garnett and his students. It is, however, so difficult to master as to discourage many from attempting to go through the details in the proof of the corona theorem.

This state of affairs has been changed by the work of T. Wolff, done in the spring of 1979. That is why we are able to present a complete proof of the corona theorem in this chapter.

The corona theorem has two equivalent formulations:

(1) If $\mathfrak{m} \in \mathfrak{M}$ there is a *net* $\{z_\alpha\}$, $|z_\alpha| < 1$, with $z_\alpha \xrightarrow[\alpha]{} \mathfrak{m}$ in $\mathfrak{M}$.

(2) If $f_1, \ldots, f_n \in H_\infty$ and

$$\sup_k |f_k(z)| \geq \text{ some } \delta > 0$$

for *all* z, $|z| < 1$, there exist functions $g_1, \ldots, g_n \in H_\infty$ such that $f_1 g_1 + f_2 g_2 + \ldots + f_n g_n \equiv 1$ on $\{|z| < 1\}$.

Let us prove the equivalence.

If (2) is *true*, let $\mathfrak{m} \in \mathfrak{M}$ and suppose there is *no* net of z, $|z| < 1$, which tends w* to $\mathfrak{m}$. By the definition of the w* topology, there exist $h_1, \ldots, h_n \in H_\infty$ and a $\delta > 0$ such that for *each* z, $|z| < 1$, *at least one of the inequalities*

$$|h_k(z) - h_k(\mathfrak{m})| \geq \delta, \quad k = 1, \ldots, n$$

must hold. Call $f_k(z) = h_k(z) - h_k(\mathfrak{m})$; $f_k \in H_\infty$. Then all the $f_k(\mathfrak{m})$ are *zero*, but for each z, $|z| < 1$, some $|f_k(z)|$ is $\geq \delta > 0$. Thence, by (2), we get $g_1, \ldots, g_n \in H_\infty$ with $g_1 f_1 + \ldots + g_n f_n \equiv 1$. So $g_1(\mathfrak{m}) f_1(\mathfrak{m}) + \ldots + g_n(\mathfrak{m}) f_n(\mathfrak{m})$ must $= 1$! But each $f_k(\mathfrak{m}) = 0$, a contradiction.

Suppose now that (1) is *true*, and let $f_1, \ldots, f_n$ satisfy the hypothesis of (2). If there are *no* $g_1, \ldots, g_n \in H_\infty$ with

$$g_1 f_1 + \ldots + g_n f_n \equiv 1,$$

the set of all sums $g_1 f_1 + \ldots + g_n f_n$ constitutes a *proper ideal* in H_∞. Since $1 \in H_\infty$, this proper ideal *must be contained in some maximal ideal*, say $\mathfrak{m}$, by the usual application of Zorn's lemma. *Then* surely $f_k(\mathfrak{m}) = 0$ for each k. By the truth of (1), we now get a net $\{z_\alpha\}$ of points, $|z_\alpha| < 1$, with $f_k(z_\alpha) \xrightarrow[\alpha]{} f_k(\mathfrak{m}) = 0$ for $k = 1, \ldots, n$. The hypothesis of (2) is *now contradicted*, so there must in fact *be* $g_1, \ldots, g_n \in H_\infty$ with $g_1 f_1 + \ldots + g_n f_n \equiv 1$.

Knowing that the corona theorem is true, mathematicians, especially Hoffman, *have indeed* been able to obtain a fairly complete description of $\mathfrak{M}$, by getting at the $\mathfrak{m} \in \mathfrak{M}$ with nets of points from $\{|z| < 1\}$. In this investigation the *interpolating sequences* studied in Chapter IX turn out to be of special importance.

B. The $\bar{\partial}$-equation

Wolff's proof of the corona theorem makes systematic use of the two differential operators

$$\partial = \frac{1}{2}\left(\frac{\partial}{\partial x} - i\frac{\partial}{\partial y}\right), \qquad \bar{\partial} = \frac{1}{2}\left(\frac{\partial}{\partial x} + i\frac{\partial}{\partial y}\right).$$

A $\mathscr{C}_\infty$ function $f(z)$ is *analytic* if and only if $\bar{\partial} f \equiv 0$, and then $f'(z) = \partial f(z)$. The *Laplacian* ∇^2 is equal to $4\partial\bar{\partial}$.

Let $g(x)$ be $\mathscr{C}_\infty$ and *of compact support*. Can we *solve the equation*

$$\bar{\partial} f(z) = g(z)$$

for $|z| < 2$, say, with a $\mathscr{C}_\infty$ *function* f (*not* necessarily of compact support)? *We can – one* solution is

$$f(z) = \frac{1}{\pi} \iint_{\mathbf{C}} \frac{g(\zeta)\, d\xi\, d\eta}{z - \zeta},$$

where $\zeta = \xi + i\eta$.

To see this, let us make the change of variable $z - \zeta = w = u + iv$, so that the above double integral goes over into

$$\frac{1}{\pi} \iint_{\mathbf{C}} \frac{g(z - w)}{w}\, du\, dv\,.$$

Since g is of *compact support*, we can, if we restrict ourselves to $|z| < 2$, rewrite the above integral as

$$\frac{1}{\pi} \iint_{|w|<R} g(z - w) \frac{du\, dv}{w}$$

with some very large R. Here g is $\mathscr{C}_\infty$ and

$$\iint_{|w|<R} \frac{du\, dv}{|w|} < \infty$$

so the previous expression – call it $f(z)$ – can be differentiated under the integral sign, yielding

$$\bar{\partial} f(z) = \frac{1}{\pi} \iint_{|w|<R} \bar{\partial} g(z - w) \frac{du\, dv}{w}.$$

I claim that the integral on the right equals $g(z)$. Without loss of generality, take $z = 0$, then we have to evaluate

$$\lim_{\rho \to 0} \frac{1}{\pi} \iint_{\rho<|w|<R} \bar{\partial} g(-w) \frac{du\, dv}{w}.$$

Here, we can replace $\bar{\partial}$ by $-\frac{1}{2}\left(\frac{\partial}{\partial u} + i\frac{\partial}{\partial v}\right)$. *Then*, since

$$\left(\frac{\partial}{\partial u} + i\frac{\partial}{\partial v}\right)\left(\frac{1}{w}\right) \equiv 0 \quad \text{for } w \neq 0,$$

the previous expression becomes

$$\frac{i}{2\pi}\lim_{\rho\to 0}\iint_{\rho<|w|<R}\left(\frac{\partial}{\partial u}\left(i\frac{g(-w)}{w}\right)-\frac{\partial}{\partial v}\left(\frac{g(-w)}{w}\right)\right)du\,dv.$$

By Green's theorem, this expression is just

$$-\frac{i}{2\pi}\lim_{\rho\to 0}\int_{|w|=\rho} g(-w)\frac{du+i\,dv}{w}+\frac{i}{2\pi}\int_{|w|=R} g(-w)\frac{du+i\,dv}{w}.$$

Now, since R is very large, and g of compact support, the line integral around $|w|=R$ is *zero*. On putting $w=\rho e^{i\phi}$, the one around $|w|=\rho$ is seen to be

$$\frac{1}{2\pi}\int_0^{2\pi} g(-\rho e^{i\phi})\,d\phi,$$

and as $\rho\to 0$, this just tends to $g(0)$. So our f *does* satisfy $\bar\partial f(0)=g(0)$, and in the same way we get $\bar\partial f(z)=g(z)$. The f given by our double integral is evidently $\mathscr{C}_\infty$, since g is.

In what follows we will be interested in solutions f of $\bar\partial f(z)=g(z)$, valid in some *slightly larger circle* than the *unit one*, for which $\sup_\theta|f(e^{i\theta})|$ *is not too large*. For this we have the

Lemma (Wolff) *Let* $h(z)$ *be* $\mathscr{C}_\infty$ *in the circle* $\{|z|<R\}$, *where* $R>1$. *Suppose that, in* $\{|z|<1\}$,

$$\left(|z|\log\frac{1}{|z|}\right)|h(z)|^2\,dx\,dy$$

and

$$\left(|z|\log\frac{1}{|z|}\right)|\partial h(z)|\,dx\,dy$$

are Carleson measures, *with 'Carleson constants' A and B respectively. Then we can find a function* $v(z)$, $\mathscr{C}_\infty$ *in some circle* $\{|z|<R'\}$ *with* $R'>1$, *such that*

$$\bar\partial v(z)=h(z),\quad |z|<1,$$

whilst $|v(e^{i\theta})|\le 9(\sqrt{A}+B)$.

Note For the definition of *Carleson measures* and their *properties*, see Chapter VIII, Section F.

Proof of lemma (As simplified by Varopoulos and Garnett) We may first suppose $h(z)$ to be *redefined* for $|z|>(1+R)/2$, say, so as to make it $\mathscr{C}_\infty$ in *all* of $\mathbb{C}$ and of *compact support*. Then the formula described above *certainly gives* us a $\mathscr{C}_\infty$ function v_0 satisfying

$$\bar\partial v_0(z)=h(z).$$

The trouble is, of course, that $|v_0(e^{i\theta})|$ may get quite large.

If $f(z)$ is *analytic* in some circle $\{|z|<R'\}$ with $R'>1$, we of course *also* have

$$\bar\partial v=h$$

with $v = v_0 - f$ in that circle, because $\bar{\partial} f \equiv 0$. The idea is to choose f so that $|v(e^{i\theta})|$ does not get too big.

In the notation of Chapter VII, Section A, $v_0(e^{i\theta})$ certainly belongs to $\mathscr{C}$. If there is an $f \in \mathscr{A}$ $(= H_\infty \cap \mathscr{C})$ such that $\|v_0 - f\|_\infty = \sup_\theta |v_0(e^{i\theta}) - f(e^{i\theta})| < d$ say, then, if $r < 1$ is *sufficiently close* to 1, $\|v_0 - f_r\|_\infty$ is *still* $< d$ with $f_r(z) = f(rz)$, and f_r *is* $\mathscr{C}_\infty$ in $\{|z| < 1/r\}$. So what we *want* to find is $\|v_0 - \mathscr{A}\|_\infty$; for any d *larger* than $\|v_0 - \mathscr{A}\|_\infty$ we will *have* a solution v, fulfilling the desired conditions and such that $|v(e^{i\theta})| \leq d$.

By the duality theory of Chapter VII, Section A,

$$\|v_0 - \mathscr{A}\|_\infty = \sup\left\{\left|\int_0^{2\pi} v_0(e^{i\theta})F(e^{i\theta})\,d\theta\right| \;;\; F \in H_1(0) \text{ and } \|F\|_1 \leq 1\right\}.$$

We may clearly *restrict* the set of $F \in H_1(0)$ over which the above sup is taken to those having an *analytic continuation* into some larger circle $\{|z| < R_F\}$, with $R_F > 1$ depending on F.

For such F, we can apply Green's theorem as in Chapter X, Subsection D.1, getting

$$\int_0^{2\pi} v_0(e^{i\theta})F(e^{i\theta})\,d\theta = \iint_{|z|<1} \left(\log\frac{1}{|z|}\right) \nabla^2(v_0(z)F(z))\,dx\,dy,$$

because $v_0(z)$ is, at any rate, $\mathscr{C}_\infty$ in the unit circle and up to its boundary. We have $\nabla^2(v_0F) = 4\partial\bar{\partial}(v_0F) = 4\bar{\partial}v_0\partial F + 4F\partial\bar{\partial}v_0$, because $\bar{\partial}F$ and $\partial\bar{\partial}F$ vanish identically. *Also*, we are supposed to have $\bar{\partial}v_0 = h$, so that $\partial\bar{\partial}v_0 = \partial h$. Therefore $\nabla^2(v_0F) = 4hF' + 4F\partial h$, and the above double integral breaks down to

$$4\iint_{|z|<1} F(z)\partial h(z)\log\frac{1}{|z|}\,dx\,dy + 4\iint_{|z|<1} F'(z)h(z)\log\frac{1}{|z|}\,dx\,dy.$$

Of these two terms, the *first* is in absolute value

$$\leq 4\iint_{|z|<1} \left|\frac{F(z)}{z}\right| |\partial h(z)||z|\log\frac{1}{|z|}\,dx\,dy.$$

Here, since $F \in H_1(0)$, $F(z)/z$ belongs to H_1 and has the same H_1-norm as F. So, by the hypothesis, and the definition of Carleson measures, our first term is in absolute value $\leq 4B\,\|F\|_1$.

Regarding the *second* term, we apply Schwarz' inequality just as in Chapter X, Subsection E.1, and see that the second term is in modulus

$$\leq 4\sqrt{\iint_{|z|<1}\left(\log\frac{1}{|z|}\right)\frac{|F'(z)|^2}{|F(z)|}\,dx\,dy \iint_{|z|<1}\left|\frac{F(z)}{z}\right| |h(z)|^2|z|\log\frac{1}{|z|}\,dx\,dy}.$$

As in Chapter X, Subsection D.3, we see that it is sufficient to estimate our second term for $F \in H_1(0)$ having only a *simple zero* at the origin, and *no others*, since *any* $F \in H_1(0)$ can be written as the sum of *two such, each* with norm no bigger than its *own*. But for *such* F,

$$\iint_{|z|<1} \left(\log\frac{1}{|z|}\right) \frac{|F'(z)|^2}{|F(z)|}\,dx\,dy \;=\; \|F\|_1$$

by a theorem of Chapter X, Subsection D.2. *Again,*

$$\iint_{|z|<1} \left|\frac{F(z)}{z}\right| |h(z)|^2 |z| \log\frac{1}{|z|}\,dx\,dy \;\le\; A\,\|F\|_1$$

for $F \in H_1(0)$, by the *hypothesis*.

We see that, for *general* $F \in H_1(0)$, analytic over the *closed* circle $\{|z| \le 1\}$,

$$\left|4\iint_{|z|<1} F'(z)h(z)\log\frac{1}{|z|}\,dx\,dy\right| \;\le\; 8\sqrt{A}\,\|F\|_1$$

(with 8 on the right and not 4 because of the restriction to the kind of F for which Schwarz' inequality was applied).

Putting the above two estimates together we see that

$$\left|\int_0^{2\pi} v_0(e^{i\theta})F(e^{i\theta})\,d\theta\right| \;\le\; 8(\sqrt{A}+B)\,\|F\|_1$$

for $F \in H_1(0)$ which are analytic over the *closed* unit circle. Therefore $\|v_0 - \mathscr{A}\|_\infty \le 8(\sqrt{A}+B)$, and the lemma is proved.

C. Proof of the corona theorem

Theorem (Carleson, 1962) *Let $f_1,\ldots,f_n \in H_\infty$; suppose $\|f_k\|_\infty \le 1$ for $k = 1,\ldots,n$ and that for some $\delta > 0$ we have*

$$\sup_k |f_k(z)| > \delta$$

for all z, $|z| < 1$. *There is a number $M(\delta, n)$ depending* only *on δ and n such that*

$$g_1f_1 + g_2f_2 + \ldots + g_nf_n \equiv 1 \quad \text{on } \{|z| < 1\}$$

with some functions $g_k \in H_\infty$ satisfying

$$\|g_k\|_\infty \le M(\delta, n).$$

Proof (Wolff, 1979) The main *analytical* idea is contained in the case $n = 2$, which we proceed to treat first.

It is sufficient to prove the theorem with functions analytic in some *slightly larger circle* than the *unit* one standing in place of the f_k. Indeed, once that is done, it will apply to the functions $f_k^{(r)}(z) = f_k(rz)$ where $r < 1$, giving us some $g_k^{(r)} \in H_\infty$ with $g_1^{(r)}f_1^{(r)}+\ldots+g_n^{(r)}f_n^{(r)} \equiv 1$ in $|z| < 1$. The bounds on the $\|g_k^{(r)}\|_\infty$ furnished by the theorem *do not depend on* r. Therefore, on making $r \to 1$, a normal family argument will give us $g_k \in H_\infty$ (with the $\|g_k\|_\infty$ admitting the *same* bounds) such that

$$g_1 f_1 + \ldots + g_n f_n = 1.$$

Here, $n = 2$, so that we have f_1 and f_2 with $\|f_1\|_\infty \le 1$, $\|f_2\|_\infty \le 1$, and, for every z, $|z| < 1$,

$$|f_1(z)| > \delta \ \text{ or } \ |f_2(z)| > \delta .$$

Let $U(w)$ be a $\mathscr{C}_\infty$ function, depending only on $|w|$, with

$$U(w) \equiv 0 \ \text{ for } \ |w| \le \delta/2,$$
$$U(w) \equiv 1 \ \text{ for } \ |w| \ge \delta,$$

and $0 \le U(w) \le 1$ elsewhere. By the hypothesis, $U(f_1(z)) + U(f_2(z)) \ge 1$ for $|z| < 1$. Put

$$\phi_k(z) \ = \ \frac{U(f_k(z))}{U(f_1(z)) + U(f_2(z))}$$

for $k = 1, 2$. The functions $\phi_k(z)$ are clearly $\mathscr{C}_\infty$ on some circle $\{|z| > R\}$ with $R > 1$, and $\phi_1(z) + \phi_2(z) \equiv 1$; in particular, $\phi_2(z) = 1$ wherever $\phi_1(z) = 0$ and vice versa. Note also that each $\phi_k(z)$ is *zero* on the set where $|f_k(z)| < \delta/2$.

Now

$$\frac{\phi_1}{f_1} \cdot f_1 \ + \ \frac{\phi_2}{f_2} \cdot f_2 \ \equiv \ 1.$$

The trouble is, of course, that ϕ_1/f_1 and ϕ_2/f_2 are *not analytic*! Here, using an idea that goes back to Hörmander, we look for some new function $v(z)$ which will *make*

$$g_1 = \frac{\phi_1}{f_1} + v f_2 \ \text{ and } \ g_2 = \frac{\phi_2}{f_2} - v f_1$$

analytic in $|z| < 1$. *Any* such v will *automatically* give us

$$g_1 f_1 \ + \ g_2 f_2 \ \equiv \ 1$$

in $|z| < 1$.

For analyticity of g_1 and g_2 we need $\bar\partial g_1 = \bar\partial g_2 = 0$ in $|z| < 1$. Since $\bar\partial f_1 = \bar\partial f_2 = 0$, we get the conditions

$$\frac{\bar\partial \phi_1}{f_1} + f_2 \bar\partial v = 0, \qquad \frac{\bar\partial \phi_2}{f_2} - f_1 \bar\partial v = 0.$$

We have $\phi_1 + \phi_2 \equiv 1$, so $\bar\partial \phi_1 + \bar\partial \phi_2 \equiv 0$, and therefore the two conditions are *compatible*, and equivalent to the single one

$$\bar\partial v \ = \ \frac{\bar\partial \phi_2}{f_1 f_2}.$$

Observe that on the open set where $|f_1(z)| < \delta/2$, $\phi_2(z) \equiv 1$ so $\bar\partial \phi_2(z) \equiv 0$; on the open set where $|f_2(z)| < \delta/2$, $\phi_2(z) \equiv 0$ so $\bar\partial \phi_2(z) \equiv 0$. Therefore

$$\left| \frac{\bar\partial \phi_2}{f_1 f_2} \right| \ \le \ \frac{4}{\delta^2} |\bar\partial \phi_2|$$

on $\{|z| < 1\}$, and

$$h(z) = \frac{\bar{\partial}\phi_2(z)}{f_1(z)f_2(z)}$$

is a nice $\mathscr{C}_\infty$ function defined on some circle $\{|z| < R\}$ with $R > 1$.

Now apply the lemma of Section B! We are looking for solutions v to $\bar{\partial}v = h$ on some circle *slightly larger* than the *unit* one; as we have just seen, for $|z| < 1$,

$$|h(z)| \le \frac{4}{\delta^2}|\bar{\partial}\phi_2(z)| = \frac{4}{\delta^2}\frac{|U(f_1(z))\,\bar{\partial}U(f_2(z)) - U(f_2(z))\,\bar{\partial}U(f_1(z))|}{|U(f_1(z)) + U(f_2(z))|^2}.$$

Since $U(f_1(z)) + U(f_2(z)) \ge 1$, this last expression is $\le K_\delta\left(|f_1'(z)| + |f_2'(z)|\right)$ by the chain rule, where $K_\delta = 4\delta^{-2}\sup_w|\text{grad } U(w)|$ *depends only on* δ. Therefore

$$|h(z)|^2|z|\log\frac{1}{|z|} \le 2K_\delta^2\left(|f_1'(z)|^2 + |f_2'(z)|^2\right)|z|\log\frac{1}{|z|}.$$

But $f_1, f_2 \in H_\infty$ are, in particular, *harmonic* in $\{|z| < 1\}$, *and* $|f_1(z)| < 1$, $|f_2(z)| < 1$ *there.*

The lemmas in Subsections E.2, E.3 of Chapter X *now show that*

$$\left(|f_1'(z)|^2 + |f_2'(z)|^2\right)|z|\log\frac{1}{|z|}\,dx\,dy$$

is a Carleson measure whose Carleson constant can be taken $\le$ *some pure number.*

Therefore

$$|h(z)|^2|z|\log\frac{1}{|z|}\,dx\,dy$$

is a Carleson measure whose Carleson constant, A_δ, can be taken to depend *only* on δ. (It is remarkable that the above boxed statement can be *proved directly, without appealing* to Chapter X, Section E or the theorem on Carleson measures! Gamelin and Davie first noticed this – see exercise at the end of this Chapter.)

Now let us look at

$$\partial h = \frac{\partial\bar{\partial}\phi_2}{f_1f_2} - \frac{\bar{\partial}\phi_2}{f_1f_2}\left(\frac{f_1'}{f_1} + \frac{f_2'}{f_2}\right).$$

Of the two terms on the right, the *second* vanishes identically on the open set where $|f_1|$ or $|f_2|$ is $< \delta/2$, and on the *complement* of that set it is in absolute value

$$\le 8\delta^{-3}\sup_w|\text{grad } U(w)|\left(|f_1'(z)| + |f_2'(z)|\right)^2.$$

The *first* term also vanishes identically on the open set just mentioned, and on its *complement* equals

$$\frac{1}{4f_1(z)f_2(z)}\nabla^2\left\{\frac{U(f_2(z))}{U(f_1(z)) + U(f_2(z))}\right\}.$$

But $\nabla^2 f_1(z) = \nabla^2 f_2(z) \equiv 0$, so the expression just written *only involves* $f_1'(z)$ and $f_2'(z)$ and is *clearly* in modulus $\leq C_\delta\,(|f_1'(z)|^2 + |f_2'(z)|^2)$, with C_δ depending only on δ. We see in this way that

$$|\partial h(z)||z|\log\frac{1}{|z|} \;\leq\; L_\delta\,(|f_1'(z)|^2 + |f_2'(z)|^2)\,|z|\log\frac{1}{|z|}\,.$$

By Subsections E.2, E.3 of Chapter X, *this makes*

$$|\partial h(z)||z|\log\frac{1}{|z|}\,dx\,dy$$

a Carleson measure with Carleson constant, B_δ, *depending only on* δ *since* $|f_1(z)| \leq 1$, $|f_2(z)| \leq 1$ *in* $\{|z| < 1\}$.

The lemma thus gives us a $v(z)$, $\mathscr{C}_\infty$ in some circle slightly larger than the unit one, with $\bar{\partial}v = h$ in $\{|z| < 1\}$ and $|v(e^{i\theta})| \leq 9(\sqrt{A_\delta} + B_\delta)$. The functions

$$g_1 = \frac{\phi_1}{f_1} + vf_2 \;\text{ and }\; g_2 = \frac{\phi_2}{f_2} - vf_1$$

will be in H_∞ – even in $\mathscr{A}$ – and satisfy $g_1f_1 + g_2f_2 \equiv 1$ in $\{|z| < 1\}$. Clearly, for $|z| \leq 1$,

$$|\phi_1(z)/f_1(z)| \leq \frac{2}{\delta}, \qquad |\phi_2(z)/f_2(z)| \leq \frac{2}{\delta},$$

so for $k = 1, 2$,

$$|g_k(e^{i\theta})| \;\leq\; \frac{2}{\delta} + 9(\sqrt{A_\delta} + B_\delta),$$

i.e.,

$$\|g_k\|_\infty \;\leq\; \frac{2}{\delta} + 9(\sqrt{A_\delta} + B_\delta),$$

proving the theorem for $n = 2$.

For $n > 2$, the situation is more complicated *algebraically*. Given $f_1, \ldots, f_n$ with $\|f_k\|_\infty \leq 1$ and $\sup_k |f_k(z)| > \delta$ for all z in the unit circle, we take the function $U(w)$ used above and put, for $k = 1, 2, \ldots, n$,

$$\phi_k(z) \;=\; \frac{U(f_k(z))}{U(f_1(z)) + U(f_2(z)) + \ldots + U(f_n(z))}\,.$$

Each ϕ_k *vanishes identically* on the set where $|f_k| < \delta/2$, and

$$\sum_1^n \phi_k(z) \equiv 1$$

for $|z| < 1$.

Assuming, as we may, each $f_k(z)$ to be analytic over the *closed* unit circle, we search for analytic functions g_k of the form

$$g_k \;=\; \frac{\phi_k}{f_k} + \sum_j v_{kj}f_j$$

with certain, as yet unknown, functions v_{kj} which we require to satisfy $v_{kj} = -v_{jk}$, $v_{kk} \equiv 0$, so as to automatically have

$$g_1 f_1 + \ldots + g_n f_n \equiv 1.$$

For analyticity of the g_k, we need $\bar{\partial} g_k \equiv 0$, which means that the v_{kj} must satisfy

$$\frac{\bar{\partial}\phi_k}{f_k} + \sum_j f_j \bar{\partial} v_{kj} = 0.$$

This *holds* if, for instance,

$$\bar{\partial} v_{kj} = \frac{\phi_k}{f_k f_j}\bar{\partial}\phi_j - \frac{\phi_j}{f_j f_k}\bar{\partial}\phi_k,$$

as may be verified directly using the relations $\phi_1 + \ldots + \phi_n \equiv 1$, $\bar{\partial}\phi_1 + \ldots + \bar{\partial}\phi_n \equiv 0$. *Here*, we *first solve each* of the equations

$$\bar{\partial} w_{kj} = \frac{\phi_k}{f_k f_j}\bar{\partial}\phi_j,$$

and *then* put $v_{kj} = w_{kj} - w_{jk}$, so that v_{kj} will *equal* $-v_{jk}$ *automatically*. The *lemma, applied in the same way as for the case* $n = 2$, shows us that we can get solutions $w_{kj}(z)$, $\mathscr{C}_\infty$ on a circle slightly larger than the unit one, and satisfying $|w_{kj}(e^{i\theta})| \le M_\delta$, a number depending *only* on δ. The (analytic) g_k obtained from these w_{kj} via the v_{kj} will satisfy

$$\|g_k\|_\infty \le \frac{2}{\delta} + 2(n-1)M_\delta,$$

and $g_1 f_1 + g_2 f_2 + \ldots + g_n f_n \equiv 1$ on $\{|z| < 1\}$.

The corona theorem is completely proved!

Exercise Let $f \in H_\infty$. Show *directly, without appealing* to the lemmas of Chapter X, Subsections E.2, E.3 or the theorem at the end of Chapter VIII, Section F, that

$$|z| \log \frac{1}{|z|} |f'(z)|^2 \, dx\, dy$$

is a Carleson measure.

Hint: One may suppose that $f(z)$ is analytic in a slightly larger circle than $\{|z| \le 1\}$, and has no zeros there. If $F \in H_1(0)$ has only a simple zero at the origin and is analytic in a circle slightly larger than $\{|z| \le 1\}$, apply the inequality

$$|(f')^2 F| \le 2|f| \frac{|(fF)'|^2}{|fF|} + 2|f|^2 \frac{|F'|^2}{|F|}$$

(idea of A.M. Davie).

Appendix I
by V.P. Havin
Jones' Interpolation Formula

1. P. Jones has proposed a surprisingly simple and direct proof of Carleson's interpolation theorem (Chapter IX, Subsection B.3). We present this proof here, having in mind the second variant of Carleson's theorem (i.e., the one for the disk and not the half plane; see the end of Subsection B.3, Chapter IX).

Let $\{z_n\}_{n=1}^{\infty}$ be a sequence of distinct points of the disk $\mathbb{D} = \{|z| < 1\}$, with

$$|z_1| \le |z_2| \le \dots \qquad \text{and} \qquad \sum_n (1 - |z_n|) < \infty.$$

We shall need the Blaschke product

$$B = \prod_{n=1}^{\infty} b_n$$

where $b_n(z) = (|z_n|/z_n)(z_n - z)/(1 - \bar{z}_n z)$ for $z_n \neq 0$ $(b_1(z) = z$ if $z_1 = 0)$, and the products

$$B_n = B/b_n.$$

Suppose that

$$\delta \overset{\text{def}}{=} \inf_n |B_n(z_n)| > 0. \tag{1}$$

Carleson's theorem asserts that under condition (1) there is, for any bounded sequence $w = \{w_n\}$ of complex numbers, a function $f \in H_\infty$ such that

$$f(z_n) = w_n, \quad n = 1, 2, \dots. \tag{2}$$

Let us suppose that we have succeeded in constructing functions Φ_n, $n = 1, 2, \dots$, analytic in the disc $\mathbb{D}$ and having the following properties:

(a) $$\Phi_n(z_k) = \begin{cases} 0, & k \neq n \\ 1, & k = n \end{cases} \qquad k, n = 1, 2, \dots ;$$

(b) $$S(z) \overset{\text{def}}{=} \sum_{n=1}^{\infty} |\Phi_n(z)| \le K < \infty$$

for all z in $\mathbb{D}$.

Then a solution to the interpolation problem (2) is given by the function

$$f(z) \overset{\text{def}}{=} \sum_{n=1}^{\infty} w_n \Phi_n(z). \tag{3}$$

Indeed, the series (3) converges everywhere in $\mathbb{D}$ by (b), and its partial sums are uniformly bounded in $\mathbb{D}$ (by the quantity $K \sup_n |w_n|$). From this it follows easily (from the theorem on normal families, for example) that $f(z)$ is analytic in $\mathbb{D}$ and belongs to H_∞. Equation (2) follows from (a). Moreover, $\|f\|_\infty \le K \sup |w_n|$, which means that the mapping $w \mapsto f$ given by formula (3) defines a linear operator, of norm not exceeding K, taking the space l_∞ of all bounded numerical sequences into the space H_∞ and putting every point $w \in l_\infty$ into correspondence with a solution f to the interpolation problem (2).

The *existence* of a sequence $\{\Phi_n\}$ having properties (a) and (b) was proved (under condition (1)) by Carleson and P. Beurling (see Chapter 7, §2 in Garnett's book). Subsequently, P. Jones pointed out a perfectly elementary procedure for the construction of functions Φ_n (Jones, 1980). Here we derive a simplified variant of *Jones' formula* (i.e., interpolation formula (3) with the Φ_n defined by equation (4) below), following the work of Vinogradov, Gorin and Hruščev (1981). Other variants and applications to interpolation problems are given in an article by Vinogradov (1983). E.A. Gorin has pointed out a modification of the Jones formula useful for interpolation by H_∞ functions continuous up to the unit circumference; regarding multiple interpolation in H_∞ see the work of Martirosian (1981).

Let us put

$$\epsilon = \left(2\log\frac{e}{\delta^2}\right)^{-1}, \qquad \alpha_n(z) = \sum_{k=n}^{\infty} \frac{1+\bar{z}_k z}{1-\bar{z}_k z}(1-|z_k|^2),$$
$$\Phi_n(z) = \left(\frac{1-|z_n|^2}{1-\bar{z}_n z}\right)^2 \frac{B_n(z)}{B_n(z_n)} \exp\left(\epsilon(\alpha_n(z_n)-\alpha_n(z))\right). \tag{4}$$

Here δ is defined by equation (1). The series defining the function α_n converges uniformly in each disk $\{|z| \le r\}$, $r < 1$, because

$$\left|\frac{1+\bar{z}_k z}{1-\bar{z}_k z}\right|(1-|z_k|^2) \le \frac{4}{1-r}(1-|z_k|) \quad \text{for } |z| \le r,\ k = 1, 2, \ldots .$$

It is clear that the functions Φ_n satisfy condition (a). Condition (b) is also fulfilled. Moreover,

$$S(z) \le \frac{2e}{\delta}\log\frac{e}{\delta^2} \quad \text{for } |z| < 1. \tag{5}$$

In order to prove the inequality (5), let us introduce the quantities

$$Z_{k,n} = \frac{(1-|z_n|^2)(1-|z_k|^2)}{|1-\bar{z}_k z_n|^2}.$$

An analogue for the half plane of the following lemma was established in the lemma of Chapter IX, Subsection B.1.

Lemma $\sum_{k=n}^{\infty} Z_{k,n} \le 1/2\epsilon$ *for* $n = 1, 2, \ldots$.

Proof Let us check that

$$1 - |b_k(z_n)|^2 = Z_{k,n}. \tag{6}$$

Indeed:

$$1 - |b_k(z_n)|^2 = 1 - \frac{|z_n - z_k|^2}{|1 - z_n\bar{z}_k|^2} = \frac{(1 - \bar{z}_n z_k)(1 - z_n\bar{z}_k) - (z_n - z_k)(\bar{z}_n - \bar{z}_k)}{|1 - z_n\bar{z}_k|^2}.$$

One need only remove parentheses and bring together the similar terms in the numerator.

Remember now that $t \le -\log(1-t)$ for $t \in (0,1)$. Therefore, by (1) and (6),

$$2\log\frac{1}{\delta} \ge -\log|B_n(z_n)|^2 = -\sum_{k \ne n} \log|b_k(z_n)|^2$$

$$\ge \sum_{k \ne n} (1 - |b_k(z_n)|^2) = \sum_{k \ne n} Z_{k,n} \ge \sum_{k=n+1}^{\infty} Z_{k,n},$$

and the lemma now follows from the definition of the quantity ϵ and the fact that $Z_{n,n} = 1$.

Corollary $\Re\alpha_n(z_n) \le 1/\epsilon$ *for* $n = 1, 2, \ldots$.

Indeed, if $k \ge n$, we have $|z_k| \ge |z_n|$ *and* $1 - |z_n|^2|z_k|^2 \le 1 - |z_n|^4$. Therefore

$$\Re\alpha_n(z_n) = \sum_{k=n}^{\infty} \frac{1 - |z_n|^2|z_k|^2}{|1 - \bar{z}_k z_n|^2}(1 - |z_k|^2)$$

$$\le \sum_{k=n}^{\infty} \frac{1 - |z_n|^4}{|1 - \bar{z}_k z_n|^2}(1 - |z_k|^2) \le 2\sum_{k=n}^{\infty} Z_{k,n} \le \frac{1}{\epsilon}.$$

We now complete the proof of (5). It follows from (1) that $|B_n(z)/B_n(z_n)| \le 1/\delta$ for $n = 1, 2, \ldots$ and $|z| < 1$. The preceding corollary shows that, for $|z| < 1$,

$$|\Phi_n(z)| \le \frac{e}{\delta}\left(\frac{1 - |z_n|^2}{|1 - \bar{z}_n z|}\right)^2 \exp(-\epsilon\Re\alpha_n(z)). \tag{7}$$

Put

$$\gamma_n(z) = \sum_{k=n}^{\infty}\left(\frac{1 - |z_k|^2}{|1 - \bar{z}_k z|}\right)^2,$$

making

$$\gamma_n(z) - \gamma_{n+1}(z) = \left(\frac{1 - |z_n|^2}{|1 - \bar{z}_n z|}\right)^2$$

and $\gamma_n(z) \downarrow_n 0$ for $|z| < 1$. Furthermore,

$$\Re\alpha_n(z) = \sum_{k=n}^{\infty} \frac{1 - |z_k|^2|z|^2}{|1 - \bar{z}_k z|^2}(1 - |z_k|^2) \ge \gamma_n(z),$$

since $1 - |z|^2|z_k|^2 \ge 1 - |z_k|^2$ for $|z| < 1$. Returning to (7), we get (for the sum $S(z)$ defined in the statement of condition (b) – translator)

$$S(z) \le \frac{e}{\delta}\sum_{n=1}^{\infty}(\gamma_n(z)-\gamma_{n+1}(z))\exp(-\epsilon\gamma_n(z))$$

when $|z|<1$. Let us use the elementary inequality $t \le e^t - 1$ with $t = \epsilon(\gamma_n(z)-\gamma_{n+1}(z))$. We get

$$S(z) \le \frac{e}{\delta\epsilon}\sum_{n=1}^{\infty}\left(e^{\epsilon(\gamma_n(z)-\gamma_{n+1}(z))}-1\right)e^{-\epsilon\gamma_n(z)} = \frac{e}{\delta\epsilon}\sum_{n=1}^{\infty}\left(e^{-\epsilon\gamma_{n+1}(z)}-e^{-\epsilon\gamma_n(z)}\right) \le \frac{e}{\delta\epsilon}.$$

Inequality (5) is proved.

2. 'A thought, once uttered, is a lie.' These words of Tiutchev apply to many important theorems in full measure. Confined within the narrow limits of its formulation, a theorem does not tell *all* of the truth about itself and is perhaps only fit for inclusion in a handbook. Its true meaning is inseparable from the proof (or proofs, if there are several of them).

Jones' proof prompts us to try to grasp once more the idea of the 'old' proof of Carleson's interpolation theorem presented in Chapter IX. It might seem that both lead to one and the same result – the description of interpolating sequences. But they actually yield the solutions to *different* problems.

In order to see this, let us consider a *finite* subset E of the disk $\mathbb{D}$. In that case the mere *possibility* of extending a function given on E to one in H_∞ is obvious on the face of it (think, for instance, of Lagrange interpolation). But the *estimation* of such extensions in terms of properties of the set E is another matter. That is what is involved here. Let us approach the question from afar.

We denote by $R_E(f)$ the trace of a function f on the set E (i.e. its restriction to E – translator). Let us agree to denote by $B_m(X)$ the set of *all* functions $\psi : X \to \mathbb{C}$ with $\sup_X |\psi| \le m$ where $m > 0$. It is clear that $R_E(B_1(\mathbb{D})) = B_1(E)$, ($E$ being, as before, a finite subset of the disk $\mathbb{D}$). In other words, *any* function ψ not exceeding 1 in modulus on E can be interpolated on E by a function f in $B_1(\mathbb{D})$; such a function can even be taken to be smooth. The fact that f coincides with ψ on the set E implies nothing about the size of $|f(z)|$ for $z \in \mathbb{D} \sim E$. The matter is completely different when it comes to *analytic* extensions. For example, the maximum modulus principle shows that if E contains more than one point, $R_E(B_1(\mathbb{D}) \cap H_\infty) \neq B_1(E)$ (a function taking the values $+1$ and -1 on E cannot be represented in the form $R_E(f)$ with $f \in B_1(\mathbb{D}) \cap H_\infty$). It follows from the simplest estimates for the derivative of a function in H_∞ that if ψ oscillates strongly on E, the modulus of any of its analytic extensions to $\mathbb{D}$ must assume quite large values. Interpolation on the set E by functions in $H_\infty \cap B_1(\mathbb{D})$ is 'not completely free' – it can only be 'more or less free'.

Let us introduce a measure for this freedom: to each function $\psi\colon E \to \mathbb{C}$ we associate the quantity

$$M(\psi, E) = \inf\{\|f\|_\infty;\ f \in H_\infty,\ R_E(f) = \psi\}$$

and then put

$$\text{Carl}\, E = \left(\sup\{M(\psi, E)\,;\ \ \psi \in B_1(E)\}\right)^{-1}.$$

The quantity Carl E will be referred to as the *Carleson index* of the set E. It is clear that Carl $E < 1$ if E contains more than one point. The closer Carl E is to unity, the 'freer' is interpolation on E by functions in $H_\infty \cap B_1(\mathbb{D})$. Indeed, Carl E is the largest number m for which $R_E(H_\infty \cap B_1(\mathbb{D})) \supseteq B_m(E)$. The condition $\sup_E |\psi| \leq$ Carl E is sufficient for solvability of the equation $\psi = R_E(f)$ for an 'unknown' $f \in H_\infty \cap B_1(\mathbb{D})$; if on the contrary Carl $E < \sup_E |\psi| < 1$, that equation is generally speaking only solvable when ψ fulfills certain special supplementary conditions difficult of survey.

Study of the Carleson indices of sets is of interest in connection with the problem of describing interpolation sequences. It is easy to see that a sequence is interpolating if and only if all of its finite subsets have Carleson indices bounded uniformly away from zero.

It is natural to expect that the Carleson index of a set is the larger, the more widely 'scattered' are its points. A fairly nice measure for such 'scattering' is furnished by the quantity

$$\delta(E) \ = \ \min\left\{\left|B_\zeta^E(\zeta)\right|;\ \ \zeta \in E\right\},$$

where

$$B^E = \prod_{\zeta\in E} b_\zeta\,, \qquad b_\zeta(z) = \frac{z-\zeta}{1-\bar{\zeta}z}\,, \qquad B_\zeta^E = B^E b_\zeta^{-1}.$$

Indeed,

$$\text{Carl}\, E \ \geq \ A\,\frac{\delta(E)}{\log(e/\delta(E))}\,, \tag{8}$$

where A is a positive absolute constant. This is just the estimate provided by both of the proofs under discussion of Carleson's theorem. It can be shown (see Chapter 7, §1 in Garnett's book) that (8) cannot be improved – by this it is meant that there are sets E with arbitrarily small $\delta(E)$ for which Carl $E \leq A_1\delta(E)/|\log\delta(E)|$.

If we look on (8) as the proof's goal, the advantages of Jones' proof are incontestable: it yields, together with the estimate (8), an explicit formula for a linear operator realizing the interpolation, besides being incomparably shorter and more elementary.

But the 'old' proof's essence is far from lying merely in the estimate (8)! The point is that the characteristic $\delta(E)$ is still too crude (as we shall see below), and can not, in principle, be used to give a *two-sided* estimate of the Carleson index. For that another quantity is needed. Put

$$Z_{\eta,\zeta} \ = \ \frac{(1-|\zeta|^2)(1-|\eta|^2)}{|1-\zeta\bar{\eta}|^2}\,, \qquad \zeta,\ \eta \in \mathbb{D}\,;$$

$$c(E) \ = \ \sup\left\{\sum_{\zeta\in E} Z_{\eta,\zeta}|B_\zeta^E(\zeta)|^{-1}\,;\ \ \eta \in E\right\}.$$

Here is the result to which the 'old' proof in fact leads:

Theorem *For every finite set $E \subseteq \mathbb{D}$,*

$$\frac{a}{c(E)} \leq \text{Carl}\, E \leq \frac{1}{c(E)}, \tag{9}$$

where $a > 0$ is an absolute constant.

This inequality captures, almost without loss, the relation of a set's Carleson index to its geometry. From it the estimate (8) follows directly – after all, $c(E) \leq (a_1/\delta(E)) \log(e/\delta(E))$ (see lemma in the first section of this appendix). On the other hand,

$$c(E) \geq \frac{1}{\delta(E)}. \tag{10}$$

Indeed, let the point $\zeta_0 \in E$ be one for which $\delta(E) = |B^E_{\zeta_0}(\zeta_0)|$. Then

$$c(E) \geq Z_{\zeta_0,\zeta_0} \left|B^E_{\zeta_0}(\zeta_0)\right|^{-1} = 1/\delta(E),$$

since $Z_{\zeta_0,\zeta_0} = 1$. From (9) and (10) we get

$$A \frac{\delta(E)}{\log(e/\delta(E))} \leq \text{Carl}\, E \leq \delta(E) \tag{11}$$

(the right-hand inequality can also be deduced directly from Schwarz' lemma).

There is some 'clearance' between the left and right-hand inequalities (the logarithm in the denominator on the left!). It is impossible to get rid of it. Indeed, we have already seen (i.e., *noted* – translator) that there are sets E which 'practically realize' the left side of (11). Consider on the other hand the 'doublet' $E_x = \{-x, x\}$, where $x \in (0, 1)$.

Simple computations show that Carl $E_x = x$ and $\delta(E_x) = 2x/(1 + x^2)$, making

$$\text{Carl}\, E_x = \frac{\delta(E_x)}{1 + \sqrt{1 - (\delta(E_x))^2}}.$$

For small values of x, the set E_x 'practically realizes' the right side of (11). Moreover, Carl $E_x \sim \delta(E_x)$ for $x \to 1$. Thus, (9) carries more information than (11).

Let us furthermore point out that the estimate (8), very useful for small values of $\delta(E)$, loses content for $\delta(E)$ close to unity. It can be shown (with the help of Earl's 'nonlinear' method – see Chapter 7, §5 in Garnett's book) that $1 - \text{Carl}\, E = O\left(\sqrt{1 - \delta(E)}\right)$ for $\delta(E) \to 1$. V.A. Tolokonnikov has observed that the order of magnitude indicated on the right is sharp; that can be seen in the doublet example just considered.

We outline a proof of inequality (9).

By the same considerations as in Subsection B.3 of Chapter IX ('Newman's procedure') it is easy to show that if $R_E(f) = \psi$ (for $f \in H_\infty$ – translator), we have*

$$M(\psi, E) = \sup\left\{\frac{1}{2\pi}\left|\int_{|z|=1} f(z)\frac{g(z)}{B^E(z)}\,dz\right|;\quad g \in H_1,\ \|g\|_1 \le 1\right\}$$

$$= \sup\left\{\left|\sum_{\zeta\in E}\psi(\zeta)\frac{g(\zeta)}{B_\zeta^E(\zeta)}(1-|\zeta|^2)\right|;\quad g \in H_1,\ \|g\|_1 \le 1\right\}.$$

Therefore

$$(\mathrm{Carl}\ E)^{-1} = \sup\{M(\psi, E);\ \psi \in B_1(E)\}$$

$$= \sup\left\{\left|\sum_{\zeta\in E}\psi(\zeta)\frac{g(\zeta)}{B_\zeta^E(\zeta)}(1-|\zeta|^2)\right|;\quad \psi \in B_1(E),\ g \in H_1,\ \|g\|_1 \le 1\right\}$$

$$= \sup\left\{\int |g|\,d\mu_E;\quad g \in H_1,\ \|g\|_1 \le 1\right\}, \tag{12}$$

where $\mu_E = \sum_{\zeta\in E}\left((1-|\zeta|^2)/|B_\zeta^E(\zeta)|\right)\delta_\zeta$ with δ_ζ the unit point mass at ζ. Here $\|g\|_1 \overset{\text{def}}{=} \frac{1}{2\pi}\int_{|z|=1} |g(z)|\,|dz|$.

Let $\eta \in \mathbb{D}$ and $g_\eta(z) \overset{\text{def}}{=} (1-|\eta|^2)/(1-\bar{\eta}z)^2$. We have $g_\eta \in H_1$ and $\|g_\eta\|_1 = 1$, and thus

$$(\mathrm{Carl}\ E)^{-1} \ge \int |g_\eta|\,d\mu_E = \sum_{\zeta\in E} Z_{\eta,\zeta}\frac{1}{|B_\zeta^E(\zeta)|}.$$

Taking the supremum for η ranging through E (or even over $\mathbb{D}$), we obtain the right-hand inequality in (9).

In order to prove the left inequality in (9) we shall proceed from the following characterization of Carleson measures (on $\mathbb{D}$), found by S.A. Vinogradov:

Proposition *Let μ be a positive Borel measure carried on a subset K of the disk $\mathbb{D}$. The following assertions are equivalent:*

(i) *There is a number $M > 0$ such that*

$$\int_{\mathbb{D}} |h|^2\,d\mu \le M\,\|h\|_{H_2}^2 \quad \text{for } h \in H_2;$$

(ii)

$$s(\mu) \overset{\text{def}}{=} \sup\left\{\int_{\mathbb{D}}\frac{1-|\eta|^2}{|1-\zeta\bar{\eta}|^2}\,d\mu(\zeta);\quad \eta \in K\right\} \quad \text{is } < \infty;$$

(iii)

$$S(\mu) \overset{\text{def}}{=} \sup\left\{\int_{\mathbb{D}}\frac{1-|\eta|^2}{|1-\zeta\bar{\eta}|^2}\,d\mu(\zeta);\quad \eta \in \mathbb{D}\right\} \quad \text{is } < \infty.$$

---

* Havin's norm $\|\ \|_1$ for $H_1(\mathbb{D})$ is $1/2\pi$ times the one used in the body of this book (see later this page) – translator

If (i) *is fulfilled, we have* $s(\mu) \leq S(\mu) \leq a_1 M \leq a_2 s(\mu)$, *where* a_1 *and* a_2 *are absolute constants.*

S.A. Vinogradov has given a direct 'Hilbert space' proof of this proposition, independent of the geometric characterization of Carleson measures (last theorem in Chapter VIII) and avoiding any use of the maximal function. His procedure is explained at the beginning of Lecture VII in Nikolskii's book. All the proof uses is a certain simple and general estimate for integral operators (the 'Vinogradov-Senichkin test') together with estimates of the Poisson kernel. The geometric characterization of Carleson measures follows easily from the proposition (and vice-versa). We emphasize, however, that for proving Carleson's interpolation theorem (or inequality (9)), that characterization is not necessary! It is, in particular, not necessary to resort to the lemma of Subsection B.2 (in Chapter IX – translator) with its 'dyadic technique'. The geometric characterization's graphic clarity is deceptive: there are situations where the apparently less effective criteria (ii) and (iii) prove to be much more practical. One such arises in the proof of the left-hand inequality in (9), to which we now return.

If $g \in H_1$ and $\|g\|_1 \leq 1$, we have $g = bh^2$ where $h \in H_2$, $\|h\|_2 = \|g\|_1 \leq 1$ and b is a Blaschke product. By the proposition, with $K = E$,

$$\int |g| \, d\mu_E \leq \int |h|^2 \, d\mu_E \leq a' s(\mu_E) = a' c(E).$$

Passing to the supremum with respect to g, we obtain, from (12), the left-hand inequality (9). The theorem is proved.

Here the almost trivial verification that the measure μ_E is Carleson has been carried out according to a scheme proposed by S.A. Vinogradov already in 1974. A far reaching development of this procedure is given in the article by Vinogradov and Rukshin. There they also consider interpolation problems (multiple interpolation with nodes of unbounded multiplicity) where the Jones proof cannot be followed (no kind of linear interpolation operator exists), but the path leading through duality and Carleson measures arrives at its goal.

Appendix II
by V.P. Havin
Weak Completeness of the Space $L_1/H_1(0)$

1. A sequence $\{x_n\}$ of points of the normed space X is said to be *weakly convergent in itself* if $\lim_{n\to\infty} x^*(x_n)$ exists for every functional $x^* \in X^*$. We shall say that such a sequence *converges weakly* if there is a point $x \in X$ for which $\lim_{n\to\infty} x^*(x_n) = x^*(x)$ whatever may be the functional $x^* \in X^*$ (and in that event we say that the sequence $\{x_n\}$ converges weakly to that point x).

A sequence weakly convergent in itself need not converge weakly – even in the case of a *Banach* space X.

Example Let $\{x_n\}$ be a sequence of functions continuous on the segment $[0, 1]$, with the property that

$$\lim_{n\to\infty} x_n(t) = \begin{cases} 1, & t = 0 \\ 0, & 0 < t \le 1 \end{cases}$$

whilst $\sup_n \|x_n\|_{\mathscr{C}([0,1])} \le 1$. Then

$$\lim_{n\to\infty} \int_{[0,1]} x_n \,\mathrm{d}\mu = \mu(\{0\}),$$

whatever may be the complex Borel measure μ on the segment $[0, 1]$. This means that the sequence $\{x_n\}$ converges weakly in itself in the space $\mathscr{C}([0, 1])$ (by F. Riesz' theorem on the general form of linear functionals on a space of continuous functions). But $\{x_n\}$ does not converge weakly. Indeed, a weak limit x of the sequence $\{x_n\}$ would have to satisfy the condition

$$\int_{[0,1]} x \,\mathrm{d}\mu = \mu(\{0\}) \tag{1}$$

for every Borel measure μ on $[0, 1]$, and that is impossible (substituting $\mu = \delta_t$ in (1) with

$t \in (0,1]$ and δ_t the unit point mass at t, we see that $x(t) \equiv 0$ on $(0,1]$; then, however, $x(0) = 0$ by continuity and (1) is violated for $\mu = \delta_0$).

Definition A normed space X is said to be *weakly complete* if every sequence of its points weakly convergent in itself does converge weakly.

Strictly speaking, our weakly complete spaces should be called *sequentially* weakly complete, but here we prefer the shorter term.

The space $\mathscr{C}([0,1])$ is not weakly complete as we have seen. A broad class of weakly complete spaces consists of the reflexive normed spaces. Indeed, let $\{x_n\}$ be a sequence weakly convergent in itself in the Banach space X. Put

$$\Phi(x^*) = \lim_{n\to\infty} x^*(x_n) \quad \text{for } x^* \in X^*. \tag{2}$$

By the Banach-Steinhaus theorem (uniform boundedness principle – translator), Φ belongs to the second adjoint space X^{**}. If X is reflexive, there is a point $x_\Phi \in X$ with

$$\Phi(x^*) = x^*(x_\Phi) \quad \text{for } x^* \in X^*.$$

To that point the sequence $\{x_n\}$ converges weakly.

All the L_p for $p \in (1,\infty)$ are in particular weakly complete, as are all of their closed subspaces (a closed subspace of a reflexive space is indeed reflexive). Therefore the Hardy spaces H_p (for the disk and the half plane) are weakly complete when $1 < p < \infty$ (reflexivity of these spaces also follows directly from the results in Chapter VII).

2. An example of a non-reflexive weakly complete space is provided by L_1 (here that symbol designates the space $L_1(\mathbb{T},m)$ with $\mathbb{T}$ the unit circumference and m normalized Lebesgue measure on $\mathbb{T}$ $(m(\mathbb{T}) = 1)$, but L_1 is also weakly complete when formed for general measure spaces). In this appendix we establish weak completeness of the factor space $L_1/H_1(0)$ (and, in passing, of the space L_1).

Let $\mathscr{Y} \subseteq L_\infty$ $(= L_\infty(\mathbb{T},m))$. We shall say that a sequence $\{f_n\}$ of functions in $L_1(\mathbb{T},m)$ is *$\mathscr{Y}$-weakly convergent in itself* if $\lim_{n\to\infty} \int f_n y$ exists for every function $y \in \mathscr{Y}$. (From now on, $\int F$ always denotes $\int_{\mathbb{T}} F\, dm$.)

A set $\mathscr{Y} \subseteq L_\infty$ will be called *rich* if, for each sequence $\{f_n\}$, $\mathscr{Y}$-weakly convergent in itself, there is some $f \in L_1$ such that

$$\lim_{n\to\infty} \int f_n y = \int f y \tag{3}$$

for every $y \in \mathscr{Y}$. We will prove that the sets L_∞ and H_∞ are rich. That will imply that the spaces L_1 and $L_1/H_1(0)$ are weakly complete (indeed, $(L_1)^* \cong L_\infty$ and $(L_1/H_1(0))^* \cong H_\infty$; the second isomorphism is established in Chapter VII, see the first table of adjoint spaces in Subsection A.1).

Let us enumerate four properties of a set $\mathscr{Y}$ ensuring, as we shall see below, its richness. Formulation of the first property involves the so-called w-convergent sequences. We say that a sequence $\{y_n\}$ of functions in L_∞ *w-converges* to a function $y \in L_\infty$ if $\sup_n \|y_n\|_\infty < \infty$

and $\lim_{n\to\infty} y_n(z) = y(z)$ for almost all (m) $z \in \mathbb{T}$. A subset of the space L_∞ will be called *w-closed* if it contains the w-limit of each w-convergent sequence of its points.

I. $\mathscr{Y}$ is a w-closed subalgebra of L_∞ (in particular, $\mathscr{Y}$ is a norm-closed subspace of L_∞).

II. Given any closed subset $E \subseteq \mathbb{T}$ of Lebesgue measure zero, one can find a uniformly bounded sequence of functions in $\mathscr{C} \cap \mathscr{Y}$ converging pointwise (i.e., *everywhere* – translator) to χ_E (the characteristic function of E) on the circumference $\mathbb{T}$; we are writing $\mathscr{C}$ for $\mathscr{C}(\mathbb{T})$.

III. Every function in $\mathscr{Y}$ is the w-limit of a sequence of functions in $\mathscr{C} \cap \mathscr{Y}$.

IV. To each Lebesgue measurable subset E of the circumference $\mathbb{T}$ one can assign functions k_E, K_E (belonging to $\mathscr{Y}$ – translator) in such fashion as to have
  (a) $|k_E(z)| + |K_E(z)| \le 1$ for $z \in \mathbb{T}$;
  (b) $\sup_E |K_E - 1| \to 0$ and $\int |K_E| \to 0$ when $m(E) \to 0$;
  (c) $\sup_E |k_E| \to 0$ and $\int |1 - k_E| \to 0$ when $m(E) \to 0$.

The pair K_E, k_E is something like a partition of unity for the circumference $\mathbb{T}$; if the Lebesgue measure of the set E is small, K_E is uniformly close to 1 on E and small 'in the mean', while k_E is uniformly small on E and close to 1 'in the mean'. Here is the simplest (and a trivial) example: $\mathscr{Y} = L_\infty$, $K_E = \chi_E$, $k_E = \chi_{\mathbb{T}\sim E}$.

Theorem *If the set $\mathscr{Y} \subseteq L_\infty$ has properties I–IV, it is rich.*

Weak completeness of L_1 (i.e., richness of the whole space L_∞) follows immediately from this theorem. Property I is obviously enjoyed by L_∞. The continuous functions

$$y_n(z) \stackrel{\text{def}}{=} \left(1 - \tfrac{1}{2}\,\text{dist}\,(z, E)\right)^n, \quad z \in \mathbb{T},$$

form a bounded sequence converging pointwise to χ_E (when $E \subseteq \mathbb{T}$ is *closed* – translator), so condition II is also fulfilled. Condition III is easy to verify, with the help of Fatou's theorem on Poisson integrals, for example, and condition IV we have already checked.

The verification of conditions I–IV for $\mathscr{Y} = H_\infty$ requires more work, and we carry it out below in Section 3. At present we proceed to the proof of the theorem. It is based on a lemma about regular functionals on L_∞.

Definition Let $\mathscr{Y} \subseteq L_\infty$. A functional $\Phi \in (L_\infty)^*$ is called *$\mathscr{Y}$-regular* if there is a function $f \in L_1$ such that

$$\Phi(y) = \int fy \tag{4}$$

for every $y \in \mathscr{Y}$.

Example Let us extend (by means of the Hahn-Banach theorem) the functional 'evaluation at the point 1' from $\mathscr{C}$ to L_∞. It is easy to see that the functional thus obtained is not $\mathscr{C}$-regular (and a fortiori not L_∞-regular).

Lemma *Let $\mathscr{Y}$ be a subspace of L_∞ enjoying properties II and III, and let $\Phi \in (L_\infty)^*$. The following assertions are equivalent:*

(i) *Φ is $\mathscr{Y}$-regular*

(ii) *Φ is continuous on $\mathscr{Y}$ with respect to w-convergence, i.e., $\lim_{n\to\infty}\Phi(y_n) = 0$ whenever the sequence $\{y_n\}$ of points of $\mathscr{Y}$ w-converges to zero.*

Proof

(i) $\Longrightarrow$ (ii) by Lebesgue's dominated convergence theorem.

(ii) $\Longrightarrow$ (i). Put $\Phi_{\mathscr{C}} = \Phi \restriction \mathscr{C} \cap \mathscr{Y}$. By the Hahn-Banach theorem, $\Phi_{\mathscr{C}}$ has a continuous extension from the subspace $\mathscr{C} \cap \mathscr{Y}$ of $\mathscr{C}$ to all of $\mathscr{C}$. Therefore, by the theorem of F. Riesz, we get a complex Borel measure ν on the circumference $\mathbb{T}$ such that $\Phi_{\mathscr{C}}(y) = \int y\, d\nu$ for $y \in \mathscr{C} \cap \mathscr{Y}$. If (ii) holds, ν must be absolutely continuous with respect to m. Indeed, let E be a closed subset of $\mathbb{T}$. It suffices to prove that $\nu(E) = 0$ if $m(E) = 0$. Let then $\{y_n\}$ be a sequence of functions in $\mathscr{Y} \cap \mathscr{C}$ corresponding as in condition II to the set E. Then, by (ii), $\lim_{n\to\infty}\Phi_{\mathscr{C}}(y_n) = \lim_{n\to\infty}\Phi(y_n) = 0$ (since y_n is w-convergent to 0 !). But by Lebesgue's bounded convergence theorem (y_n tending *everywhere* to χ_E – translator),

$$\lim_{n\to\infty}\Phi_{\mathscr{C}}(y_n) = \lim_{n\to\infty}\int y_n\, d\nu = \int \chi_E\, d\nu = \nu(E).$$

Thus, ν is absolutely continuous with respect to m, making $d\nu = f dm$ for a certain function f in L_1, and (4) is true for each $y \in \mathscr{Y} \cap \mathscr{C}$. From condition (ii) and property III of the space $\mathscr{Y}$ it now follows that (4) holds for any $y \in \mathscr{Y}$. Q.E.D.

Let us continue with the proof of our theorem. Let $\{f_n\}$ be a sequence $\mathscr{Y}$-weakly convergent in itself. Put (cf. (2))

$$\Phi_n(y) = \int f_n y \quad \text{and} \quad \Phi(y) = \lim_{n\to\infty}\Phi_n(y)$$

for $y \in \mathscr{Y}$. According to condition I, $\mathscr{Y}$ is a *closed* subspace of L_∞. Therefore the functional Φ is continuous on $\mathscr{Y}$ by the Banach-Steinhaus theorem (uniform boundedness principle – translator). Let us extend it to all of L_∞ with preservation of continuity and linearity – the extended functional we continue to denote by the same letter Φ. The theorem will be proved if we can verify that Φ is $\mathscr{Y}$-regular.

Put $\mathscr{B}_{\mathscr{Y}} = \{y \in \mathscr{Y}\,;\ \|y\|_\infty \le 1\}$. On $\mathscr{B}_{\mathscr{Y}}$ we define a new metric, setting

$$\rho(y_1, y_2) \overset{\text{def}}{=} \int |y_1 - y_2| \quad \text{for } y_1, y_2 \in \mathscr{B}_{\mathscr{Y}}.$$

The *sphere* $\mathscr{B}_{\mathscr{Y}}$ equipped with the metric ρ turns out to be a *complete* metric space. Indeed, if the sequence $\{y_n\}$ of points of $\mathscr{B}_{\mathscr{Y}}$ is Cauchy for the metric ρ, completeness of the space L_1 (in norm $\|\ \|_1$ – translator) gives us a $y \in L_1$ with

$$\lim_{n\to\infty}\int |y_n - y| = 0. \tag{5}$$

In such case we can extract from $\{y_n\}$ a subsequence w-convergent to y so, by condition I, $y \in \mathscr{Y}$; (5) now shows us that $y \in \mathscr{B}_{\mathscr{Y}}$ and that $\lim_{n\to\infty}\rho(y_n, y) = 0$. Completeness of the sphere $\mathscr{B}_{\mathscr{Y}}$ with respect to the metric ρ is established.

Denote by φ_n the restriction of the functional Φ_n to the sphere $\mathscr{B}_{\mathscr{Y}}$. Put $\varphi = \Phi \restriction \mathscr{B}_{\mathscr{Y}}$. It follows from Lebesgue's dominated convergence theorem that each of the functions φ_n is continuous with respect to convergence in the metric ρ. (*Translator's note*: Suppose that for some $\delta > 0$ and some n we have $|\varphi_n(y_k) - \varphi_n(y)| \geq \delta$ with y and the y_k in $\mathscr{B}_{\mathscr{Y}}$ and $\rho(y_k, y) \xrightarrow[k]{} 0$. A suitable subsequence $\{y_{k_j}\}$ of $\{y_k\}$ is then w-convergent to y – see above – but that makes $\varphi_n(y_{k_j}) - \varphi_n(y) = \int f_n(y_{k_j} - y) \xrightarrow[j]{} 0$ by the theorem of Lebesgue referred to, and we have a contradiction.) The function φ, equal to the pointwise limit of the φ_n on the *complete* metric space $\mathscr{B}_{\mathscr{Y}}, \rho$, is therefore ρ-continuous at some point b of the sphere $\mathscr{B}_{\mathscr{Y}}$ (according to the well-known theorem of Baire whose proof can, for instance, be found in Yosida's book).

It is clear (by the bounded convergence theorem – translator) that w-convergence of points in $\mathscr{B}_{\mathscr{Y}}$ implies their convergence in the metric ρ. Therefore our point $b \in \mathscr{B}_{\mathscr{Y}}$ has the property that

$$\lim_{n\to\infty} \Phi(z_n) = \Phi(b) \quad \text{for any sequence of } \textit{points } z_n \textit{ in the sphere}$$
$$\mathscr{B}_{\mathscr{Y}} \ \rho\text{-convergent (or } w\text{-convergent – translator) to } b\,. \tag{6}$$

The italicized words of this statement are most vital. If we could dispense with them, the proof of our theorem could be completed immediately: taking a sequence $\{x_n\}$ of points in $\mathscr{Y}$, w-convergent to zero, we could write the identity

$$-\Phi(x_n) \;=\; -\Phi(b) + \Phi(z_n)$$

with $z_n = b - x_n$ and thus convince ourselves that $\lim_{n\to\infty} \Phi(x_n) = 0$ (the z_n being w-convergent to b). The functional Φ would turn out to be regular (by the lemma). The trouble is, however, that the points z_n w-convergent to b may leave the sphere $\mathscr{B}_{\mathscr{Y}}$; on account of that some work still lies ahead of us.

Two properties of the functional Φ are at our disposal: property (6) and Φ's continuity with respect to the *norm* for L_∞. Therefore the quantity $|\Phi(x)|$ is small if the vector $x \in \mathscr{Y}$ is the sum of a fixed number of others, each of which is either very small in L_∞ norm or of the form $\pm(b - v)$ with v *lying in* $\mathscr{B}_{\mathscr{Y}}$ and close in the metric ρ to b.

Let then $\{x_n\}$ be a sequence of elements of the space $\mathscr{Y}$ w-convergent to zero. We verify that $\lim_{n\to\infty} \Phi(x_n) = 0$. Without loss of generality we may suppose that $x_n \in \mathscr{B}_{\mathscr{Y}}$ for $n = 1, 2, \ldots$, for otherwise we could go over to the sequence of points $x_n / \sup_k \|x_k\|_\infty$. Taking a small number $\sigma > 0$, we put

$$E_n \;=\; E_n(\sigma) \;=\; \{z \in \mathbb{T};\ |x_n(z)| > \sigma\}.$$

It is clear that $\lim_{n\to\infty} m(E_n) = 0$. Let us write the following identity:

$$x_n = \left(K_{E_n} x_n + k_{E_n} b - b\right) + \left(b - k_{E_n} b\right) + \left(1 - K_{E_n}\right) x_n \stackrel{\text{def}}{=} \left(v_n^{(1)} - b\right) + \left(b - v_n^{(2)}\right) + v_n^{(3)}. \tag{7}$$

Here K_E, k_E is the 'partition of unity' from condition IV. We shall see in a moment that if n is large, the quantities $\|v_n^{(3)}\|_\infty$, $\rho(v_n^{(1)}, b)$ and $\rho(v_n^{(2)}, b)$ are small, and that $v_n^{(1)}$ and $v_n^{(2)}$ belong to $\mathscr{B}_{\mathscr{Y}}$. That will enable us to prove that the quantity $|\Phi(x_n)|$ is small.

The functions $K_{E_n} x_n$, $k_{E_n} b$ (and thus the $v_n^{(j)}$, $j = 1, 2, 3$) belong to $\mathscr{Y}$ since that space is a subalgebra of L_∞. Let us estimate $\|v_n^{(3)}\|_\infty$. If $z \in E_n$, $|v_n^{(3)}(z)| \leq \sup_{E_n} |1 - K_{E_n}| \xrightarrow[n]{} 0$

(since $m(E_n) \xrightarrow[n]{} 0$); if, on the other hand, $z \in \mathbb{T} \sim E_n$, then $|v_n^{(3)}(z)| \leq 2|x_n(z)| \leq 2\sigma$. Thus,

$$\|v_n^{(3)}\|_\infty \leq 2\sigma \text{ for all sufficiently large } n. \tag{8}$$

Inclusion of the $v_n^{(j)}$ in $\mathscr{B}_\mathscr{Y}$ for $j = 1,2$ follows from property IV(a) of the functions K_E, k_E and from the relations $x_n \in \mathscr{B}_\mathscr{Y}$, $b \in \mathscr{B}_\mathscr{Y}$.

Furthermore,

$$\rho\left(v_n^{(1)}, b\right) \leq \int |K_{E_n}| + \int |k_{E_n} - 1| \xrightarrow[n]{} 0,$$

$$\rho\left(v_n^{(2)}, b\right) \leq \int |1 - k_{E_n}| \xrightarrow[n]{} 0.$$

Therefore, by (6), $\lim_{n\to\infty} \Phi\left(v_n^{(1)} - b\right) = \lim_{n\to\infty} \Phi\left(v_n^{(2)} - b\right) = 0$, and by (7) and (8),

$$\limsup_{n\to\infty} |\Phi(x_n)| \leq \|\Phi\| \limsup_{n\to\infty} \|v_n^{(3)}\|_\infty \leq 2\sigma\|\Phi\|,$$

whatever may be the value of $\sigma > 0$. Thus, $\lim_{n\to\infty} \Phi(x_n) = 0$, and the theorem is proved.

3. It remains to verify that $\mathscr{Y} = H_\infty$ enjoys properties I–IV.

Property I holds for H_∞ because (by the bounded convergence theorem – translator) Fourier coefficients are w-continuous functionals, and membership of a function $f \in L_\infty$ in H_∞ is equivalent to the vanishing of its coefficients with negative index.

Property II is easy to establish with help of the Fatou function ϕ for the (*closed* – translator) set E constructed in Subsection A.1 of Chapter II; the required sequence is $\{\phi^n\}_{n=1}^\infty$.

Property III. If $y \in H_\infty$, Fatou's theorem makes $y(e^{i\theta}) = \lim_{r\to 1} y(re^{i\theta})$ almost everywhere (m), where

$$y(re^{i\theta}) = \frac{1}{2\pi}\int_0^{2\pi} y(e^{it})\frac{1-r^2}{1-2r\cos(\theta-t)+r^2}\,dt.$$

It is clear that $\sup|y(re^{i\theta})| \leq \|y\|_\infty$, while the function $z \longmapsto y(rz)$, $z \in \mathbb{T}$, belongs to $\mathscr{C} \cap H_\infty$ (for $r < 1$ – translator).

Property IV. Verification of this condition is somewhat more involved. The requirements placed by IV(c), for example, on the function k_E (analytic in $\mathbb{D}$ and not exceeding 1 in modulus there!) are not so easy to impose simultaneously. On the one hand it is to be close to 1 everywhere on $\mathbb{T}$ in the metric ρ (and that makes it close to 1, in a known sense, everywhere in $\mathbb{D}$). On the other hand it is to be uniformly small on a set of small but nevertheless strictly positive measure, and that tends to make it *small* in $\mathbb{D}$.

Let $m(E) > 0$. Put $f_E = \left(\chi_E + i\widetilde{\chi_E}\right)/\sqrt{m(E)}$, where χ_E is the characteristic function of the set E and $\widetilde{\chi_E}$ its harmonic conjugate (see Section E of Chapter I). The function f_E extends from $\mathbb{T}$ into the unit disk $\mathbb{D}$ as an analytic function with positive real part, and $f_E(0) = \sqrt{m(E)}$, $\Re f_E = 1/\sqrt{m(E)}$ a.e. (m) on E. Putting $\lambda(x) = |\log x|^{-1/2}$ for $x \in (0,1)$, let us introduce the following function g_E, analytic in $\mathbb{D}$:

$$g_E \stackrel{\text{def}}{=} \left(\frac{1}{1+f_E}\right)^{\lambda(m(E))}$$

The function g_E (more precisely, its boundary value on $\mathbb{T}$) is already almost our k_E, and $1-g_E$ almost K_E. Indeed, $\Re(1+f_E)^{-1}$ is >0 in $\mathbb{D}$ (which permits us to define a single-valued branch there of $(1+f_E)^{-\lambda(m(E))}$). Furthermore, $|1+f_E| \geq 1+\Re f_E > 1$ in $\mathbb{D}$, so all the values $(1+f_E(z))^{-1}$, $z \in \mathbb{D}$, lie in the right half disk $\mathbb{D} \cap \{\Re w > 0\}$. If the measure $m(E)$ is small, the exponent $\lambda(m(E))$ is also small, and the set $g_E(\mathbb{D})$ is located within a narrow sector of the disk $\mathbb{D}$, with bisector $[0,1]$ (and of opening $\pi\lambda(m(E))$). This means that for $m(E)$ tending to zero the set $g_E(\mathbb{D})$ is contained in an ellipse $\mathscr{E}(m(E))$ with foci at 0, 1, semimajor axis $\frac{1}{2}+o(1)$ and semiminor axis $o(1)$, i.e., that

$$|g_E(z)| + |1-g_E(z)| \;\leq\; 1+\mu(m(E)) \quad \text{for } z \in \mathbb{D}, \tag{9}$$

where μ is a certain positive function on $(0,1)$ with $\lim_{x\to 0}\mu(x) = 0$. Let us observe moreover that when $m(E) \longrightarrow 0$,

$$\begin{aligned} g_E(0) &= (1+f_E(0))^{-\lambda(m(E))} \longrightarrow 1\,, \\ \sup_E |g_E| &\leq (m(E))^{\lambda(m(E))/2} \longrightarrow 0\,. \end{aligned} \tag{10}$$

Finally, let us put

$$k_E = \frac{g_E}{1+\mu(m(E))}\,, \qquad K_E = \frac{1-g_E}{1+\mu(m(E))}\,.$$

Condition IV(a) is fulfilled by virtue of (9). The first of conditions IV(c) follows from (10). To verify the second one, it suffices to show that

$$\int |g_E - 1| \longrightarrow 0 \quad \text{when} \quad m(E) \longrightarrow 0\,. \tag{11}$$

The function $|\Im g_E|$ is uniformly bounded by the semiminor axis of the ellipse $\mathscr{E}(m(E))$ and $\Re g_E(z) \in [0,1]$ for $z \in \mathbb{D}$, so (11) is equivalent to the relation

$$\int_{\mathbb{T}} \Re g_E \, dm \longrightarrow \int_{\mathbb{T}} 1 \, dm = 1\,,$$

a consequence of (10) (since $g_E(0) = \Re g_E(0) = \int \Re g_E$). Property IV(b) of the function K_E follows in evident fashion from what has already been proven.

Our theorem can thus be applied with $\mathscr{Y} = H_\infty$, showing that space to be rich and $L_1/H_1(0)$ weakly complete. This fact was established independently and by different methods in the papers of Mooney (1972) and Havin (1973); see also Amar (1973). Subsequently, the proof was simplified (Havin, 1974), but the construction of partitions of unity remained comparatively complicated. Here we have reproduced the very simple construction proposed by Garnett.

In connection with this appendix see Garnett's book. The theorem on weak completeness of the space L_1 is a classical result (see the theorem of Vitali, Hahn and Saks in Yosida's book).

Postscript

The translations of the above two appendices were submitted to Professor Havin, who then went through them. Besides suggesting some improvements, he has asked that two more recent proofs of the result on weak completeness of $L_1/H_1(0)$ be mentioned. These are in papers by G. Godefroy and J.L. Fernández; exact references can be found in the bibliography.

Bibliography

Books

Bari, N.K. *Trigonometricheskiie riady.* State Pub. Hse. for Physico-Math. Lit., Moscow, 1961.

Bari, N.K. *A Treatise on Trigonometric Series.* (English translation of previous item.) Pergamon, New York, 1964.

Browder, A. *Introduction to Function Algebras.* Benjamin, New York, 1969.

Collingwood, E.F. & Lohwater, A.J. *The Theory of Cluster Sets.* Cambridge University Press, London and New York, (1966).

Duren, P. *Theory of H_p Spaces.* Academic Press, New York, 1970.

Gamelin, T. *Uniform Algebras.* Prentice-Hall, Englewood Cliffs, 1969.

Garnett, J. *Bounded Analytic Functions.* Academic Press, New York, 1981.

Golusin, G.M. *Geometrische Funktionentheorie.* (German translation of first edition of following item.) Deutscher V. der Wiss., Berlin, 1957.

Goluzin, G.M. *Geometricheskaia teoriia funkstiĭ kompleksnovo peremennovo.* Second edition, Nauka, Moscow, 1966.

Goluzin, G.M. *Geometric Theory of Functions of A Complex Variable.* (English translation of above item.) Amer. Math. Soc., Providence, 1969.

Guzman, M. de. *Differentiation of Integrals in* $\mathbb{R}^n$. Lecture Notes in Math., **481**. Springer, Berlin, 1975.

Helson, H. *Lectures on Invariant Subspaces.* Academic Press, New York, 1964.

Hoffman, K. *Banach Spaces of Analytic Functions.* Prentice-Hall, Englewood Cliffs, 1962.

Katznelson, Y. *An Introduction to Harmonic Analysis.* Wiley, New York, 1968.

Nikolskii, N.K. *Treatise on the Shift Operator. Spectral Function Theory.* With appendix by S.V. Hruščev and V.V. Peller. Springer, Berlin, 1986.

Privalov, I.I. *Granichnyie svoĭstva analiticheskikh funktsiĭ.* State Pub. Hse. for Tech. and Theor. Lit., Moscow, 1950.

Priwalow, I.I. *Randeigenschaften analytischer Funktionen.* (German translation of above item.) Deutscher, V. der Wiss., Berlin, 1956.

Sarason, D. *Function Theory on the Unit Circle.* Department of Math., Virginia Poly. Inst. and State Univ., Blacksburg, 1978.

Stein, E.M. *Singular Integrals and Differentiability Properties of Functions.* Princeton University Press, 1970.

Titchmarsh, E.C. *Introduction to the Theory of Fourier Integrals.* Oxford University Press, Second edition, 1948.

Tsuji, M. *Potential Theory in Modern Function Theory.* Maruzen, Tokyo, 1959.

Yosida, K. *Functional Analysis.* Springer, Berlin, 1965.

Zygmund, A. *Trigonmetrical Series.* First edition, Monografje Matematyczne, Warsaw, 1935. Reprinted by Chelsea, New York, 1952.

Zygmund, A. *Trigonometric Series.* (Greatly augmented and revised re-edition of above item.) Two volumes, Cambridge University Press, 1959.

Research Papers and Articles

Adamian, V. Nievyrozhdennyie unitarnyie stsepleniia poluunitarnykh operatorov. *Functs. Analiz i Prilozh.* 7: 4 (1973), pp. 1–16.

Adamian, V., Arov, D. & Krein, M.G. O beskonechnykh Gankelevykh matritsakh i obobshchennykh zadachakh Karateodori-Feiera i F. Rissa. *Funkst. Analiz i Prilozh* 2: 1 (1968), pp. 1–19.

Adamian, V., Arov, D. & Krein, M.G. Beskonechnyie Gankelevy matritsakh i obobshchennyie zadachakh Karateodori-Feiera i Shura. *Funkst. Analiz i Prilozh* 2: 4 (1968), pp. 1–17.

Adamian, V., Arov, D. & Krein, M.G. Analytic properties of Schmidt pairs for a Hankel operator and the generalized Schur-Takagi problem. *Math. USSR Sbornik* **15** (1971), pp. 31–73. (English translation of article appearing in *Mat. Sbornik* **86** (128), (1971), pp. 31–73.)

Adamian, V., Arov, D. & Krein, Infinite Hankel block matrices and related extension problems. *A.M.S. Translations* 2: 111 (1978), pp. 133–156. (English translation of article appearing in *Izv. Akad. Nauk Armian. SSR., Ser. Mat.* **6** (1971), pp. 87–112.)

Amar, E. Sur un théorème de Mooney relatif aux fonctions analytiques bornées. *Pacific J. Math.* **49** (1973), pp. 311–314.

Behrens, M. The maximal ideal space of algebras of bounded analytic functions on infinitely connected domains. *Trans. A.M.S.* **161** (1971), pp. 359–380.

Behrens, M. Interpolation and Gleason parts in *L*-domains. *Trans. A.M.S.* **286** (1984), pp. 203–225.

Bernard, A., Garnett, J. & Marshall, D. Algebras generated by inner functions. *J. Funct. Analysis* **25** (1977), pp. 275–285.
Beurling, A. On two problems concerning linear transformations in Hilbert space. *Acta Math.* **81** (1949), pp. 239–255.
Burkholder, D., Gundy, R. & Silverstein, M. A maximal function characterization of the class H_p. *Trans. A.M.S.* **157** (1971), pp. 137–153.
Carleson, L. On bounded analytic functions and closure problems. *Arkiv för Mat.* **2** (1952), pp. 283–291.
Carleson, L. An interpolation problem for bounded analytic functions. *Amer. J. Math.* **80** (1958), pp. 921–930.
Carleson, L. A representation formula for the Dirichlet integral. *Math. Zeitsch.* **73** (1960), pp. 190–196.
Carleson, L. Interpolation by bounded analytic functions and the corona problem. *Annals of Math.* **76** (1962), pp. 547–559.
Carleson, L. Maximal functions and capacities. *Annales Inst. Fourier, Grenoble* **15** (1965), pp. 59–64.
Carleson, L. On convergence and growth of partial sums of Fourier series. *Acta Math.* **116** (1966), pp. 135–157.
Carleson, L. Two remarks on H_1 and BMO. *Advances in Math.* **22** (1976), pp. 269–277.
Carleson, L. & Garnett, J. Interpolation sequences and separation properties. *J. d'Analyse Math.* **28** (1975), pp. 273–299.
Carleson, L. & Jacobs, S. Best uniform approximation by analytic functions. *Arkiv för Math.* **10** (1972), pp. 219–229.
Chang, S.-Y. A characterization of Douglas subalgebras. *Acta Math.* **137** (1976), pp. 81–89.
Chang, S.-Y. Structure of subalgebras between L_∞ and H_∞. *Trans. A.M.S.* **227** (1977), pp. 319–332.
Chang, S.-Y. & Garnett, J. Analyticity of functions and subalgebras of L_∞ containing H_∞. *Proc. A.M.S.* **72** (1978), pp. 41–45.
Chang, S.-Y. & Marshall, D. Some algebras of bounded analytic functions containing the disk algebra, in *Banach Spaces of Analytic Functions, Lecture Notes in Math.* **604**. Springer, Berlin, (1977), pp. 12–20.
Coifman, R. A real variable characterization of H^p. *Studia Math.* **51** (1974), p. 269–274.
Coifman, R. & Fefferman, C. Weighted norm inequalities for maximal functions and singular integrals. *Studia Math.* **51** (1974), pp. 241–250.
Davie, A., Gamelin, T. & Garnett, J. Distance estimates and pointwise bounded density. *Trans. A.M.S.* **175** (1973), pp. 37–68.
Douglas, R.G. Toeplitz and Wiener-Hopf operations in $H_\infty + \mathscr{C}$. *Bull. A.M.S.* **74** (1968), pp. 895–899.
Douglas, R.G. On the spectrum of a class of Toeplitz operators. *J. Math. and Mechanics* **18** (1968), pp. 433–435.
Douglas, R.G. On the spectrum of Toeplitz and Wiener–Hopf operators, in *Proceedings of the Conference on Abstract Spaces and Approximation, Oberwohlfach, 1968.* I.S.N.M. 10, Birkhäuser, Basel (1969).

Douglas, R.G. & Rudin, W. Approximation by inner functions. *Pacific J. Math.* **31** (1969), pp. 313–320.

Douglas, R.G. & Sarason, D. A class of Toeplitz operators. *Indiana Univ. Math. J.* **20** (1971), pp. 891–895.

Earl, J. On the interpolation of bounded sequences by bounded functions. *J. London Math. Soc.* **(2)** 2 (1970), pp. 544–548.

Fatou, P. Séries trigonométriques et séries de Taylor. *Acta Math.* **30** (1906), pp. 335–400.

Fefferman, C. & Stein, E.M. H_p spaces of several variables. *Acta Math.* **129** (1972), pp. 137–193.

Fernandez, J.L. A boundedness theorem for L^1/H_0^1. *Michigan Math. J.* **35** (1988), pp. 227–231.

Gamelin, T. Localization of the corona problem. *Pacific J. Math.* **34** (1970), pp. 73–81.

Gamelin, T. The Shilov boundary of $H_\infty(U)$. *Amer. J. Math.* **96** (1974), pp. 79–103.

Garnett, J. Interpolating sequences for bounded harmonic functions. *Indiana Univ. Math. J.* **21** (1971), pp. 187–192.

Garnett, J. Two remarks on interpolation by bounded analytic functions, in *Banach Spaces of Analytic Functions. Lecture Notes in Math.* **604**. Springer, Berlin, (1977), pp. 32–40.

Garnett, J. Harmonic interpolating sequences, L_p, and *BMO*. *Annales Inst. Fourier Grenoble* **28** (1978), pp. 215–228.

Garnett, J. & Jones, P. The distance in *BMO* to L_∞. *Annals of Math.* **108** (1978), pp. 373–393.

Garnett, J. & Latter, R. The atomic decomposition for Hardy spaces in several complex variables. *Duke Math. J.* **45** (1978), pp. 815–845.

Gleason, A. & Whitney, H. The extension of linear functionals on H_∞. *Pacific J. Math.* **12** (1962), pp. 163–182.

Godefroy, G. Espaces H_1 sur des domaines généraux, in *Analyse harmonique: groupe de travail sur les espaces de Banach invariants par translation.* Université de Paris-Sud, Département de mathématique, Orsay, (1984). Exposé No. 5, 18 pp.

Godefroy, G. Sous-espaces bien disposés de L^1 – applications. *Trans. A.M.S.* **286** (1984), pp. 227–249.

Havin, V.P. Obobshchenie teoremy Privalova-Zigmunda o module nîepreryvnosti sopriâzhennoĭ funktsii. *Izv. Akad. Nauk Armiân. S.S.R.* **6** (1971), pp. 252–258 and pp. 265–287.

Havin, V.P. Slabaîa polnota prostranstva $L^1/H^1(0)$. *Vestnik Leningrad Gosudarst. Univ.* **13** (1973), pp. 77–81.

Havin, V.P. Prostranstva H^∞ i L^1/H_0^1. *Zapiski nauchn. semin. LOMI* **39** (1974), p. 120–148.

Havinson, S. On some extremal problems of the theory of analytic functions. *A.M.S. Translations* **(2)** 32 (1963), pp. 139–154. (English translation of article appearing in *Moskov. Univ. Uchen. Zapiski,* **148** *Matem.* **4** (1951), pp. 133–143, with historical appendix by the translator, A. Shields.)

Havinson, S. Extremal problems for certain classes of analytic functions in finitely connected regions. *A.M.S. Translations* **(2)** 5 (1957), pp. 1–33. (English translation of article appearing in *Mat. Sbornik* **36**, (78) (1955), pp. 445–478.

Helson, H. & Lowdenslager, D. Prediction theory and Fourier series in several variables. *Acta Math.* **99** (1958), pp. 165–202.

Helson, H. & Lowdenslager, D. Prediction theory and Fourier series in several variables, II. *Acta Math.* **106** (1961), pp. 175–213.
Helson, H. & Sarason, D. Past and future. *Math. Scand.* **21** (1967), pp. 5–16.
Helson, H. & Szegö, G. A problem in prediction theory. *Annali di Mat. Pura ed Applicata* **4**, 51 (1960), pp. 107–138.
Herz, C. Bounded mean oscillation and regulated martingales. *Trans. A.M.S.* **193** (1974), pp. 199–215.
Hoffman, K. Bounded analytic functions and Gleason parts. *Annals of Math.* **86** (1967), pp. 74–111.
Hunt, R., Muckenhoupt, B. & Wheeden, R. Weighted norm inequalities for the conjugate function and Hilbert transform. *Trans. A.M.S.* **176** (1973), pp. 227–251.
John, F. & Nirenberg, L. On functions of bounded mean oscillation. *Comm. Pure Appl. Math.* **14** (1961), pp. 415–426.
Jones, P. A complete bounded complex submanifold of $\mathbb{C}^3$. *Proc. A.M.S.* **76** (1979), pp. 305–306.
Jones, P. Extension theorems for *BMO*. *Indiana Univ. Math. J.* **29** (1979), pp. 41–46.
Jones, P. Factorization of A_p weights. *Annals of Math.* **111** (1980), pp. 511–530.
Jones, P. Bounded holomorphic functions with all level sets of infinite length. *Michigan Math. J.* **27** (1980), pp. 75–79.
Jones, P. Carleson measures and the Fefferman-Stein decomposition of $BMO(\mathbb{R})$. *Annals of Math.* **111** (1980), pp. 197–208.
Jones, P. L_∞ estimates for the $\bar{\partial}$-problem. *Preprint, Math. Dept., Univ. of Chigago* (1980).
Jones, P. Ratios of interpolating Blaschke products. *Pacific J. Math.* **95** (1981), pp. 311–321.
Jones, P. L^∞ estimates for the $\bar{\partial}$-problem in a half-plane. *Acta Math.* **150** (1983), pp. 137–152.
Kahane, J.-P. Another theorem on bounded analytic functions. *Proc. A.M.S.* **18** (1967), pp. 827–831.
Khavin, V.P.: *see under* Havin, V.P.
Khavinson, V.P.: *see under* Havinson, V.P.
Koosis, P. On functions which are mean periodic on a half-line. *Comm. Pure Appl. Math.* **10** (1957), pp. 133–149.
Koosis, P. Interior compact spaces of functions on a half-line. *Comm. Pure Appl. Math.* **10** (1957), pp. 583–615.
Koosis, P. Weighted quadratic means of Hilbert transforms. *Duke Math. J.* **38** (1971), pp. 609–634.
Koosis, P. Moyennes quadratiques de transformées de Hilbert et fonctions de type exponentiel. *C.R. Acad. Sci. Paris* **276** (1973), pp. 1201–1204.
Koosis, P. Sommabilité de la fonction maximale et appartenance à H_1. *C.R. Acad. Sci. Paris,* **286** Sér. A (1978), pp. 1041–1043.
Koosis, P. Sommabilité de la fonction maximale et appartenance à H_1. Cas de plusieurs variables. *C.R. Acad. Sci. Paris, Sér. A,* **288** (1979), pp. 489–492.
Latter, R.H. A decomposition of $H_p(\mathbb{R}^n)$ in terms of atoms. *Studia Math.* **62** (1979), pp. 92–101.
Lax, P. Remarks on the preceding paper. *Comm. Pure Appl. Math.* **10** (1957), pp. 617–622.

Lax, P. Translation invariant subspaces. *Acta Math.* **101** (1959), pp. 163–178.

Lee, M. & Sarason, D. The spectra of some Toeplitz operators. *J. Math. Anal. Appl.* **33** (1971), pp. 529–543.

de Leeuw, K. & Rudin, W. Extreme points and extremum problems in H_1. *Pacific J. Math.* **8** (1958), pp. 467–485.

Lindelöf, E. Sur la represéntation conforme d'une aire simplement connexe sur l'aire d'un cercle, in *Quatrième Congrès des Mathematiciens Scandinaves, Stockholm* (1916), pp. 59–90.

Marshall, D. Blaschke products generate H_∞. *Bull. A.M.S.* **82** (1976), pp. 494–496.

Marshall, D. Subalgebras of L_∞ containing H_∞. *Acta Math.* **137** (1976), pp. 91–98.

Martirosian, V.M. Effektivnoie resheniie zadachi kratnoi interpoliatsii v H^∞ primeneniiem metoda biortogonalizatsii M.M. Dzhrbashiana. *Izvestiia Akad. Nauk. Armian. S.S.R. ser. matem.* **16** (1981), pp. 339–357.

Mooney, M. A theorem on bounded analytic functions. *Pacific J. Math.* **43** (1972), pp. 457–463.

Muckenhoupt, B. Weighted norm inequalities for the Hardy maximal function. *Trans. A.M.S.* **165** (1972), pp. 207–226.

Muckenhoupt, B. Hardy's inequality with weights. *Studia Math.* **44** (1972), pp. 31–38.

Nevanlinna, R. Über beschränkte Funktionen die in gegebenen Punkten vorgeschriebene Werte annehmen. *Ann. Acad. Sci. Fenn.* **13** (1919), No. 1.

Nevanlinna, R. Über beschränkte analytische Funktionen. *Ann. Acad. Sci. Fenn.* **32** (1929), No. 7.

Newman, D.J. Pseudo-uniform convexity in H_1. *Proc. A.M.S.* **14** (1963), pp. 676–679.

Riesz, F. & M. Über die Randwerte einer analytischen Funktion, in *Quatrième Congrès des Mathematiciens Scandinaves, Stockholm* (1916), pp. 27–44.

Rogosinski, W. & Shapiro, H.S. On certain extremum problems for analytic functions. *Acta Math.* **90** (1953), pp. 287–318.

Sarason, D. Generalized interpolation in H_∞. *Trans. A.M.S.* **127** (1967), pp. 179–203.

Sarason, D. Approximation of piecewise continuous functions by quotients of bounded analytic functions. *Canad. J. Math.* **24** (1972), pp. 642–657.

Sarason, D. An addendum to 'Past and future'. *Math. Scand.* **30** (1972), pp. 62–64.

Sarason, D. Algebras of functions on the unit circle. *Bull. A.M.S.* **79** (1973), pp. 286–299.

Sarason, D. Functions of vanishing mean oscillation. *Trans. A.M.S.* **207** (1975), pp. 391–405.

Sarason, D. Algebras between L_∞ and H_∞, in *Lecture Notes in Math.* **512**. Springer, Berlin, 1976, pp. 117–129.

Sarason, D. Toeplitz operators with semi almost-periodic symbols. *Duke Math. J.* **44** (1977), pp. 357–364.

Sarason, D. Toeplitz operators with piecewise quasicontinuous symbols. *Indiana Univ. Math. J.* **26** (1977), pp. 817–838.

Shapiro, H.S. & Shields, A. On some interpolation problems for analytic functions. *Amer. J. Math.* **83** (1961), pp. 513–532.

Srinivasan, T. & Wang, J.-K. On closed ideals of analytic functions. *Proc. A.M.S.* **16** (1965), pp. 49–52.

Stein, E.M. A note on the class $L\log L$. *Studia Math.* **32** (1969), pp. 305–310.

Uchiyama, A. The construction of certain *BMO* functions and the corona problem. *Pacific J. Math.* **99** (1982), pp. 183–204.

Varopoulos, N. Sur un problème d'interpolation. *C.R. Acad. Sci. Paris* **274** Sér. A (1972), pp. 1539–1542.

Varopoulos, N. *BMO* functions and the $\bar{\partial}$-equation. *Pacific J. Math.* **71** (1977), pp. 221–273.

Varopoulos, N. A remark on *BMO* and bounded harmonic functions. *Pacific J. Math.* **73** (1977), pp. 257–259.

Vinogradov, S.A. Nĭeskol'ko zamechaniĭ o svobodnoĭ interpoliatsii ogranichennymi i medlenno rastushchimi analiticheskimi funktsiiami. *Zapiski nauchn. semin. LOMI* **126** (1983), pp. 35–46.

Vinogradov, S.A., Gorin, E.A. & Hruščev, S.V. Svobodnaia interpoliatisiia v H^∞ po P. Dzhonsu. *Zapiski nauchn. semin. LOMI* **133** (1981), pp. 212–214.

Vinogradov, S.V. & Rukshin, S.E. O svobodnoĭ interpoliatsii rostkov analiticheskikh funktsiĭ v prostranstvakh Khardi. *Zapiski nauchn. semin. LOMI* **107** (1982), pp. 36–45.

Widom, H. Toeplitz matrices, in *Topics in Real and Complex Analysis*, edited by I.I. Hirschman. Mathematical Association of America (1965), pp. 197–209.

Widom, H. Toeplitz operators on H_p. *Pacific J. Math.* **19** (1966), pp. 573–582.

Wilson, J.M. A simple proof of the atomic decomposition for $H^p(\mathbb{R}^n)$, $0 < p \leq 1$. *Studia Math.* **74** (1982), pp. 25–33.

Ziskind, S. Interpolating sequences and the Shilov boundary of $H_\infty(\Delta)$. *J. Funct. Analysis* **21** (1976), pp. 380–388.

Index